**CONSILIENT MAPPING
NINE PROBES FOR ARCHITECTURE IN KOREA**

통섭지도: 한국건축을 위한 아홉 개의 탐침

국민대학교 건축대학 엮음
COMPILED BY SCHOOL OF ARCHITECTURE, KOOKMIN UNIVERSITY

공간사

- Sungyong Joh
- H-Sang Seung
- Kilyong Park
- Itami Jun
- Anouk Legendre + Nicolas Desmazieres
- Hyunsik Min
- Byungyoon Kim
- ELEPHANT + GANSAM Partners
- Jean Nouvel
- Youngsun Chung
- Andres Perea Ortega
- MVRDV + Youngwook Joung
- Gonghee Lee
- Kerl Yoo
- David-Pierre Jalicon
- Woongwon Yoon + Jeongjoo Kim
- Dominique Perrault
- Jungkyu Kim
- Ilburm Bong
- Satoshi Matsuoka + Yuki Tamura
- Daniel Valle + KIOHUN
- JUNGLIM Architecture
- HEERIM Architects
- HNA ONGODANG + DOGMA
- Peter Eisenman + HAEAHN Architecture
- Yoongyoo Jang
- Moongyu Choi
- Minsuk Cho
- FOA
- Leemjong Jang
- Youngjoon Kim
- Tesoc Hah
- Ciro Najle
- Young Kweon
- Jooryung Kim
- Shaun Murray
- Haewon Shin + Liu Yuyang
- Haewon Shin
- Changjoong Kim
- Sooin Yang
- Kihong Kim
- Kyunghoon Lee
- Paul Preissner
- Lonn Combs + Rona Easton
- Marcos Novak
- Junsung Kim + Hailim Suh
- Hailim Suh + Junsung Kim
- Hailim Suh + Nader Tehrani
- Ilburm Bong
- Hoon Moon
- OMA + SAMOO Architects
- Ken Min Sungjin
- Zaha Hadid
- Alberto Francini + Andrea Boschetti
- SPACE Group + BEOM Architects
- Wangdon Choi
- JINA Architects
- KUNWON Architects + DA Group
- SAMOO Architects + SOM
- MOOYOUNG Architects + DESTEFANO
- GANSAM Partners
- WONDOSHI Architects
- WONYANG Architectural Design + HAEAHN
- TOMOON Engineering + UNSANGDONG
- CHANJO Architects + HILLER Architecture

MAPPING
ARCHITECTURE IN KOREA

- Haeinsa Temple Culture Complex
- Hyehwa Cultural Center for Daejeon University
- Architecture as an Allegory
- Pinx Museum
- Jeongok Prehistory Museum
- National Library of Czech Republic
- Pyoukhoa Nuri Peace Park — Youth Training Center
- Asia Publication Culture and Information Center
- A-Municipal Library
- Seoul Performing Arts Center
- Seoul Performing Arts Center
- Sonyudo Park Landscape
- Seoul Performing Arts Center
- New Multi-functional Administrative City
- Anyang Peak
- Busan City Sofa
- Universal Creativity
- Becoming from Relationship, the Neo-Organism
- Asian Culture Complex
- Millennium Community Center
- Graduate School of Architecture Kyunghee University
- Aqua-Art Bridge
- Nungpyung-ri House
- Seoul Performing Arts Center
- Myung Film Company Building
- Ewha Campus Center
- Design Manual for Heyri Art Valley
- Rethinking Prototype
- Jeongok Prehistory Museum
- Seoul Performing Arts Center
- Imjingak Memorial
- Cheonggyecheon Culture Center
- Iran Oil Industry Headquarters
- The First Town
- Sewoon District #4 Urban Redevelopment Project
- Asian Culture Complex
- Gallery Yeh
- Program-Diagram : Process Concept
- M^3_Questions on the Space
- Handsome Hotel
- Ewha Campus Center
- ⊞ + Fish & Fish
- Jeongok Prehistory Museum
- Heryoojae Women's Hospital
- Hamburg Architectural Olympiad
- Systematized Prototype
- Protostructures
- PARAscape : City, Organization and Architecture
- Namjune Paik Museum
- Eyes of Pordenone
- Disturbing Territories
- Chichi Earthquake Memorial
- Catalog City
- Eagon Window Image Shop
- The Last House-House for the Dead
- Better, Cheaper, Faster
- Living Glass
- Residential Complex Plan in Catalunya
- Chichi Earthquake Memorial
- Digital Technology and Cyberspace
- Jeongok Prehistory Museum
- Maarun, South Bank Planning Development
- Jeongok Prehistory Museum
- New National Library Czech Republic
- Allo Series
- House of Open Books
- Borim Publishing House and Marionette Theater
- Obzee Fashion
- Why Not? A Question to the Law of Causality about Form
- New-Type-Body-Architecture 1.0/2.0
- Ssalon de Sson 1.0
- Museum of Art, Seoul National University
- Leeum Samsung Museum of Art
- S-Gallery
- Hilton Namhae Golf & Spa Resort
- Ewha Campus Center
- Asian Culture Complex
- New Headquarters of Lombardia District
- Speed Dome
- Architectural Production as a System
- Underground Campus and International Dormitory, Sogang University
- The First Town
- Tower Palace III
- Danang Administration Center
- Boheon Building
- Gangdong Culture Art Center
- Osong Bio-Health Science Technopolis
- Design Center of Gwangju
- LG Electronics Seocho R&D Campus

머리말

한국 현재건축의 지형도

한국 건축이 모더니즘을 시작한 지 100년, 광복 후 우리 스스로의 환경에서 새로운 꽃을 피운 지 60년이다. 그리고 우리 건축사회는 이미 제4세대를 넘어선다. 한국 현대건축은 불끈거리는 그 생명력에도 불구하고 가끔 주인의 행방을 찾을 수 없다. 아직 아무도 우리 상황의 지도를 확연히 그려보지 않았기 때문이다. 활달한 담론이 벌어지고 있으며, 유의할 만한 작품들이 생산되고 있지만 시대의 국면으로는 낱알의 사실이다. 세계 건축과의 교류가 어느 시기보다도 활발하지만 정작 무엇이 교차되고 있는지는 흐릿하다.

시야視野

혼돈처럼 보이는 한 시대의 단면이며 개인 차이가 확연하지만, 그 건축 태도 중에서 동조, 변이, 차이의 형색을 알아볼 수 있을 줄 믿는다. 보통 복잡한 대상을 한꺼번에 보고자 할 때 시선을 높일 필요가 있다. 바로 지도를 그리는 일이다. 이 지도를 그리는 시기는 현재이며, 한국과 그 근린을 범위로 모두 9개 가상의 그물망을 만들었다.

　　　현대적 패러다임에서 어떤 문화의 태도를 구분하고 유형을 가른다는 생각에 저항이 있을 수도 있겠다. 그 규칙이라는 것에는 이종변이가 그득한 현상에서 소수의 가치가 은닉될 여지가 없지는 않다. 그러나 드러난 사실만으로도 시대의 지시성을 알 수 있을 것으로 믿는다.

그것이 국계된 것이건 단순한 영향관계이건 동조의 요인이 연쇄 행태로 나타나고, 그것이 맥으로 읽힐 때 지도가 그려진다. 산봉우리의 연쇄로서 산맥이 만들어지는 것은 계곡과 강줄기처럼 상대적인 준위準位가 있기 때문이다. 다시 말해 양으로서의 가치와 음으로서의 가치가 함께있다. 산맥은 계곡을 만들며 또는 분지를 이루며 달린다. 물론 거기에는 수많은 우열의 차, 이단, 불평등이 있어 산수가 그리는 미적 쾌감을 돋운다. 한 시대사회가 만드는 이 맥의 계보는 넓을 수도 있고, 얕은 채 주변으로 잦아들고 말 수도 있다.

사회는 누적된 시간의 현상이기에 전후의 맥락이 있다. 관자의 위치에 따라 앞선 것이 흐릿할 수도 있고 뒤에 선 것이 모호할 수도 있다. 시대 상황을 앞에서 보느냐 뒤에서 보느냐의 차이다. 무엇보다 기대되는 것은 산맥의 형상에서 지배적인 세력이 떠오르냐 또는 다양성의 밋밋함-으로 나타날 것이냐다.

가급적 우리는 이 한국 현대건축의 지형도를 동시대성으로 자르려 했으나, 정확히 '현재'라는 경도에서 대상 작품의 범위를 자르지 못하였다. 불가피하게 CT촬영처럼 몇 개의 단층을 복층으로 자르게 된 이유다. 문화의 진척은 좀처럼 시간의 관성에서 벗어나기 힘들다. 그래서 조금 앞서가는 사람과 조금 뒤에 선 사람을 함께 보게 될 것이다.

건축 개념의 유형적 가설로서 9개의 탐침探針

한국 현대건축이라는 산덩이는 많은 분자와 여러 지층으로 구성된 복잡한 질료이다. 이를 분석해 정체를 밝히려는 입장에서는 몇 개의 유형적 가설이 필요했다. 그것은 보편성으로서 건축의 뜻이기도 하지만, 한국적 상황이 갖는 특수해가 포함된다. 여러 단계의 논의를 거쳐 우리는 이 가설을 9개의 탐침으로 정리하였다. 이 시대적인 관점은 9개의 그물로 비유할 수도 있다. 이 그물을 한국 현대건축의 현재성에 투망한다.

- 알레고리로서 건축
- 보편적 창의
- 관계로부터의 생성
- 프로토타입 다시 생각하기
- 프로그램 다이어그램
- 도시조직의 패러스케이프
- 디지털 테크놀로지와 사이버공간
- Why not? 형태의 인과율에 던지는 물음
- 시스템으로서의 건축 생산

물론 이러한 태도를 구분해보는 것은 유형화와 경계를 쌓는 일이 아니라 교환, 공유 그리고 교집합交集合의 힘을 그리기 위한 것이다. 이 모색은 시대사회를 보는 거시적 시선과 동시에 감춰져 있을지도 모르는 가치를 위해 미시적이어야 한다. 다만 우리의 시선을 들어울릴 필요가 있는 것은 여기를 향해 오고 있는 대상을 응시해야 하기 때문이다.

통섭通涉

한국 현대건축은 이제 환갑이다. 한국 건축의 60년이 이룬 종의 다양성은 이제 그 분파구조를 읽기 어려울 만큼 복잡해졌다. 한국은 20세기 중반에야 모더니즘을 받아들이며 혹독한 사회적 곤란을 경과했다. 1970년대까지 근대성에는 한국성이라는 명제가 상대하며 한국 건축의 대계大系를 이루었다. 1980년대 이후 세계의 창이 넓어지며, 포스트모더니즘을 우회하고 해체를 시선 너머에 둔다. 세대구조로 보아도 지난 60년 사이에 우리는 다섯 세대를 쌓았다.

　　　　문화에서 무엇이 중요한가는 관점에 따라 달라지지만, 지난 시기에 우리가 끌어안고 있던 것은 통합의 보자기였다. 1990년대에 이르러 차이의 가치를 긍정하게 된다. 차이는 분별을 만들고 다양성의 근거가 된다. 그 차이의 가치관이 밋밋하였던 한국 건축의 지평에서 봉우리를 만든다.

　　　　한국 건축의 종 다양성은 지난 반세기의 축적된 경험과 새로운 건축가들이 만든 유전자의 변이 결과이다. 근대성과 낭만성이 혼재되던 일제강점기의 제1세대, 국가의 재건과 근대문화를 엮던 제2세대, 1970년대를 이끌던 중견의 제3세대, 4.3그룹을 중심으로 하던 제4세대 그리고 현재이다. 현재의 단면으로 보아 여기에는 국제적인 수학 경험을 가지고 한국에 재귀한 이종異種의 경험이 상당수 포함된다. 특히 미국과 유럽에서의 학습경험 또는 서구의 건축 방법들이 이접移接되는 현상이다. 다른 종의 이입 경로는 한국이 설계시장을 개방하면서 외국 건축가들과의 문화 교접인데, 대부분 설계경기에서 우승하거나 컨소시엄을 통하거나 독자적으로 초대된 건축가들이다. 이들 중 상당 부분은 포장된 채 들어온 것이 포함되어 있다.

　　　　이러한 구도에서 이룬 한국 현대건축의 다양성은 얼핏 풍요의 시대로 보이나, 문제는 그 건강함이다. 이접이 야기할 아토피도 염려된다. 타자성과 문화충돌은 서구와 아시아적 가치 사이에서 흔히 벌어졌던 일이다. 문화의 교차가 이를 잡종강세雜種强勢의 뜻은 분명하지만, 지배와 피지배의 흥건한 기억은 문화의식을 네거티브하게 한다. 그러나 이러한 섞임의 불안정에도 불구하고 문화는 섞이면서 성장한다.

　　　　우리는 이 섞임의 문제를 '통섭'으로 이해하고, 이 시대에 작동할 운韻의 생동을 그리려 한다. 한국이라는 그릇에 담긴 건축적 방법과 사유가 알갱이의 상태가 아니라는 뜻이다. 요체들이 응고되지 않기 위해서는 기운이 필요하고, 또한 서로 건네고 통하며 이치를 이룬다는 뜻이다.

그릇은 테두리 외연과 내용을 담을 내역으로 구성되며 접시처럼 평퍼짐하거나 사발처럼 깊기도 하다. 한국이라는 그릇은 세계라는 큰 물 위에 떠 있다. 그 그릇의 외연은 자꾸 넓어지며, 나아가서는 경계를 흐린다. 그만큼 넘나듦이 쉬워지고 활발해질 것이다. 우선 이 활발한 기운의 줄기를 9가지로 나누어 상황을 헤아리지만, 상황이라는 것이 그러하듯 물결의 구분은 고착적인 것이 아니다. 서로가 결을 나누거나 공유하며 새로운 자극에 의해 거래를 계속할 것이다.

지난 100년 동안 한국 건축은 5세기 동안 벌어졌던 서양의 경험을 받아들였고, 지난 30년 동안 엄청난 종의 다양성을 일구었다. 이제 한국 현대건축은 세계성을 동시성으로 말하고 있다. 우리의 상황은 다양성에 충일하지만, 분명한 것은 다양함 자체가 중요한 것은 아니다. 이번 프로그램은 우리 당대의 건축가들이 고민하는 생각을 한꺼번에 볼 기회이다. 그리고 그것이 통섭되며 자기화에 이르는 것을 보고자 한다.

한국 건축의 장면들

흐리지만 넓은 것, 진하지만 좁은 것, 흐리면서 좁은 것, 진하면서도 넓은 것. 물론 이 유형 간에는 인력이 작용하는데, 공격적인 영향 성향과 수세적인 수용 태도이다.

이제 한국의 현재성은 포만감을 느낄 만큼 종의 다원성을 이루었다. 방법론 시장에서는 데이터, 다이어그램, 어반 호크, 언어학, 프로토타입과 어깨를 부대끼며 만난다. 그리고 이 다원성은 세계와 동시성에 있다. 이 종의 다원성과 동시성의 동태에는 원천적으로 컴퓨터의 지배력이 작용한다. 그중에서 디지털건축은 따로 장르를 형성하기도 하지만, 그 수단이 자꾸 새로운 유전자를 만들 동력을 제공한다.

세계와의 동시성은 다시 말해 국제주의이다. 1950년대 우리는 국제주의를 에스페란토로 말했지만, 이제 컴퓨터 알고리즘과 마야 언어로 말한다. 그로피우스의 국제주의가 격렬히 부정되던 것과 지금 겪고 있는 디지털 문화에서의 국제주의는 어떤 차이가 있는가. 한국적 아이텐티티를 위한 정서는 한국 현대건축에서 벌어지는 현상을 보기 위해 이번 기획에 57작가를 초대하였으며, 거기에서 어떤 기운의 운동韻動을 읽고자 한다. 이 운동의 해독으로서 매핑이 유효하리라 생각하며, 여기에 9개의 탐침을 꽂는다.

박길룡
국민대학교 건축대학 학장

PREFACE

Map of Consilience,
the Ranges of the Mountain of the Korean Architecture

It has been 100 years since Korean architecture started Modernism, and 60 years since it bloomed again in our own environment after the liberation from the Japanese occupation. The Korean architectural society has already entered into the 4th generation.

Although the Korean modern architecture is dynamic and vital, it sometimes is lost with no whereabouts of the owner. It is because nobody has drawn a clear map of our situation yet. Even though there are lively discourses and significant works that are being produced, they just remain to be individual facts that are scattered around. Despite the most active exchanges ever with the international architectural community, it is still cloudy and hard to capture what is really being achieved.

Field of Vision

Although it is just a slice of an era that looks like a chaos, its architectural attitudes reveal the different colors of alignment, variation and disparity.

Some people might want to refuse to be involved in this structure, but it is not easy to be completely free from the norm of a society. It is true that there may be resistance against the idea of classifying and dividing the modern paradigm. It might not be the case that the norm does not have room for concealment behind the phenomenon of the bocage of heterozygous.

Be it an implicit agreement or a simple cause and effect relation, a map is drawn when the alignment factors appear to have a chain reaction and it is read as an artery. It is similar to a mountain range that is created by the linkage of mountain peaks. It happens because of the relative level such as the valley and the course of the river. In other words, the Yang value and the Yin value coexist. The mountain range runs side by side or branching out with the valley. The numerous disparities, heresies and inequalities of course add to the beauty of the mountains and the water. The genealogy of this range that is created by one society of one era could be deep, or a shallow one that ends up submerging in the surrounding environment.

As a society is a phenomenon of accumulated time, it has a before-and-after context. Depending on where the viewer stands, what is in front can be unclear while what is behind can be vague. It is a difference that occurs whether the situation of the era is viewed from the front or back. What is expected the most is whether a dominant power will emerge from the mountain range or it will be just a boring but diverse situations.

Although we try to cut the map of the Korean modern architecture along the line of what is contemporary, we could not have a clear cut scope for the works on the measurement of the "present." This is because we tried to cut several slices into smaller ones like a CT scan. The evolution of the culture is hard to go beyond the inertia of time. So, the person who is a little up ahead and the person who lags a little behind will be viewed together. This is the premise of the map.

THE 9 PROBES AS THE HYPOTHESES OF THE CONCEPT OF ARCHITECTURE

The Korean modern architecture is a substance that is composed of many molecules and diverse layers. To analyze and identify it, several hypotheses were needed. They were universal architectural thoughts, but also covered the uniqueness of the situation of Korea.

Through a number of phases of discussions, we have narrowed the hypotheses down to directors. This point of view can be compared to nets that we throw to the contemporaneity of the Korean modern architecture.

- Architecture as an Allegory
- Generated by Relations
- Universal Consent
- Rethinking Prototype
- Program Diagram
- Parascape of Urban Structure
- Digital Technology and Cyber Space
- Why not? Form out of Causality
- Architectural Production as a System

Classifying the attitudes as above is certainly not to draw the line against typicalization, but to show the power of exchanges, sharing and intersection. This should be done in a macro gaze that looks at the society and the era, and at the same time in a micro way for the value that might be hidden. It's just that we need to lift our eyes up because we need to stare at the object that is coming toward here.

Consilience

The Korean architecture has just celebrated its 60th birthday. The diversity that has been achieved for the past 60 years is one that is even hard to read the structure of. It was not until the mid-20th century that Modernism was adopted and had to go through severe social tribulation. Until the 1970s, modernity was confronted by the proposition of being Korean, which were the two great pillars of the Korean architecture. Since the 1980s, as the window to the outside world became larger, Postmodernism has been circumvented and Deconstructivism has been put right over the gaze. Even from the perspective of the structure of the generations, we have gone through 5 of them for the past 60 years.

What is important in culture varies according to the viewpoint, but what we embraced in the past was a bundle of integration. It was in the 1990s when we came to have a positive attitude toward the value of difference. Difference creates senses and becomes the ground of the diversity of species. The value of the difference creates the peaks on the horizon of the Korean architecture. The peaks will draw a new topography along with the valleys and the ranges.

The diversity of species in the Korean architecture is the result of the accumulated experiences for the past half-century and the mutation of the genes conducted by new architects. The first generation was the period of the Japanese occupation, when modernity was mixed with Romanticism. It was followed by the second generation that went through the reconstruction of the country and the modern culture, and the third generation that pushed the country through the 1970s. More recently, we had the fourth generation that was centered around the Group 4.3, and we have come to the present situation. The present slice tells us that it includes the exotic species who have received international education and come back to Korea. The learning experience and architectural methods especially from the United States and Europe are being mixed with each other in Korea. The road that those different species take to penetrate Korea is the country's opening up of the architecture market, which results in the cultural exchanges with foreign architects. This is mostly done by winning international design contests, through consortia, or by independently invited architects. In other words, they are the species that have entered this country already packaged.

The diversity of the Korean modern architecture that has been achieved in this structure might seem rich at a glance, but how healthy the diversity is remains to be a problem. Concerns over atopy are being raised. Cultural clashes have commonly been witnessed between the Western and the Asian values. Cultural exchanges strengthen it, but the vivid memory of dominating and being dominated makes the cultural awareness a negative thing. Despite the instability due to such mixture, cultures grow through the mixtures.

We intend to understand this issue of mixture as consilience, and to paint the picture of vividness of the rhyme that will work in this era. The purport of this is to let the Korean way of architecture and thinking not remain in the status of granules. For cardinal points not to solidify, they need energy, and to interact and communicate with each other to make sense.

A container should have the external appearance and the internal content. It can be in the form of a f at plate or a deep bowl. The container in the name of Korea is floating in the great sea of the world. The external appearance of the container is increasing in size, and goes on to blur the boundary. It means crossing the boundary will become easier and more active. Such active energy has been classified into 9 so that we can understand the situation. Just like the situation, the classification of the flow will not stay the same. The different classes will share their contacts and be stimulated to continue to do so.

The situation we are in is overflowing with diversity, but it is clear that the diversity itself is not important. This program presents itself as an opportunity to look at what our contemporary architects worry about. We will also take this opportunity to view the consilience of all of them.

Scenes in the Korean Architecture

Faint yet wide; distinct yet narrow; faint and narrow; dark and wide… Needless to say, gravity is in force among these types and there exist attitudes of acceptance which are aggressive or defensive.

Today, the modernity of Korea has achieved the Pluralism of species to the point of saturation. In the markets of methodology, one can witness data, diagrams, Urban Hawks, linguistics, and methodology as a prototype. And this diversity has simultaneity with the world. In the dynamics of Pluralism and simultaneity of the species, in principle, the dominance of the computer is in force. Among them, digital architecture constitutes a separate genre, yet its means continuously providing the driving force for making new genres.

The simultaneity with the world is, in other words, Internationalism. In the 1950s we dabbled with Internationalism through Esperanto, yet today we pass through computer algorithms and Maya language. What difference would then exist between the time when Internationalism advocated by Walter Gropius was strongly denied and the Internationalism experienced am d the digital culture of today? What are the sentiments for Korean identity?

This project has invited 57 architects to investigate phenomena taking place in contemporary Korean architecture and intends to interpret some movements of certain kinds of energy out cf them. It is deemed that the mapping will be effective in deciphering the movements, so we establish the nine probes here.

Kilyong Park
Dean, School of Architecture, Kookmin University

CONTENTS

	한국 현재건축의 지형도_박길룡	004
	키워드 맵 작도법	016
	지도를 읽는 일의 효용에 관하여_봉일범	018
조성룡	해인사 신행 문화 도량	032
승효상	대전대학교 혜화문화관	036
박길룡	알레고리로서 건축	042
이타미 준	핀크스미술관	052
아눅 리정드르 + 니콜라 데마지에르	전곡선사박물관	058
	체코국립도서관	062
민현식	평화누리 + 청소년수련원	064
김병윤	아시아출판문화정보센터	070
	A-시립도서관	074
엘레펀트 + 간삼파트너스	서울공연예술센터	076
장 누벨	서울공연예술센터	080
정영선	선유도공원 조경	084
안드레 페레아 오르테가	서울공연예술센터	090
	행정중심복합도시	092
MVRDV + 정영욱	안양 피크	096
	부산국제영상센터	098
이공희	보편적 창의	100
이공희	관계로부터 생성, 그 신유기	108
유걸	아시아문화전당	114
	밀레니엄 커뮤니티센터	116
	경희대학교 건축전문대학원	118
다비드 피에르 잘리콩	아쿠아 아트 브리지	120
	능평리 주택	122
윤웅원 + 김정주	서울공연예술센터	124
	명필름 사옥	126
도미니크 페로	이화여대 캠퍼스센터	128
김종규	헤이리아트밸리 건축설계 지침	134
봉일범	프로토타입 다시 생각하기	140
마츠오카 사토시 + 타무라 유키	전곡선사박물관	144
	서울공연예술센터	146
다니엘 바예 + 기오헌	임진각 기념관	148
정림건축	청계천문화센터	152
희림건축	이란석유성	156
온고당 + 도그마	행정중심복합도시, 첫마을	160
피터 아이젠만 + 해안건축	세운상가4구역 도시환경정비사업	164
장윤규	아시아문화전당	168
	예화랑	172
장윤규	프로그램 다이어그램 - 프로세스 컨셉	174
최문규	M³ 공간에의 질문	180
조민석	한섬호텔	186
FOA	이화여대 캠퍼스센터	192
장림종	田 + 피쉬 & 피쉬	196
	전곡선사박물관	198
김영준	허유재병원	200
	함부르크 건축 올림피아드	204

하태석	시스템적 프로토타입	206
시로 나흘레	프로토스트럭처	212
권영	패러-스케이프 : 도시, 조직 그리고 건축	218
김주령	백남준미술관	224
	포르데노네의 눈	226
숀 머레이	영역 교란	228
신혜원 + 유양 리우	치치 대지진 추모공원	236
신혜원	카탈로그 시티	240
김찬중	이건창호 쇼룸	242
	최후의 집, 납골당	244
양수인	더 나은, 더 싸게, 더 빠르게	246
	리빙 글래스	248
김기홍	카탈루냐 주거단지 계획	250
	치치 대지진 추모공원	252
이경훈	디지털 테크놀로지와 사이버공간	254
폴 프라이스너	전곡선사박물관	262
	사우스뱅크 개발 계획	264
론 콤 + 로나 이스턴	전곡선사박물관	266
	체코 신국립도서관	268
마르코스 노박	Allo 시리즈	270
김준성 + 서혜림	열린책들 사옥	274
서혜림 + 김준성	보림출판사	278
서혜림 + 나데어 테라니	오브제 사옥	280
봉일범	Why Not? 형태의 인과율에 던지는 물음	284
문훈	신몸건축 1.0/2.0	294
	싸롱 드 쏜 1.0	298
OMA + 삼우설계	서울대학교 미술관	300
	삼성미술관 리움	304
켄 민 성진	S 갤러리	308
	힐튼 남해 골프 & 스파 리조트	310
자하 하디드	이화여대 캠퍼스센터	314
알베르토 프란치니 + 안드레아 보체티	아시아문화전당	318
	신 롬바르디아 정부 청사	320
공간그룹 + 범건축	스피드 돔	322
최왕돈	시스템으로서의 건축 생산	326
진아건축	서강대학교 지하캠퍼스 및 국제학사	336
건원건축 + DA그룹	행정중심복합도시, 첫마을	340
삼우설계 + OMA	타워팰리스 III	344
무영건축 + 데스테파노	베트남 다낭시청사	348
간삼파트너스	보헌빌딩	352
원도시건축	강동문화예술센터	356
원양건축 + 해안건축	오송생명과학단지	360
토문건축 + 운생동	광주디자인센터	364
창조건축 + 힐러건축	LG전자 서초 R&D 캠퍼스	368
	전시	372
	심포지엄 1_한국 현대건축의 프로세스	384
	심포지엄 2_신종을 위한 패러다임	396
	프로필	408

CONTENTS

	Map of Consilience, the Ranges of the Mountain of the Korean Architecture_Kilyong Park	004
	Cartography of Keyword Map	016
	About the Operation of Map-Reading_Ilburm Bong	018
Sungyong Joh	Haeinsa Temple Culture Complex	032
H-Sang Seung	Hyehwa Cultural Center for Daejeon University	036
Kilyong Park	Architecture as an Allegory	042
Itami Jun	Pinx Museum	052
Anouk Legendre + Nicolas Desmazieres	Jeongok Prehistory Museum	058
	National Library of Czech Republic	062
Hyunsik Min	PyoungHoa Nuri Peace Park + Youth Training Center	064
Byungyoon Kim	Asia Publication Culture and Information Center	070
	A-Municipal Library	074
Elephant + Gansam Partners	Seoul Performing Arts Center	076
Jean Nouvel	Seoul Performing Arts Center	080
Youngsun Chung	Seonyudo Park Landscape	084
Andrés Perea Ortega	Seoul Performing Arts Center	090
	New Multi-functional Administrative City	092
MVRDV + Youngwook Joung	Anyang Peak	096
	Busan City Sofa	098
Gonghee Lee	Universal Creativity	100
Gonghee Lee	Becoming from Relationship, the Neo-Organism	108
Kerl Yoo	Asian Culture Complex	114
	Millennium Community Center	116
	Graduate School of Architecture Kyunghee University	118
David-Pierre Jalicon	Aqua-Art Bridge	120
	Nungpyung-ri House	122
Woongwon Yoon + Jeongjoo Kim	Seoul Performing Arts Center	124
	Myung Film Company Building	126
Dominique Perrault	Ewha Campus Center	128
Jongkyu Kim	Design Manual for Heyri Art Valley	134
Ilburm Bong	Rethinking Prototype	140
Satoshi Matsuoka + Yuki Tamura	Jeongok Prehistory Museum	144
	Seoul Performing Arts Center	146
Daniel Valle + KIOHUN	Imjingak Memorial	148
Junglim Architecture	Cheonggyecheon Culture Center	152
Heerim Architects	Iran Oil Industry Headquarters	156
HNA Ongodang + Dogma	The First Town	160
Peter Eisenman + Haeahn Architecture	Sewoon District #4 Urban Redevelopment Project	164
Yoongyoo Jang	Asian Culture Complex	168
	Gallery Yeh	172
Yoongyoo Jang	Program-Diagram : Process Concept	174
Moongyu Choi	M³_Questions on the Space	180
Minsuk Cho	Handsome Hotel	186
FOA	Ewha Campus Center	192
Leemjong Jang	田 + Fish & Fish	196
	Jeongok Prehistory Museum	198
Youngjoon Kim	Heryoojae Women's Hospital	200
	Hamburg Architectural Olympiad	204

Tesoc Hah	Systematized Prototype	206
Ciro Najle	Protostructures	212
Young Kweon	PARAscape : City, Organization and Architecture	218
Jooryung Kim	Namjune Paik Museum	224
	Eyes of Pordenone	226
Shaun Murray	Disturbing Territories	228
Haewon Shin + Liu Yuyang	Chichi Earthquake Memorial	236
Haewon Shin	Catalog City	240
Changjoong Kim	Eagon Window Image Shop	242
	The Last House-House for the Dead	244
Sooin Yang	Better, Cheaper, Faster	246
	Living Glass	248
Kihong Kim	Residential Complex Plan in Catalunya	250
	Chichi Earthquake Memorial	252
Kyunghoon Lee	Digital Technology and Cyberspace	254
Paul Preissner	Jeongok Prehistory Museum	262
	Maarun, South Bank Planning Development	264
Lonn Combs + Rona Easton	Jeongok Prehistory Museum	266
	New National Library Czech Republic	268
Marcos Novak	Allo Series	270
Junsung Kim + Hailim Suh	House of Open Books	274
Hailim Suh + Junsung Kim	Borim Publishing House and Marionette Theater	278
Hailim Suh + Nader Tehrani	Obzee Fashion	280
Ilburm Bong	Why Not? a Question to the Law of Causality about Form	284
Hoon Moon	New-Type-Body-Architecture 1.0/2.0	294
	Ssalon de Sson 1.0	298
OMA + SAMOO Architects	Museum of Art, Seoul National University	300
	Leeum Samsung Museum of Art	304
Ken Min Sungjin	S-Gallery	308
	Hilton Namhae Golf & Spa Resort	310
Zaha Hadid	Ewha Campus Center	314
Alberto Francini + Andrea Boschetti	Asian Culture Complex	318
	New Headquarters of Lombardia District	320
SPACE Group + BEOM Architects	Speed Dome	322
Wangdon Choi	Architectural Production as a System	326
JINA Architects	Underground Campus and International Dormitory, Sogang University	336
KUNWON Architects + DA Group	The First Town	340
SAMOO Architects + SOM	Tower Palace III	344
MOOYOUNG Architects + DESTEFANO	Danang Administration Center	348
GANSAM Partners	Boheon Building	352
WONDOSHI Architects	Gangdong Culture Art Center	356
WONYANG Architectural Design + HAEAHN	Osong Bio-Health Science Technopolis	360
TOMOON Engineering + UNSANGDONG	Design Center of Gwangju	364
CHANJO Architects + HILLER Architecture	LG Electronics Seocho R&D Campus	368
	Exhibition	372
	Symposium #1 Design Processes of Korean Contemporary Architecture	384
	Symposium #2 Paradigms for New Species	396
	Profile	408

키워드 맵 작도법

CARTOGRAPHY OF KEYWORD MAP

키워드 맵은 두 개의 레이어를 갖는다.
바탕이 되는 베이스 맵은 국민대학교 건축대학 교수진이 집필한 아홉 개의 에세이로부터 추출된 이론적인 키워드들로 이루어져 있으며 그 위에 덧그려지는 쓰레드 맵은 수록 작가들의 글에서 추출된 키워드들로 이루어져 있다.

베이스 맵의 키워드들은 아홉 개의 '탐침'들로 범주화되고, 수평적으로는 작동의 전략, 원칙, 방법론으로부터 그것이 만들어내는 효과 또는 결과적인 현상으로(왼쪽에서 오른쪽으로), 수직적으로는 주관적이며 개인적이고 시적인, 정서적인 것으로부터 객관적이고 집단적이며 엄밀하고 체계적인 것으로(위에서 아래로) 각각 개념적인 정성적 스펙트럼에 따라 배열되어 있다.

작가들로부터 연유한 키워드들은 베이스 맵의 층위에 놓인 주제어들과의 개념적 유사성에 따라 분포되며, 각각 아홉 개의 탐침들과 동일하게 표지된 하나의 색선으로 하나의 작품 또는 프로젝트로부터 파생된 키워드들이 연결됨으로써 쓰레드 맵을 형성한다.

그 결과 키워드 맵은 한국 건축의 현황을 드러내 보여주는 일종의 엑스레이 투사 사진, 또는 언제나 끊임없이 변화하는 한국 건축 전체의 사유 양상을 어느 한순간에 고정시켜 놓은 것과도 같은 지형적인 윤곽을 보여준다.

The KEYWORD MAP consists of two layers: the BASE MAP of theoretical main keywords abstracted from the essays written by the faculty members of School of Architecture, Kookmin University (SAKU) and the THREAD MAP of keywords abstracted from the writings by the contributors.

Main keywords in BASE MAP are categorized into the group of 9 Probes and aligned basically in accordance with the qualitative spectrum from left [operating strategy, principles, methodologies] to right [its effects or resultant phenomena, etc]; from top [subjective, individual, poetic, emotional] to bottom [objective, collective, rigorous, systematic].
Keywords in THREAD MAP are distributed according to conceptual affinity with the main keywords of BASE MAP layer; then a batch of keywords from a project or a work is threaded by a zig-zagged line which is color-coded the same as 9 Probes.

As a result, the KEYWORD MAP shows a sort of x-ray projection of the current status of Korean Architecture, or a captured topographical profile of ever-changing the whole of architectural thought, in Korea.

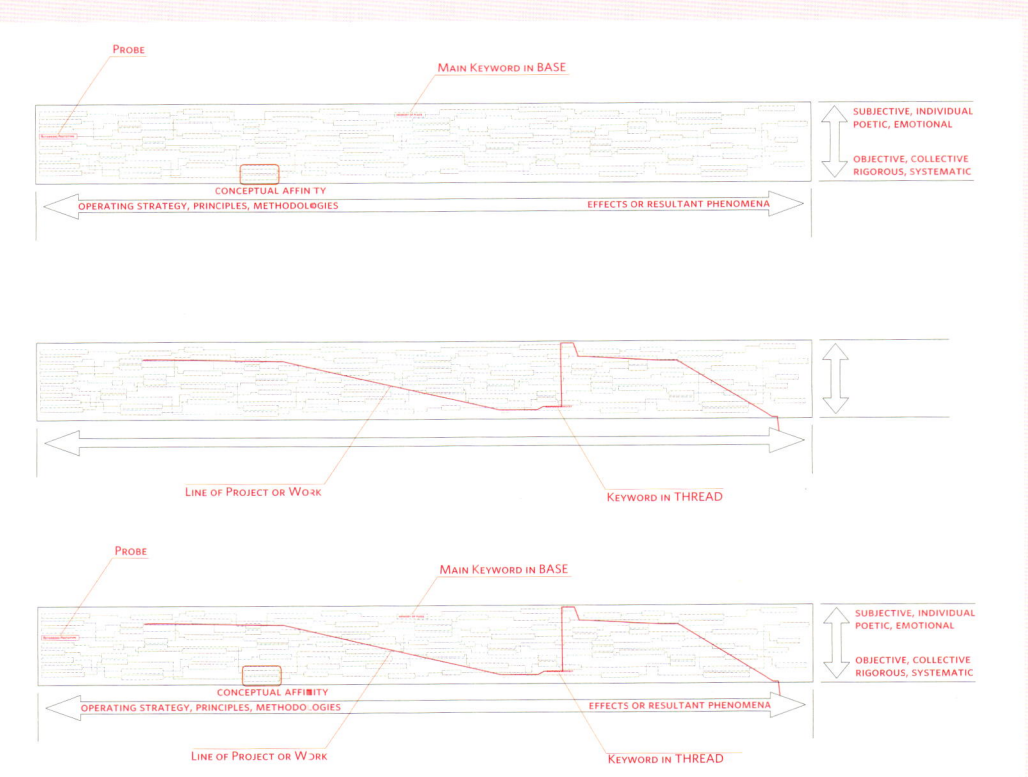

지도를 읽는 일의 효용에 관하여

봉일범
국민대학교 건축대학

이 책에 실려 있는 키워드 맵의 의의는 그것을 작성한 과정이 '컨텍스트를 벗어난 텍스트' 사이의 관계를 새로 맺어주는 추상적인 구축작업이었다는 데서 출발한다. 달리 말하면 건축가들의 글로부터 추출됨으로써 본래 그것이 속해 있던 문맥을 상실한 낱알의 키워드들을 재료 삼아 하나의 관계망을 이루는 지도를 그려낸 작업의 과정이 일반적인 분석이나 해석, 의미화, 이론화의 과정은 아니었다는 것이다. 이 말은 곧 키워드 맵의 작성이라는 결코 간단치도 수월치도 않았던 작업이 이끌어낸 '결론'이 과연 무엇이냐의 당연한 물음에 대해 결코 단정적이거나 권위적인 즉답을 내놓지는 않겠다는 뜻이지만, 그렇다고 해서 우리가 도달한 궁극의 지점이 무기력한 허무주의나 교묘한 지적 회피가 아님은 물론이다.

지도를 읽는 수많은 방법들, 그렇게 읽어낸 수없이 다양한 해석과 잠정적인 결론들을 일일이 열거하는 것은 지도를 작성한 주체의 소관이 아니라는 생각이다. 그보다는 오히려 완성된 지도를 읽어내는 것과는 전혀 다른 차원에서 감지할 수 있었던 한국 건축의 '의미 있는' 지향성을 이야기하는 것이 먼저일 듯하다.

I
단정적인 결론을 내는 일에 대한 의도적인 주저와 판단의 유보가 가져올 가능성들에 대한 믿음에도 불구하고, 그러나 지도 작업의 재료가 되었던 표집된 키워드 전체의 집합에는 일정 정도의 유의미한 경향성이 엄존하는 것 또한 사실이다. 그것은 지도 작성에 앞서 이미 작가와 작품을 불문하고 수다히 등장하는 특정한 키워드들의 빈도수라는 지극히 정량적인 기준만으로도 명료하게 드러나는데, 모두 10회 이상 등장하는 키워드들을 단순 빈도수로 나열해보면 다음과 같다.

URBAN 28 NATURE 24 SPACE 23 LANDSCAPE 22 CONNECT(ION/IVITY) 20
PROGRAM(MATIC) 13 SYSTEM 13 NETWORK 12 STRUCTURE 12 CITY 12 SKIN 11
ECO- 10 TOPOGRAPHY 10

여기서 우리는 한국 건축의 현재와 어떤 식으로든 관련을 맺고 있는 건축가들이 건축의 영역을 도시와 자연, 경관으로 확대하고 있음을 (또는 확대하기를 갈망하고 있음을) 알 수 있으며 자족적인 단일체로서의 건물을 넘어 무엇인가와 연결하고 관계 맺는 일을 진지하게 사고하고 있음을 엿보게 된다.

그런데, 전체의 키워드를 망라하고 보면 단순한 빈도수를 넘어 고려해야 할 또 다른 요인들을 무시할 수 없을 듯한데, 일례로 space라는 키워드의 경우 그 자체를 요체로 하는 건축의 특수성에 비추어 과연 여타의 키워드들과 같은 정도로 특정 의미를 지시하는 말인가를 생각해보면 오히려 논외로 하는 것이 마땅할 것이고 skin과 surface, urban과 city 등의 유의어들이 만들어내는 집합적인 의미에 대해서도 다시 생각해야 마땅하리라는 것이다. 그런 점에서 다시 정리해보면, 다소의 견해차는 있을 수 있다 해도 아마도 다음에 나열하는 네 개의 주제, 여덟 개의 개념들이 한국 건축의 현재가 겨냥하고 있는 사유의 큰 구조를 만들고 있다고 보아도 무방할 듯하다.

도시 스케일의 연결과 확장, 관계성
(1) **도시와 도시를 이루는 기반시설로 확장되는 건축**이라는 개념. 도시는 건축을 사고하기 위한 참조항으르서 이해되기도 하고 도시=건축이라는 동일시의 대상으로서 등장하기도 한다. 이는 곧 직능의 한계를 넘어서려는 무언의 집단적 합의에 기반하는 시도들일 수도 있으며 건축 프로젝트의 규모가 확대되고 그 배경이 되는 대지가 이미 인공의 도시인 경우가 대부분이라는 현실적인 문제들이 배후에 자리하고 있는 것으로도 생각된다.

URBAN 28 CITY 12 INFRASTRUCTURE 7

그리고 도시적 확장이라는 이 첫 번째 개념은 다시 (2) **연결과 관계, 집합적 총체로서의 건축**이라는 두 번째 개념과도 일부 상통하는 의미들을 내포하는 것으로 보인다. 연결과 공동체, 소통, 집합체, 맥락이라는 키워드들 모두가 공히 담고 있는 '함께(con-/com-)'라는 의미는 곧 하나의 건물이 이루는 폐쇄적인 세계의 확장과 외파外破를 의미하기 때문이다.

CONNECT(ION/IVITY) 20 COMMUNITY 8 COMMUNICATION 6
RELATION 7 COLLECTIVITY 5 CONTEXT 8

연속적인 포괄로서의 자연과 경관

한편 단일체로서의 건축 또는 단절된 공간단위들의 집합이라는 고정된 건축관을 벗어나 **(3) 건축보다 넓고 연속적이며 포괄적이라는 의미에서의 지형과 경관 언어의 차용 또는 그 은유**라는 개념 또한 주요한 경향을 이루고 있는 것으로 보인다.

LANDSCAPE 22 (XXX+SCAPE 3)   TOPOGRAPHY 10

이것은 좀 더 느슨한 분류기준을 적용해보자면 다음과 같은 두 개의 개념들과도 유사성을 갖는다고 생각되는데, 그 하나가 **(4) 자연과 생태, 건축의 인공성과 대비되는 또는 그것을 보완하고 교정하기 위한 장치로서의 자연관**인 듯하다. 이 네 번째의 개념은 지난 수년간에 걸쳐 점차 논의가 확대되고 심화되어가는 환경친화적 건축 또는 지속가능한 개발이라는 이상과도 불가분이다.

NATURE 24   ECO- 10   ENVIRONMENT 8   GREEN 3   SUSTAINABLE 2

그리고 또 다른 하나는 바로 **(5) 고정된 물리적 단위로서의 건축을 넘어서는 역동적이고 유동적인 개념**으로 보인다. 살아 움직이는 자연 상태의 생명체 또는 집합적인 사회적 양태 등과 마찬가지로 건축을 고립되고 동결된 하나의 덩어리로 보는 것이 아니라 지속적으로 변화하며 움직이는, 나아가 외부의 환경과 유기적인 입출入出의 대사작용을 하는 의사擬似 생명체로 바라보는 시각 등이 그것이다.

MOVEMENT 9   FLOW 6   TIME 5

관습적인 건축관에 대한 재사고

한편, 가장 두드러지게 건축의 관습적인 결정자들에 대해 의문을 제기하고 재정의하려는 시도들은 주로 이종異種간의 결합이나 다중화를 의도하는 **(6) 프로그램에 관한 독자적인 재해석 또는 재조합**으로 나타난다. 건축의 물리화 이전에 놓여 있는 '미리 지시되고 작성된 것'으로서의 프로그램을 수동적으로 수용하던 단계를 벗어나려는 근대 이후의 시도는 이미 그 연원이 30여 년을 훌쩍 넘어서고 있다지만 그럼에도 여전히 새로운 시도로서의 유효성을 지니고 있는 것으로 보이며, 고전적인 분류 체계로 본 건축의 요소들 중 **(7) 구조와 표면의 부상** 또한 간과할 수 없는 신개념의 한 축으로 보인다. 구조와 표면의 분리라는 근대건축의 도그마를 뒤집어 다시금 고딕으로 회귀하려는 시도처럼 보이기까지 하는 구조화된 표면이라는 개념은 특히 주목할 만하다.

PROGRAM(MATIC) 13   STRUCTURE 12   SKIN 11   SURFACE 4

컴퓨터 환경에서 통합적으로 사고되는 프로세스-디자인

(8) 건축의 디자인 프로세스와 그 결과물을 동일한 시스템 환경의 일부로 파악하는 개념. 이것은 어쩌면 가장 새로운, 가장 최근에 등장한 건축 분야에서의 혁신적인 사고방식일 터인데, 이것은 건축가의 자의적, 직관적, 경험적 판단을 넘어 객관화된 자료의 처리 방식을 통해 디자인 프로세스를 의식적으로 통제하는 한편 다부문간 인자들의 네트워킹을 통해 프로세스의 자동성과 확장성을 보장함으로써 리서치-디자인-생산이 통합되는 건축 환경을 목표로 하고 있는 개념이라 정의할 수 있을 것이다.

SYSTEM 13   NETWORK 12   TECHNOLOGY 6

위의 네 가지 주제, 여덟 가지 개념들로 정리된 내용들 중에서도 특히 건축 내부의 논리에 대한 재사고 외에 가장 두드러지는 것은 역시 고립된 단위체로서의 건축이라는 관념을 벗어나 도시, 자연 등의 외적 요소들 사이에서 모색하는 관계 맺기와 연결, 흐름, 확장, 이종결합과 같은 개념들일 것이다. 그리고 이와 같은 큰 흐름이 바로 다음의 두 접두어가 개념어 속에 포함되는 빈도수를 비약적으로 증가시킨 것으로 보인다.

INTER- 18   MULTI- 15

이는 '건축이 곧 사고의 방식' 이라는 극단적인 선언이 결코 낯설어 보이지 않는 지경에 이를 정도로 건축에 관한 사유가 점점 더 건축의 외부를 향해 관념적으로나 물질적으로나 확대일로에 놓여 있는 최신의 경향과도 일맥상통한다. 그리고 그 대극으로서, 일례로 composition과 같은 지극히 고전적인 주제어가 단 2회 등장하는 것으로 그치고 말았다는 사실만 보아도 건축 개념의 진화라는 대강의 줄기는 명확히 짚어볼 수 있을 것으로 보인다.

한편, 이와는 별개로 이 책 자체가 '한국 건축' 을 그 소재로 삼고 있다는 점에서 보면 건축가들의 글 속에 한국 건축이라는 특수성을 반영하고 있는 키워드들이 매우 드물다는 사실이 눈에 띈다.

TRADITION 6   HISTORY 6   KOREA(NESS) 5

역사적인 맥락, 지역적인 특수성을 대신하여 전지구적 보편성과 현재성이 건축가들에게는 훨씬 더 민감하게 받아들여지고 있다는 뜻으로 이해되는데, 이것을 적잖은 한국 건축가들이 해외에서의 수학과 실무를 경험했고 대다수 대규모 프로젝트들이 외국 건축가들과의 합작으로 이루어지고 있다는 맥락에서 세계화의 당연하고도 긍정적인 결과로 볼 것인지, 아니면 가깝게는 일본 건축과 비교해볼 때 한국 건축이 여전히 세계 속에서 하나의 독립적인 분파로서 공정하게 자리매김되지 못하고 있는 결정적인 요인인지... 심각하게 고민해야 할 문제라 생각된다.

　　　　이상에서 개괄한 바와 같이 총체적이고도 집합적인 확연한 경향성을 짚어보는 외에 지엽적인 독해와 해석, 이론화의 결과들에 권위를 부여하기는 쉽지 않을 뿐더러 반드시 그러해야 할 일로도 보이지 않는다. 각 작품을 나타내는 선들을 따라 키워드들 사이의 인접성에 주목해 작품들의 개념적 경로를 추적해보든, 지도 전체에서 읽히는 키워드 분포의 소밀疏密관계에 주목하든... 키워드 맵을 읽는 방법은 무수히 많을 수 있겠으나 그 각각이 확정적으로 진위를 논할 대상은 아닐 뿐더러 '지도를 읽는 일' 자체가 이와는 전혀 별개의 과정이 될 것이고, 또 되어야 하기 때문이다.

II

실상 지도라는 것에는 세 가지의 인자들이 결부되어 있다고 생각되는데, 그 첫째가 지도를 있게 한 구체적이며 사실적인 대상일 것이고 둘째는 그것을 한눈에 볼 수 있도록 축약하거나 기호로 바꾸어 표기하는 작도의 방식 또는 그 방식에 따라 지도를 그리는 행위일 것이며 세 번째는 바로 그렇게 작성된 지도를 읽어 길을 찾는다는 참조 또는 수행의 과정일 것이다. 첫 번째의 구체적이고도 사실적인 대상이 이 경우에 한국 현대건축을 있게 한 사고의 총화 또는 좀 더 정확히 말하자면 그것을 드러내 보이려는 의도로 수집한 글들이고 두 번째의 작도법이 앞서 '키워드 맵 작도법'에서 설명한 바와 같다면, 이 글이 초점을 두고 있는 부분은 바로 세 번째의 '지도를 읽어 길을 찾는' 후행적인 수행이다. 그러므로 이 복잡한 '지도'로부터 무엇을 얻을 것인가의 문제는 (일반적인 지도 활용의 경우들과 마찬가지로) 이 지도가 의미하는 바가 무엇인가의 문제가 아니라 그것을 가지고 과연 어디에 도달할 수 있을 것인가의 현실적이고도 유용한 문제가 되어야 한다.

　　　　그런데, 모두冒頭에서 밝힌 바와 같이 지도를 작성하는 행위가 추상적인 구축 작업일 뿐 아니라 지도를 둘러싸고 일어나는 여타의 행위들 역시 대부분 구축의 행위라는 사실에 주목할 필요가 있어 보인다. 지도를 작성하는 작업은 본질적으로 환원의 작업이지만, 존재하지 않는 종류의 지도를 만들기 위해 그 환원의 원칙을 '정해야 하는' 입장에 놓인다면 이 또한 명확한 원칙에 따라 요소들을 관계 맺어야 한다는 점에서 재현의 한계를 넘어서는 구축의 작업이다. 지도를 읽어내는 일 역시 수다한 요소와 관계들 속에 주관적 판단을 투영해 특수한 종류의 목적을 성취해낸다는 점에서 엄연한 구축의 행위이자 생산의 행위이다. 지도를 참조하는 일의 의의는 결국 그 실질적 효용에 있는 것이지 그것이 내포한 의미의 해석에 있는 것은 아니라는 뜻이다.

지도라는 이름이 붙은 것들은 대개 특수한 목적에 부합하는 특수한 효용을 갖기 마련이다. 일례로 배를 운항하면서 참조하는 항로도는 그것을 읽는 방법을 알든 모르든 일반인들에게는 아무 '의미'가 없는 물건이지만 그것을 필요로 하는 특수한 부류의 사람들에게는 불가결의 도구로서 지대한 '효용'을 갖지 않는가. 그런 점에서 한국 건축의 현재를 있게 한 근원으로서의 사고들이 이루고 있는 집단적 총화를 대상으로 하는 키워드 맵은 그 효용 역시 어떤 특수한 종류의 건축적 사고를 구축해가기 위한, 또는 제각각 관심있는 목적점을 향해 길을 찾아가기 위한 용도로 사용될 때 그 효용가치를 득하게 되는 것이지 결코 한국 건축을 재단하기 위해, 대단히 환원적이고 추상적인 말로 성급히 그것을 사조화하기 위해 쓰이는 것은 아니다.

지도의 작성이나 그 독해, 참조는 고고학도 역사학도 아니다. 진위 여부를 떠나 지도 읽기는 해석의 문제가 아니라는 것이다. 더욱이 계속해서 변화하고 있는 한국 건축 내부의 사고들을 대상으로 하는 이 지도는 현재진행형의 유동적인 실체를 따라가는 동시적인 트레이싱의 과정이자 그 트레이싱을 통한 예측 불허의 프로젝션을 의도하고 있는 것이다. 그러므로 건축가들로부터 연유한 키워드들과는 별도로 선정된 아홉 개의 이론적인 '탐침'들이 키워드 빈도수에 드러나는 여덟 개의 개념들과 어떻게 결부되는가의 문제에 대해서도, 유사성에 주목하든 불일치에 주목하든 해석에 무게를 둘 일이 아니라 이 두 층위를 하나의 지도 위에 겹쳐 그림으로써 얻게 되는 미지의 충돌이 드러낼 새로운 개념의 단서들을 포착하는 일에 집중해야 할 것이다. 지도를 작성한 우리의 시선은 과거지향이 아니라 현재와 미래에 그 초점을 두고 있기 때문이다. 어떻든 주어진 대상에 선고를 내리고 싶은 것이 아니라 예상과 추론에 속하지 않는 새로운 촉발을 기대하는 것이 이 생경스런 '지도'를 용감히 드러내놓는 취지이다.

이와 같은 의도를 갖는 '건축적 사고의 지도 그리기'는 아마도 처음이 아닌가 싶은데, 그 작도법과 관련하여 제기될 수 있는 몇 가지 치명적일지도 모를 문제들에도 불구하고, 우리의 결론은 이 지도가 결코 종착점이나 수렴점이 아니라 시발점이자 확산의 기점이 되리라고 기대한다. 이와 유사한 '건축사유의 지도'들이 다양한 형식을 빌어 양산되는 경우를 상상해보면, 어느 시점엔가 우리는 새로운 개념을 구축하기 위해 다양한 '건축 개념도'들을 펼쳐보며 그 누구도 가지 않았던 새 길을 궁리하는 모험적인 건축가들의 모습을 당연시하게 될지도 모를 일이다.

About the Operation of Map-Reading

Ilburm Bong
School of Architecture, Kookmin University

The implication of the cartography of Keyword Map is first of all that making the map is a construction process to define the new relationship among "texts without context". In other words the process of map-drawing, in which key-words as raw material were firstly extracted from the architects' writings and lost their original context, is not that of analysis, interpretation, or theorization in general. So we are not to answer directly to the question of what the conclusion is. In spite of this hesitation, however, we were not plunged into intellectual nihilism, either.

It would not be our task to draw up a list of numerous ways of map-reading, diverse interpretations, or tentative conclusions. It would rather be helpful to show "meaningful" conceptual bias which we could capture on a different level from map-drawing.

I
Regardless of our intentional resistance to explicit conclusion and belief in possibilities from reserved judgement, it is also true that there is clear conceptual direction in the total set of key-words. we can find it easily - even before map-drawing - just from the quantitative criterion how often a certain keyword appears. The frequency list of the words more than 10 times is:

URBAN 28   NATURE 24   SPACE 23   LANDSCAPE 22   CONNECT(ION/IVITY) 20
PROGRAM(MATIC) 13   SYSTEM 13   NETWORK 12   STRUCTURE 12   CITY 12   SKIN 11
ECO- 10   TOPOGRAPHY 10

From this simple list, we can immediately be aware of the overt intention of architects to extend the realm of architecture toward city, nature, and landscape. The architects at least on this book seem to consider the idea of connection and relation between architecture and the other as a new possibility.

But on the other hand, we cannot help thinking over other implications than simple frequency. For example, key-word "space" can be doubted whether it signifies so specific meaning as the other words in that it is the very essence of architecture and so would rather be out of consideration. We should also consider the collective meaning consisting of synonyms such as skin and surface, or urban and city. As a result, here we define 4 themes and 8 ideas as conceptual points on which Korean architects as a collective whole concentrate. Although there may be opposite opinions, it can be regarded as the current streams of architectural thoughts in and of Korea today.

EXTENSION TO THE URBAN SCALE AND RELATIONSHIP WITH THE OTHER

(1) The notion of architecture which extends to the city and urban infrastructure: the "city" is interpreted as a reference to re-think architecture, or as the other but identified with architecture itself. There seems to be two implicit backgrounds: the silent conspiracy to get over the limit of profession and the real problem that the scale of architectural projects has been increased with its location in the middle of artificial city instead of rural settings.

URBAN 28   CITY 12   INFRASTRUCTURE 7

This first idea of urban extension and permeation has in common with **(2) the notion of architecture as connected, related, and collective whole**. For the prefix "con-/com-" in such key-words as connection, community, communication, and context means, in a sense, extension and explosion from a closed unity of an individual building.

CONNECT(ION/IVITY) 20   COMMUNITY 8   COMMUNICATION 6   RELATION 7   COLLECTIVITY 5   CONTEXT 8

NATURE AND LANDSCAPE AS CONTINUITY AND COMPREHENSION

In terms of the other sort of experimental attempts to break with convention of singular architecture or architecture composed with a set of separated spatial units, there emerges the third idea or conceptual trend: **(3) borrowing form and metaphor from the language of topography and landscape** in the sense that they are larger, more continuous, and more comprehensive than architecture.

LANDSCAPE 22 (XXX+SCAPE 3)   TOPOGRAPHY 10

The idea related with topography and landscape seems to inclined toward the similar direction with the other two. The first is **(4) nature and ecology, or the idea of nature contrasted with but at the same time complementing and enriching the artificiality of**

architecture. Needless to say, this idea is not far from the most up-to-date ideal of environmental or sustainable architecture.

NATURE 24   ECO- 10   ENVIRONMENT 8   GREEN 3   SUSTAINABLE 2

And the second is **(5) the notion of dynamism and liquidity to overcome the conception of passive and frozen architecture**. It is, put in other way, not about isolated and monolithic architecture but about ever changing, ever moving architecture, and moreover about architecture as a pseudo bio-creature dealing in metabolic input and output with its surroundings.

MOVEMENT 9   FLOW 6   TIME 5

Re-thinking Conventional inner Logic of Architecture

The attempts to questioning and re-defining the methodology in architectural design is embodied mainly as **(6) subjective re-interpretation or re-organization of given program** aiming at the interbreeding or multiplying. The after-modern attempt to overturn the program as "pre-dictated guideline" has been more than 30 years; nevertheless, it has still operative as a "new" idea. Similarly, **(7) re-attention to structure and surface** among the architectural elements from traditional taxonomy seems not to be a negligible spoke in a large wheel of new ideas. Structured surface, especially, as a sort of attempt to return to Gothic in that it is to deny the Modern dogma of separation between structure and surface is worth while to take into account.

PROGRAM(MATIC) 13   STRUCTURE 12   SKIN 11   SURFACE 4

Process-design Integrated into Computer Environment

(8) The notion of identifying design process and its results as a part of a larger system: this is, in a certain sense, most recently emerging innovative thought in architectural field. We can define this as a mode of thought in architectural design in which the trinity of research-design-production is integrated through the conscious control over the process with objective data operation beyond the arbitrary, intuitive, and experiential decision of individual architects on one hand; and the extended and self-sufficient design process by networking among multidisciplinary factors on the other.

SYSTEM 13   NETWORK 12   TECHNOLOGY 6

In short, the most distinctive idea inferred so far is one that aims connection, relation, interchange, extension, and hybridization with the factors out of limited boundary of architecture, let alone re-thinking about inner conventional logic of architecture, which explains the frequency of those two prefixes:

INTER- 18   MULTI- 15

This tendency accords with the recent atmosphere in which architectural thought is growing extended toward the outside of the profession in both physical and abstract ways to the extent that the daring argument that "architecture is a way of thinking" is not heard strange. We can be also aware of the clear line of evolution from the fact that the very traditional key-word "composition", for example, appears only 2 times.

In addition, considering that this book is about "Korean Architecture" in itself, it is noticeable that there exist very few key-words reflecting the regional specificity.

TRADITION 6   HISTORY 6   KOREA(NESS) 5

It can be understood as the most of architects on this book are more susceptible to the global universality and condition of being present than historical context or regional speciality. To accept it positively as a natural result of globalization in that not a few Korean architects have experienced study and practice in abroad and not a few projects in Korea are executed in collaboration with foreign architects, or to accept it negatively as the critical cause of the fact that Korean architecture is not appreciated justly as a unique branch in comparison with near Japan; that is the question to be answered seriously.

Regardless of the overview on the clear but somewhat general tendencies so far, it is neither easy nor necessary to give authorities to all the detailed and partial readings, interpretations, and theorizations. Despite there can be innumerable ways to read the map - whether you trace the conceptual path of a work by the zig-zagged line through the key-words on the layer of Thread Map or you take notice of the density or sparsity of key-words' distribution-, the individual readings are not subject to the judgement in terms of true or false. The act of "map-reading" can and should have the other operations indifferent to judgement or conclusion.

II
There are three levels involving the suggestive term "map". The first one is concrete and real target, i.e. the *raison d'être* of map; the second is the cartography or the act of map-drawing to transpose the real into the abstract and down-scaled signs; and the third is the effectuating act of referring the map to find unknown ways. While in this case the first level is the whole set of architectural thoughts enabling Korean Architecture to emerge -or more exactly the collected writings to see it- and the second is the cartography inscribed next to Keyword Map, our focus is on the third level of effectuation *a posteriori* "to find ways by referring the map". Therefore, the question of what we can get from this complicated map should be not about what it means but (like in practical use of map) about useful and real problem of where we can reach with it.

Then we need to notice that almost all the other acts with the map as well as the map-drawing process which we already noted are virtually acts of construction. If we had to "define" the principles of reduction to make an inexistent kind of map, the process to draw map is that of construction transcending the limit of representation in that we should make relations among the elements in accordance with clear principles. Referring to map is also the act of construction and production because we should achieve a certain purpose by projecting the subjectiveness onto the numerous elements and relations among them. The operation of map-reading is not revelation of its implicit meaning but substantial effects.

Maps in general has the specific operation according to its purpose. Navigation map, for instance, does not have any "meaning" to people on the ground whether they can read it or not, but has enormous and real "effects" to people on the ocean. In the same way, the operation of Keyword Map reflecting the whole set of architectural thoughts is way-finding to a specific kind of conceptual goal in order to construct the unknown yet and so the "new" architectural thoughts, which are all the different to architect by architect; it will never operate in order to cut out Korean Architecture into the hasty conclusion with abstract and even empty words.

Map-drawing or its referring is neither archaeology nor historicism. Regardless of its authenticity, the act of referring the map is not that of interpretation. Moreover, to see the Keyword Map reflecting the ever-changing spectrum of thoughts is not only to trace the liquid substance in simultaneous and multiple ways but also to intend unexpected projection through those tracings. About how the nine theoretical "probes" irrelevant directly to the key-words from architects' writings can be related to eight ideas listed above, therefore, we should concentrate on the new conceptual seeds which the unexpected collision between these two dimensions will germinate. Be it consistent or not, again the interpretation of the meaning is out of our interest. As our view directs toward not the past but the present and the future, so we dare to present Keyword Map not for a decisive sentence to Korean Architecture but for a new detonation of ideas beyond mere anticipation or reasoning.

"Mapping the architectural thoughts" which we have performed by Keyword Map may be the first case in our profession at least in Korea. In spite of the crucial problems in terms of the cartography, our argument is not far from the expectation that this Map will be not terminal or converging but starting or diverging point. Supposing the proliferation of different kind of "maps of architectural thoughts" in different forms, we can imagine easily the moment when we see brave architects refer to the "maps" in order to find a way of constructing the new concept of architecture.

029

HAEINSA TEMPLE CULTURE COMPLEX
HYEHWA CULTURAL CENTER FOR DAEJEON UNIVERSITY
ARCHITECTURE AS AN ALLEGORY
PINX MUSEUM
JEONGOK PREHISTORY MUSEUM
NATIONAL LIBRARY OF CZECH REPUBLIC
PyoungHoa Nuri Peace Park + Youth Training Center
ASIA PUBLICATION CULTURE AND INFORMATION CENTER
A-MUNICIPAL LIBRARY
SEOUL PERFORMING ARTS CENTER
SEOUL PERFORMING ARTS CENTER
SEONYUDO PARK LANDSCAPE
SEOUL PERFORMING ARTS CENTER
NEW MULTI-FUNCTIONAL ADMINISTRATIVE CITY
ANYANG PEAK
BUSAN CITY SOFA
UNIVERSAL CREATIVITY?
BECOMING FROM RELATIONSHIP, THE NEO-ORGANISM
ASIAN CULTURE COMPLEX
MILLENNIUM COMMUNITY CENTER
GRADUATE SCHOOL OF ARCHITECTURE KYUNGHEE UNIVERSITY
AQUA-ART BRIDGE
NUNGPYUNG-RI HOUSE
SEOUL PERFORMING ARTS CENTER
MYUNG FILM COMPANY BUILDING
EHWA CAMPUS CENTER
DESIGN MANUAL FOR HEYRI ART VALLEY
RETHINKING PROTOTYPE
JEONGOK PREHISTORY MUSEUM
SEOUL PERFORMING ARTS CENTER
IMJINGAK MEMORIAL
CHEONGGYECHEON CULTURE CENTER
IRAN OIL INDUSTRY HEADQUARTERS
THE FIRST TOWN
SEWOON DISTRICT #4 URBAN REDEVELOPMENT PROJECT
ASIAN CULTURE COMPLEX
GALLERY YEH
PROGRAM-DIAGRAM : PROCESS CONCEPT
M³_QUESTIONS ON THE SPACE
HANDSOME HOTEL
EWHA CAMPUS CENTER
⊞ + FISH & FISH
JEONGOK PREHISTORY MUSEUM
HERYOOJAE WOMEN'S HOSPITAL
HAMBURG ARCHITECTURAL OLYMPIAD
SYSTEMATIZED PROTOTYPE
PROTOSTRUCTURES
PARAscape : CITY, ORGANIZATION AND ARCHITECTURE
NAMJUNE PAIK MUSEUM
EYES OF PORDENONE
DISTURBING TERRITORIES
CHICHI EARTHQUAKE MEMORIAL
CATALOG CITY
EAGON WINDOW IMAGE SHOP
THE LAST HOUSE-HOUSE FOR THE DEAD
BETTER, CHEAPER, FASTER
LIVING GLASS
RESIDENTIAL COMPLEX PLAN IN CATALUNYA
CHICHI EARTHQUAKE MEMORIAL
DIGITAL TECHNOLOGY AND CYBERSPACE
JEONGOK PREHISTORY MUSEUM
MAARUN, SOUTH BANK PLANNING DEVELOPMENT
JEONGOK PREHISTORY MUSEUM
NEW NATIONAL LIBRARY CZECH REPUBLIC
ALLO SERIES
HOUSE OF OPEN BOOKS
BORIM PUBLISHING HOUSE AND MARIONETTE THEATER
OBZEE FASHION
WHY NOT? A QUESTION TO THE LAW OF CAUSALITY ABOUT FORM
NEW-TYPE-BODY-ARCHITECTURE 1.0/2.0
SSALON DE SSON 1.0
MUSEUM OF ART, SEOUL NATIONAL UNIVERSITY
LEEUM SAMSUNG MUSEUM OF ART
S-GALLERY
HILTON NAMHAE GOLF & SPA RESORT
EWHA CAMPUS CENTER
ASIAN CULTURE COMPLEX
NEW HEADQUARTERS OF LOMBARDIA DISTRICT
SPEED DOME
ARCHITECTURAL PRODUCTION AS A SYSTEM
UNDERGROUND CAMPUS AND INTERNATIONAL DORMITORY, SOGANG UNIVERSITY
THE FIRST TOWN
TOWER PALACE III
DANANG ADMINISTRATION CENTER
BOHEON BUILDING
GANGDONG CULTURE ART CENTER
OSONG BIO-HEALTH SCIENCE TECHNOPOLIS
DESIGN CENTER OF GWANGJU
LG ELECTRONICS SEOCHO R&D CAMPUS

ARCHITECTURE
GENERATED BY
UNIVERSAL CON
RETHINKING PRO
PROGRAM DIAG
PARASCAPE
DIGITAL AND CY
FORMS OUT OF
PRODUCTION A

S AN ALLEGORY
ELATIONS
ENT
TOTYPE
RAM

ERSPACE
AUSALITY
A SYSTEM

Sungyong Joh
Johsungyong Architect office, Korea
Haeinsa Temple Culture Complex
Hapcheon, Gyeongsangnam-do, Korea
2004-

가야산국립공원에 자리잡은 합천 해인사는 한국의 3대 대찰 중 하나이며 법보사찰이다. 특히 13세기에 제작된 팔만대장경 목판본과 그것을 소장하고 있는 장판각 건물로 유명하며 유네스코 세계문화유산에 등재되어 있다.

이 프로젝트는 해인사의 부대시설 계획에 그치지 않고 이 시대 사찰건축의 새로운 방향을 제안한다. 해인사의 공간 구성과 스님, 신자, 방문자의 움직임과 활동을 분석하여 기단과 건물, 담으로 구성되는 단순한 아이디어에 주목하였다. 화엄사상의 사찰공간은 대개 3개의 동심원이 이루는 영역으로 나뉘며 각 영역은 중심마당과 그것을 둘러싸는 건물로 구성되어 경내를 이루며 방문자들은 주 접근장소에서 길과 마당을 통해 박물관과 도서관, 세미나실과 대강당(법당)에 이르는 동안 문화적 프롬나드를 경험한다.

건물 크기는 외부 공간과 주변 자연을 고려하면서 주의 깊게 조율되어, 땅과 주변의 경관과 관계를 맺는 건물의 '집합적' 개념을 제안한다. 이를 통해 원래 이 땅의 풍경이 드러나고, 새로운 '풍경의 집합체'가 되어갈 것이다.

COLLECTIVITY

TRADITION

LANDSCAPE

NETWORK

URBAN VOID

SPATIAL COMPOSITION

SENSE OF PLACE

TOPOGRAPHICAL RESTORATION

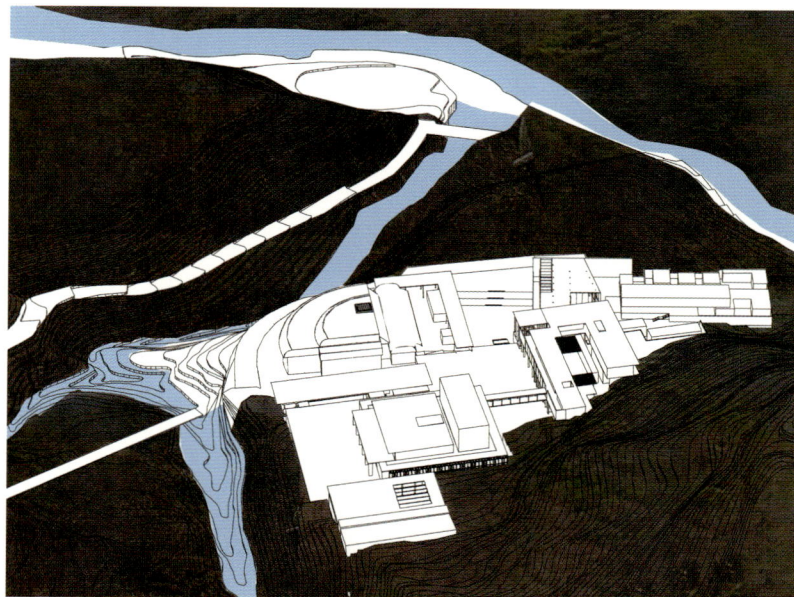

가야산 자락의 대지와 신행 문화 도량 SITTING OF CULTURE COMPLEX IN THE MIDDLE OF MT. GAYA

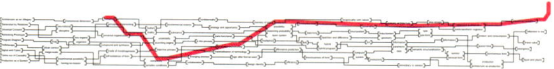

해인사 신행 문화 도량 　　　조성룡
　　　　　　　　　　　　　조성룡도시건축

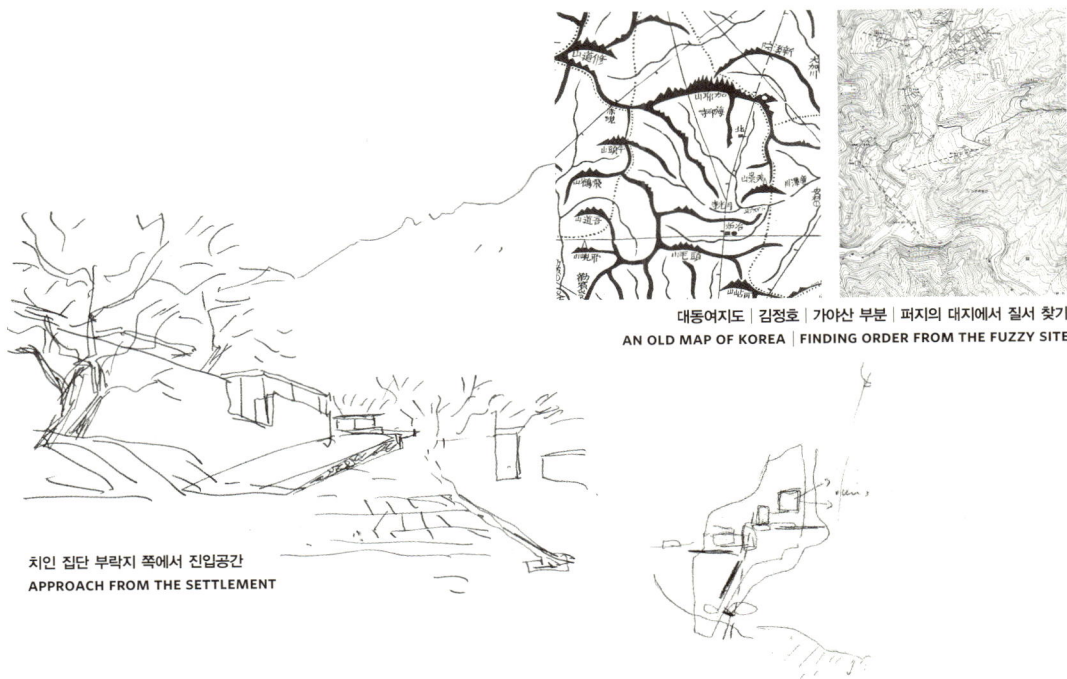

대동여지도 | 김정호 | 가야산 부분 | 퍼지의 대지에서 질서 찾기
AN OLD MAP OF KOREA | FINDING ORDER FROM THE FUZZY SITE

치인 집단 부락지 쪽에서 진입공간
APPROACH FROM THE SETTLEMENT

두 군데의 참선실과
산쪽의 작은 마당

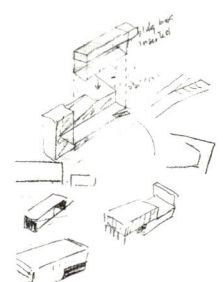

문화동(자료-도서관/다용도실)

This is a new Buddhist temple and affiliated cultural facilities complex in the middle of deep in Mt. Gaya National Park in southeastern region of the Korean Peninsula. Haeinsa Temple is one of the three "jewels" of Korean Buddhism, very old and distinguished religious space, also with the famous collection of Tripitaka Koreana, 80,000-plus wooden printing blocks of Buddhist scripture since 1,251 A.D., and with buildings where the blocks are stored, which were put on the World Culture Heritage List by UNESCO.

Analyzing the old temples regarding the theory and phenomenon of spatial composition refer to the activities of Buddhists and visitors through the religious and cultural programs, we had found some ideas from the spatial setting of existing buildings, the concept of precinct of spatial field with its very unique but natural platform and stone walls dispatched. It's very simple idea but linked strongly with the topographical conditions considering the surrounding landscape.

Visitors would experience the cultural promenade beginning with the main access space through paths and courtyards, cultural space including the museum and library to seminary and main hall.

Adjusting the volume of each building considering with the outdoor space and surrounding nature, the proposed scheme shows the notion of building complex relation with the land and surrounding *revealing* the landscape and will become a *collectivity* of landscape totally.

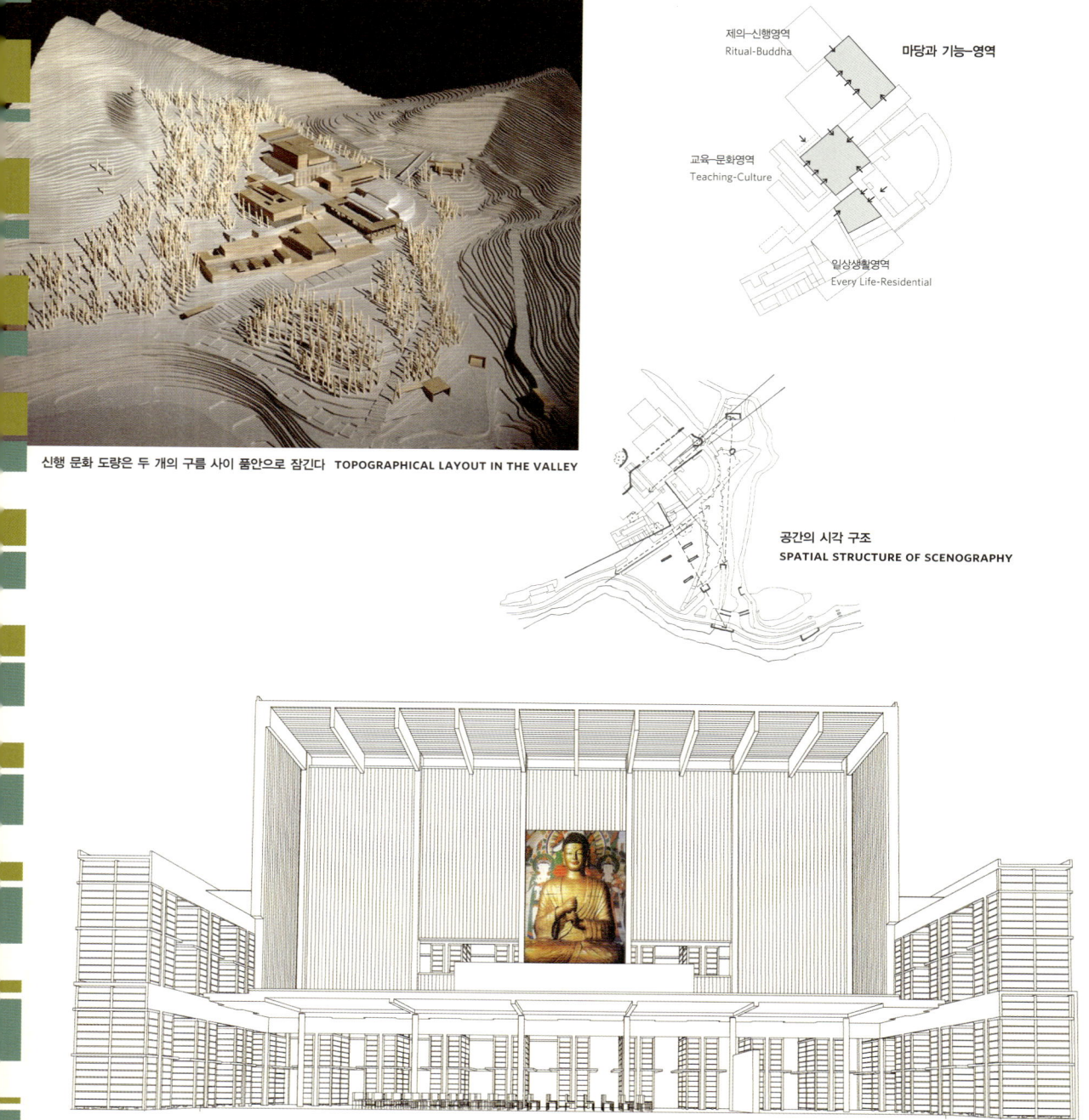

신행 문화 도량은 두 개의 구름 사이 품안으로 잠긴다 TOPOGRAPHICAL LAYOUT IN THE VALLEY

제의-신행영역 Ritual-Buddha
마당과 기능-영역
교육-문화영역 Teaching-Culture
일상생활영역 Every Life-Residential

공간의 시각 구조
SPATIAL STRUCTURE OF SCENOGRAPHY

법보공간 | 팔만대장경으로 둘러싸인 법당 "JEWELRY SPACE" OF BUDDHISM | MAIN HALL FOR THE RITUAL

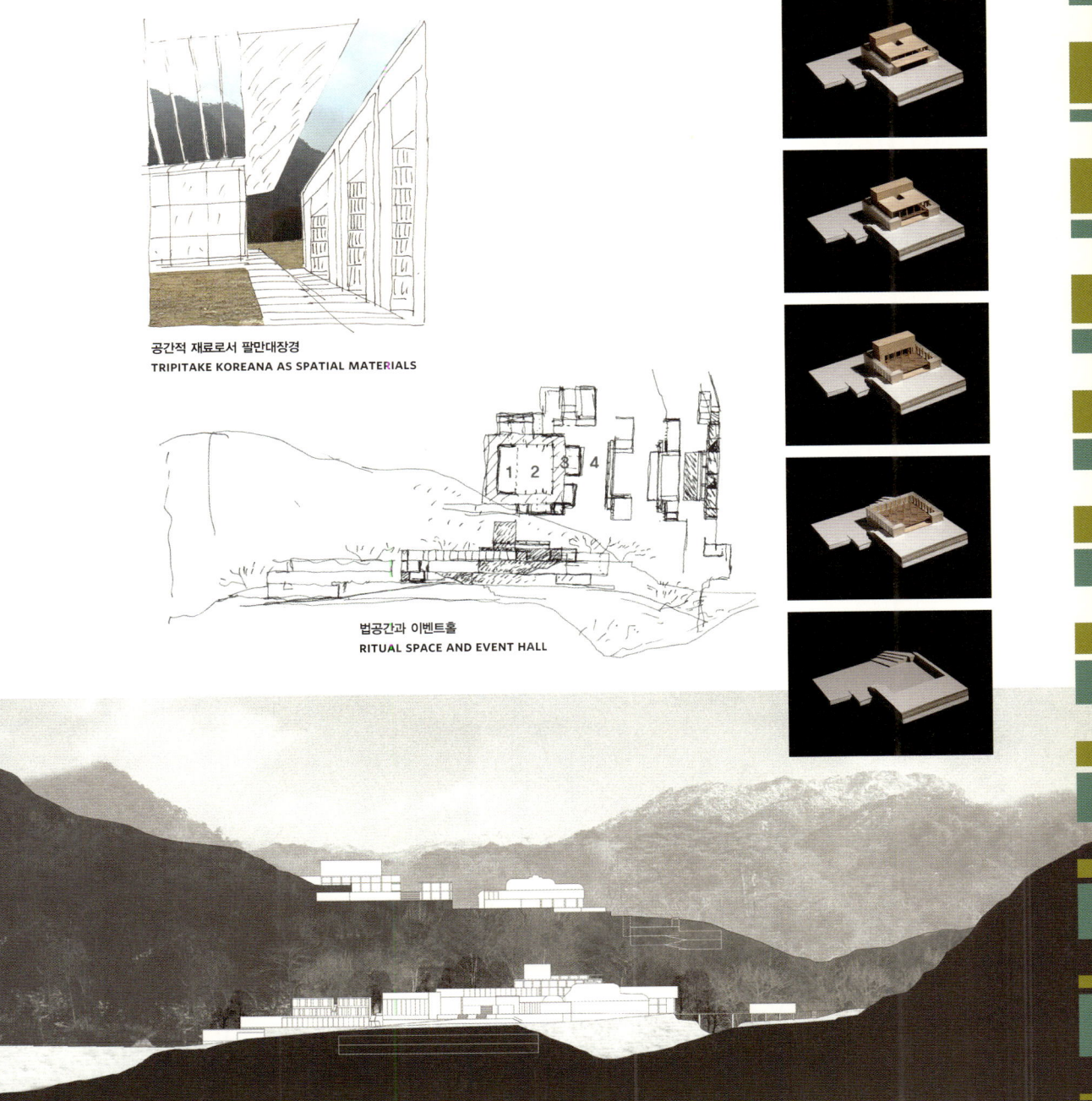

공간적 재료로서 팔만대장경
TRIPITAKE KOREANA AS SPATIAL MATERIALS

법공간과 이벤트홀
RITUAL SPACE AND EVENT HALL

다시 풍경으로 MERGING INTO THE LANDSCAPE

H-Sang Seung
Iroje Architects & Planners, Korea
Hyehwa Cultural Center for Daejeon University
Daejeon, Korea
2001-2003

주어진 프로그램은 다목적 공연장과 동아리 회합실, 식당과 회의장, 전시실, 학생상담실과 어학실습실 등 하나의 건물로 묶기에는 너무 다양하고 독립적인 시설들이었다. 이 복잡다단한 시설들을 하나의 출입동선으로 제어하는 것은 그 출입의 성격과 기능으로 보아 불가능할 뿐 아니라 불필요하였다. 오히려 여기저기서 서로 다른 이벤트가 동시에 일어나는 것을 허용하는 것이 더욱 유효한 것으로 판단되었다. 학생들을 실내 공간에서만 수용한다는 것은 자유분방한 이들의 문화 행위를 특수 분야로만 한정짓는 것이 분명하였다.

옥외공간의 필요성을 절감하고 이곳에 새로운 땅을 만들기로 했다. 즉 진입도로에 면하여 평탄면을 하나 만드는 것이다. 토목공사 없이 새로운 땅을 만드는 일은 건물을 데크로 만들어서 사용하는 것이다. 이 깊이 54m, 폭 62m의 데크는 가운데 커다란 개구부가 있다. 개구부 아래는 진입도로에서 10m 아래로 경사져서 내려간다. 이 경사진 오픈스페이스는 집회장이 될 수 있어 실내의 다목적 공연장이 수용하지 못하는 1,000명이 함께하는 대규모 집회를 가능하게 한다. 물론 이 야외집회장은 아래쪽 무대 부분에서 실내 다목적 공연장의 무대와 일치된 레벨이어서 두 집회공간을 연계하여 이용하면 훨씬 활발한 이벤트도 가능할 것이다. 야외집회장은 평소에는 휴식이나 소규모 회합 혹은 전시가 가능한 구조이며, 무엇보다도 아래의 기숙사 시설과 연결되는 외부 길로도 쓰이는 공간이다.

1,000평 면적의 새로운 땅-데크는 좌우에 유리로 싸인 두 개의 매스가 올려져 있다. 그 사이에 조그만 집회공간도 있고 벤치가 있기도 하며 나무도 심어져 있다. 마치 하나의 공원이기도 하고 광장이기도 하며 마당이기도 하다. 심지어는 하늘로 올라가는 길처럼 보이기도 할 것이다. 이 데크를 오르는 경로는 다양하다. 가운데 야외집회장을 한정시키는 경사진 길을 따라 식당 앞 발코니를 거쳐 오르거나 전시장과 회의시설이 있는 곳의 중정을 통해 오르기도 하고 혹은 좌우의 유리박스를 통해 접근할 수도 있으며, 진입도로에서 직접 계단을 통해 오르기도 한다.

이곳에 오르면 이 땅은 모든 지면에서 떠 있는 특별한 장소로 느껴질 것이다. 이 특별한 장소는 특별한 프로그램을 가지고 있지는 않으나, 학생이나 교직원 혹은 방문자는 저마다 특별한 추억을 이곳에서 만들 것이며 그것은 그들의 새로운 삶이 된다.

MULTIPURPOSE

CONNECTION

OPEN SPACE

ZONING

ORGANIC

RELATIONSHIPS

TRADITIONAL

앞쪽의 혜화문화관과 뒤쪽의 기숙사가 이루는 공간적 맥락
SPATIAL CONTEXT CONSISTING OF CULTURAL CENTER AND DORMITORY

The project required putting together a complex amalgam of facilities. A multi-purpose concert hall, club rooms, dining rooms, conference rooms, exhibition hall, counselors' office, language learning center - they seemed too diverse and independent to be placed together in one building. Controlling the different facilities with a single access route seemed both impossible (due to the different purposes and functions) and unnecessary. Rather, the more efficient solution was to allow spontaneous occurrence of different events at different places.

In addition, confining the students only to the indoors would highly limit their cultural activities, which by nature pursue freedom. This is why I felt that an outdoor space is also needed, and decided to create a new "land". It was to be a "plateau" adjacent to the road leading into the Center. Given the school's shortage of flat areas, it also seemed very much needed. Creating a new space without civil engineering meant that we had to build a deck to the building.

The resulting 54m-deep, 62m-wide deck has a huge opening in the middle. The opening has a 10m-slope leading down to the road. The entire deck is a sloped open space, that serves as an excellent assembly hall with a capacity of 1,000, enabling the types of performance that cannot be contained in the indoor multi-purpose concert hall. The lower stage of the outdoor hall is level with the indoor hall. If used jointly, they will undoubtedly provide for much bigger and livelier events.

The outdoor hall is designed to generally support rest and relaxation, small gatherings or exhibitions. It also hosts the passage to outside, with connection to the dormitory below. The 3,300m^2 new land - deck - is ended at each side by a glass-covered mass. In the middle sits a small assembly space along with benches and trees. It feels like a park, a square and a backyard all at once. It may also look like a passage to heaven. There are different ways of going up the deck. One is to take the sloped path that limits the outdoor gathering area, through the balcony in front of the dining hall. The other is to pass through the central courtyard (with the exhibition hall and conference facilities), or through the glass box at each end, or up the stairs directly from the entry road.

However one reaches it, once up on the deck, it will feel like a special zone that seems to hover above everything earthly. The place itself does not offer special programs. But each visitor, be it a student, professor, school administrator, or an outsider, will make special memories up on this deck, and for that moment, live a new life, however ephemeral that may be.

Campus Plateau 전경, 새로운 대지의 형성과 비운 마당 VIEW OF CAMPUS PLATEAU, NEW TOPOGRAPHY AND COURTYARD

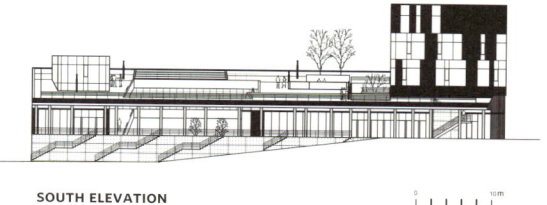

SOUTH ELEVATION

대전대학교 혜화문화관　승효상
이로재

기숙사에서 바라본 혜화문화관　VIEW FROM THE DORMITORY

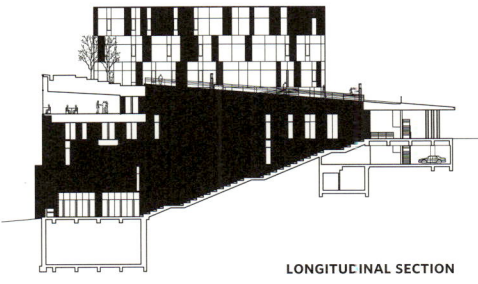

LONGITUDINAL SECTION
경사대지의 지형을 읽는다
SHOWING THE SLOPED TOPOGRAPHY

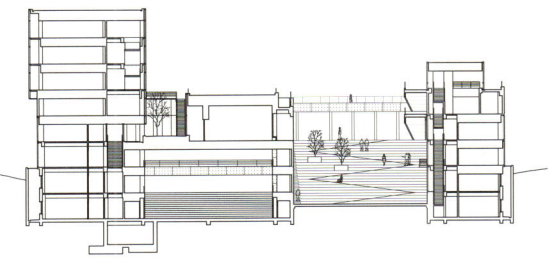

CROSS SECTION
두 건물과 계단 마당의 얼개
COMPOSITION OF TWO BUILDINGS AND STEPPED COURTYARD

포치 PORCH

포치 후 중정 전경 VIEW OF STEPPED COURTYARD FROM THE PORCH

계단 마당 STEPPED COURTYARD

옥상의 새 지반 NEW PLATEAU ON THE ROOF

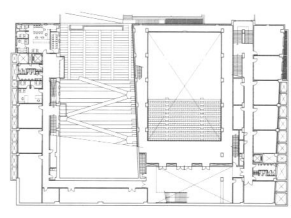
1F PLAN

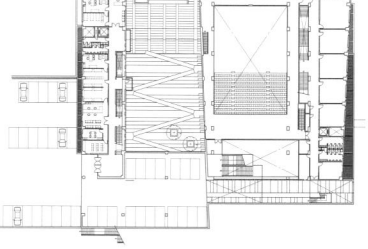
3F PLAN

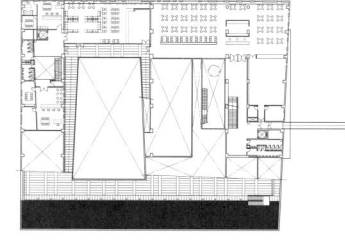
4F PLAN

대전대학교 혜화문화관 | 승효상 이로재

대지의 경사를 받아내는 계단 코트 하단 | 계단 코트의 목적성과 리듬
STEPPED COURTYARD COINCIDES WITH SLOPPED SITE | RHYTHM AND PURPOSEFULNESS

알레고리로서 건축

박길룡
국민대학교 건축대학

모더니즘 이후 현대건축은 두 가지 믿음을 청산하여야 했다. 첫째는 이지주의의 도그마에 의해 마비된 감성을 푸는 것이고, 둘째는 건축이 물상의 입장을 떠나 더 넓은 미의 향수를 위해 자신을 여는 것이다. 여기에서 건축을 이룰 사상에 대해 두 가지 시선을 서론으로 대신한다. 첫째, 상황과 본질에 대해서 동시적同時的 시야여야 하며, 사상의 통섭성을 강조한다. 둘째, 건축은 대상에 대해 훨씬 더 확장된 이해로 시작되어야 하며, 그것은 물상을 넘어 현상의 차원까지다.

건축을 분석의 결말로 기다리지 않고, 은유로 시작하는 것은 대상을 필연에서 우연까지 확장하는 일이다. 그것이 건축을 논리구조나 계층적 방법론으로 시작하는 것과의 차이다. 그것은 문학적 감성처럼 이루어지지만, 그대로 시작詩作처럼 되는 것은 아니다. 다만 의미와 묘법이 분리되지 않고, 이해와 감성이 하나이면서 문제를 통합성으로 접근한다면 문제 사이의 간극까지 통섭할 수 있다.

상황으로서 조건

조선의 지도는 도면식과 회화식 지도가 있다. 회화식 지도는 서양과 동양에서 18~19세기에 걸쳐 공통적으로 나타나는 양식이지만, 서양은 지리상 발견 이후 보다 과학적인 지도를 만들기 위해 평면화된다. 조선과 서양의 회화 지도는 의미 정보와 심미 묘사에서도 차이가 있어서, 서양은 다분히 서사적이고 동양은 서정적이다. 다만 대상이 지닌 지리 정보와 거시 풍경의 감정을 포괄하려는 뜻은 비슷하다.

PHENOMENAL DIMENSION

METAPHOR

SCENERY ARCHITECTURE

SITUATIONAL CONDITION

RELATIONAL AESTHETICISM

REALISM

TIME-DESIGN

CONFIDENTIALITY

MEMORY OF PLACE

RECIPROCALITY WITH NATURE

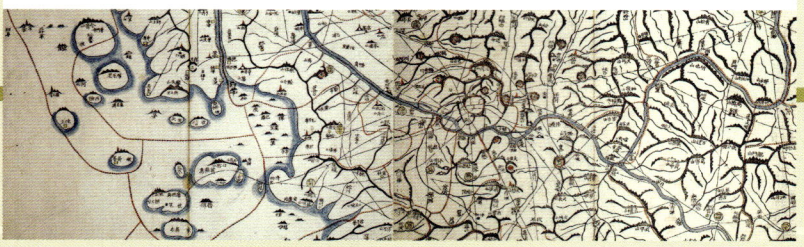

대동여지도 | 김정호 | OLD KOREAN MAP, DAEDONGYOJIDO | JUNGHO KIM

Architecture as an Allegory

Kilyong Park
School of Architecture, Kookmin University

The architecture after Modernism has to abandon its two beliefs; it has to recover "emotion" which was suppressed by intellectual dogma. And, it also has to open itself for more diverse types of beauties, leaving materialistic standpoint. Two views towards the material and the image co-exist as essential elements in forming architecture. First view is that the circumstance and the nature have to be dealt with, equally and simultaneously, while recognizing their consilience. Second view asserts that architecture start with more extended views, which goes beyond the materialistic dimension to reach the phenomenal one.

By placing object in the level of inadvertency, beyond inevitable logics, architecture starts with metaphor rather than analytic conclusion, which differentiate from the attitude that derives from a logical system or a progressive method. It is attained same as literary emotion, but not as poem writing. By not separating meaning and its illustration, by integrating recognition and emotion, and by comprehensive approaching to the problems, the consilience of even the gaps between the problems is achievable.

Condition as a Circumstance

There are two types of maps in Joseon Dynasty (1392~1910); picturesque and diagrammatic. Although they are similar in styles in the West, the Western maps are flattened to be more scientific after new geographical discoveries. Comparing two styles, the Western maps are more narrative while the Eastern maps are more lyrical. Both of two, however, shares the intention to include geographical information and illustration on the actual scene, as well.

The map of Joseon has been produced under the cooperation of painter, geographer, mathematician and scholar to delineate actual topology emotionally as well as realistically. It was a painter not a technician who has drawn map, meaning that map integrates geographical information and visual emotion as a circumstance. Thus, the picturesque delineation becomes not less important as the density of information.

It is problematic how to understand the given conditions in architecture;

조선의 회화식 지도를 그리기 위해서는 화원畵員과 상지相地가 합작해야 하며, 산사算士와 학자들이 함께 산형山形과 수파水波를 심정審定한다. 아니면 김정호와 같이 판각과 회화 그리고 지리와 산수의 능력을 두루 갖춘 재능인이 통합적인 작업을 이루기도 한다. 지도기술자가 아닌 화가가 지도를 만든다는 뜻은 지리가 가진 시각 감성과 정보를 하나의 상황으로 거둔다는 것이다. 따라서 정보의 밀도와 미술로서 묘법이 함께 중요해진다. 특히 조선 초기부터 한국의 옛 지도는 실경산수實景山水의 영향이 지배적이며, 조선 후기 겸재 정선에 이르러서는 진경산수眞景山水에 이른다. - 우리 옛지도와 그 아름다움 | 효형출판 | 2005 | 옛 지도와 회화 | 안휘준 | p.2005

건축에 주어진 조건을 이해하는 태도가 계층적이냐 통합적이냐의 문제이다. 구조주의 관점에서 조건은 분리된 정보들이 수직적으로 쌓이며 하나로 정리되어간다. 계층적 방법론은 조건마다 면밀함을 보장하지만, 대신 켜와 켜 사이의 이해를 놓칠 수 있다. 그것이 사상의 융합 작용을 위해 간극을 강조하는 이유이기도 하다. 그렇게 해서 요소들은 개별 사실로서 현전되는 것이 아니라, 어떠하든 '관계의 미학'을 통해 섭렵된다.

상황으로서 조건이란 이해되어야 할 조건들이 켜로 분리된 사실이 아니라, 그것은 처음부터 한통자의 사실이라는 것이다. 대지와 시각구조가 분리되지 않고, 시간과 환경이 따로가 아니다. 직관까지 포함하여 현상의 조건을 포괄할 수 있을 때, '분석' 하지 않는 이해에 이를 수 있다. 문제는 '상황' 이라는 것의 애매함이다. 그러기에 상황으로서 접근은 대상에 대한 건축가의 더 깊은 관조를 필요로 한다. 그것은 대상을 향한 자기 내밀의 긴장 다음에 가능한 의사소통의 일이다. 사상은 고착된 물상이 아니라 증변症變을 이루며 다른 사실로 전이하는 현상성現象性에 더 주목함이다.

통합적 사유

에즈라 파운드의 〈칸토스Cantos〉는 전통적인 시 형식을 뛰어넘어 하나의 시 형식을 포괄하는 문미학文美學이다. 시작詩作 자체가 그의 사고와 행동, 그의 시적 재능의 총화인 이 장시는 너무나 많은 것을 포함하고 있다. 그러기에 우리가 그 실체를 이해하기 위해서는 관자 자신의 다중적인 시선이 필요하다. 파운드는 공자의 서경과 서양문화사를 결합시키는데, 한자와 영어를 섞어 구문한다. 보다 더 인상적인

progressive or integrative? In the Structuralists' view, each condition is organized, and being stacked different layers of information vertically. It is the view of the Progressive to examine each condition accurately, but with the danger of losing the relationship between layers. These are the reason to emphasis on the gaps for the fusion of views. By the fusion, each element is fully understood thorough "the aesthetics of relationship", rather than presented individually.

Condition as a circumstance refers to the fact that the conditions are integrated entities from the beginning, not separable factual layers to be understood. Site and visual structure are not divided, and time is indissoluble from environment. It is possible to reach "the understanding without analysis" when the conditions of phenomena are incorporated, including intuition.

The problem is the ambiguous nature of "Circumstance". The approach to circumstance, therefore, demands architects more contemplation on the object. The communication gets to be possible only after a private tension towards the object. It is noticeable that material and image is not fixed, rather they are transferred to factual events.

INTEGRATED THINKING

The Cantos by Ezra Pound is a literature aesthetics, suggesting new style beyond conventional poem style. This long poem includes too many things while being the sum of his thinking, action and his talent on poem. It is required to observe the poem with multi-layered views to fully understand it. Pound combines Confucian Book of Odes with the Western cultural history by blending of ancient Chinese character with English in his sentence structure. It is more impressive that his poem is spatialized by the obvious differences of size, location and coordinates of each letter and spacing of lines. Ancient Chinese books and history are combined with his knowledge to broad meanings and to deepen sensitivity.

Architecture is achieved not only by logical process. Modernism architecture separate concepts form program, emphasize on the orderly process of analysis and solutions to limit itself within its own ideological framework. In Modernism architecture, a methodical system is required as more complicated facts are involved. In the progressive method, the reason for everything has to be proved continuously. It creates architecture in the method of casting information in a mold.

t is vertical to think progressively. It develops through the process of clue-evidence-concept-solution-design, vertically. If it is possible, however, to grasp entire process simultaneously, which may be call synthetic thinking, this is horizontal. In

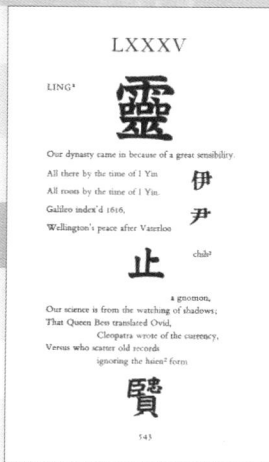

에즈라 파운드 | 서경書經
EZRA POUND | THE CANTOS LXXXV

것은 시문의 공간화이다. 두드러지는 글자 크기차와 위치좌표 그리고 행간거리를 포함하는 시각 형식으로 구성된다. 중국의 대학, 도덕경, 고대왕조사가 그의 지식과 결합되고 시간, 공간, 의미체계의 통합구조는 그만큼 의미 대역을 넓게 하고 의식의 수심을 깊이 한다. -이일환 | 에즈라 파운드 시집 《칸토스》 | 문학과지성사 | 1990 | p.306~307

상황에서 단서를 잡고, 증거를 확보하고, 심증을 굳히며, 움직일 수 없는 논증으로 건축이 이루어지는 것만은 아니다. 모더니즘은 프로그램과 개념의 작업을 분리하고, 분석과 해결이 계층화되면서 스스로 사유의 범위를 한정해갔다. 모더니즘건축은 규합될 성질이 복잡해지면서, 작업은 분화되고 시스템으로서 방법이 긴요해진다. 계층적 방법에서는 끊임없이 왜 그래야 하는가를 증명해야 한다. 대상으로부터 채굴한 정보를 모아, 자신의 방법론이라고 믿는 틀에 부어 건축을 주조한다.

보통 방법론으로 결과를 얻는 생각의 구조는 수직적이다. 즉 단서-증거-개념-해결-설계가 수직적으로 전개된다. 최초 단서에서부터 증거와 뜻, 미적 쾌감까지를 일거에 이룬다 하고 이를 통합적 사유라 하자. 이 사유의 방법은 횡적이다. 그것은 서예에서처럼 수단과 표현이 동시에 일어난다. 대상을 상황으로 보며 사실들의 작용을 통합적으로 사유하면서, 건축은 메타구조가 된다.

현상으로서 건축

모던건축은 기본적으로 물상이며 고정체였다. 모더니즘은 건축을 붙들어매야 사진이 뚜렷해지는 것처럼 알았다. 물론 우리는 건축을 고정태로 만들지만, 그것이 완성되는 순간부터는 '현상으로서 나타남'이 된다. 건축이 현상태가 되는 것은 그 동안 건축에서 주변 것이라고 간주되던 것을 포섭하는 일이며, 동시에 그 이면의 것과 관계들을 더 구체적으로 엮는 일이다. 우리가 어떻게 나부끼는 건축을 그릴 수 있는지, 건축이 시간을 흔든다든지. 사실 건축은 훨씬 흐릿한 일이다.

여기에서 건축이 현상으로 나타난다 하지만, '현상으로서 디자인'과 '현상학'은 구분된다. 그것은 사념의 미학이 아니라 구체적인 사실이어야 하며 어떤 지각으로건 쥐어져야 한다. 물질이 떠올리는 감성을 붙잡을 지각의 그물이 필요하다. 감성은 휘발성이 강하기 때문에 이를 사실화하기 위해서는 해부학적 이해가 유용하다. 시간이 흐르는 방법, 기억의 메커니즘, 빛의 동작 습성, 바람의 행동 유

horizontal thinking, expression and mean occur in the same time, just as in the Oriental calligraphy. Architecture becomes a meta-structure, when the objects are viewed as a circumstance, and the actions of factual elements are thought synthetically.

Architecture as a Phenomenon

Basically, modern architecture has been materialistic and static, and has fixed architecture just as photography. Although architecture exists in static state, it becomes a phenomenal appearance as soon as it is built. Becoming phenomenal implies that the elements that were considered to be in perimeters are grasped, and in concurrence with organizing the elements in background. In fact, architecture can be vague, being fluttering or shaking time and so on.

Architecture as a phenomenon is distinct form "Phenomenology". It has to be a concrete commitment to be apprehended in any manner, rather than speculated aesthetics. It requires the network of perceptions to grasp the emotions deriving from the object. In order to actualize emotions, an anatomical understanding is useful since they are volatile, such as the way time flows, the Mechanism of memory, the behavior of light and the habit of mud.

The notions of nature and urban are resulted from the interaction with our feelings. Therefore, its phenomenal quality of material to be becomes richer, as the limit of the object is stretched. Time is an axis of phenomenon as material and image are not fixed. Architecture can be called "time design" as it deals with the object with a biological time. It becomes a literature when the focus of allegory is being phenomenal rather than materialistic. Metaphor expands its power beyond visual object, and its emotional action reaches the level of five senses.

Architecture as an Allegory

Architecture is more than mere illustration, as it is ultimately phenomenal body. Architecture with the synthetic thinking, not believing in obvious analysis, reaches broader image and materiality which reflect phenomena. In large measure, therefore, architecture is an allegory. There are number of attitudes, such as metaphor, analogy, allegory, simulacra and so on, stimulate the allegory in between inclusion, expansion, synthesis and phenomenal body. It is not possible to render the differences between the attitudes; subtle differences are found in terms of degree of being either narrative or emotional, either more logical or literature, either more with denotation of impression or with intention.

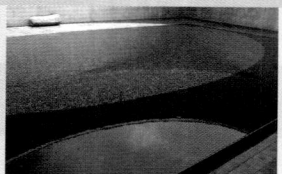
METAPHOR OF NATURE | ITAMI JUN
PINX MUSEUM

형, 땅의 습속 등이 그러하다.

결국 자연이나 도시에서 우리의 느낌이란 반응과 작용의 결과들이다. 그러기에 포획하려는 대상을 넓히거나 깊이 할수록 그물질될 현상이 훨씬 풍부해진다. 사상은 고정체가 아니기에 항상 시간은 현상의 기축이 된다. 건축적 대상들은 어떻든 생체시간을 갖기 때문에 시간디자인이라고 할 만하다. 알레고리의 대상을 물상에 두지 않고 현상에서 볼 때 다분히 문예적으로 된다. 은유의 힘은 시각 대상 너머로 확대되기에 그 감정이입은 시각 이상의 오감에 이른다.

알레고리로서 건축

건축은 궁극적으로 현상체이기에 물상의 묘사력 이상이 된다. '통합적 사유' 로서 건축은 '명명백백한 분석'을 믿지 않지만, 훨씬 포괄적으로 사상에 이른다. 사상들은 현상하기에 건축은 다분히 알레고리이다. 이 포괄, 확장, 통합성, 현상체 사이에 알레고리를 준동시키는 데에는 은유Metaphor, 유추Analogy, 알레고리Allegory, 시뮬라크르Simulacre 등의 몇 가지 태도가 있다. 이 태도들을 구분하는 차이를 명확히 말할 수는 없지만 좀 더 서사적이거나 감상적이거나, 좀 더 논리체이거나 문학체이거나, 좀 더 인상의 외연이거나 자의의 개입 차이로 이해된다.

은유 또는 문학성 비유, 직유, 은유, 환유 등은 모두 상징적 수단이지만, 사유의 표현화 단계가 다르다. 우리가 표현하고자 하는 것을 이웃의 것과 비추어 말하되, 직유는 상징성의 제기가 더 구체적이고, 비유는 우리가 표현하고자 하는 것을 이웃의 것에 빗대어 말하는 것이고, 은유는 아예 그것을 다른 대상으로 대체해 버리는 것이다. 그래서 은유는 말을 던지는 사람의 관점이 깊이 작용한다. 작가가 작용까지를 지배하려는 상징에 비해, 은유는 관자의 자리를 더 많이 배려한다. 은유란 어떤 것을 다른 것의 관점으로 보는 계획이다.

유추 또는 구조적 감성 우리가 미적 체험으로서 흔히 견주는 아날로지와 상징은 구분된다. 유추를 작동시키기 위해서는 단서가 필요한데, 그것이 '그러할 수 있는' 피할 수 없는 단서도 있으나, 좀 더 막연한 심증이 모두 가능하다. 어느 한 사실에서 그러할 수 있는 다른 사상을 끄집어내는 것이기에 유추는 단서 대 표현이라는 일대일의 상대성이기 쉽다. 유추는 관자와의 소통이 일방적이지 않지만, 작가 나름대로의 논리구조가 따른다. 그래서 은유보다는 직접적이며 작가가 준비한 체험의 경로를 공유한다.

METAPHOR OR LITERATURE Metaphor, simile, analogy and metonymy are symbolic means, but their expression of thinking varies. A simile symbolizes the object more direct way, an analogy refers it to neighboring idea and a metaphor replaces it with other object. A metaphor, thus, viewpoint of the author is more active. In a metaphor, spectator is more attentive than a symbolic system in which author try to dominate even the interaction. Metaphor is a way of seeing with other point of view.

ANALOGY OR STRUCTURAL EMOTION As an aesthetic experience, an analogy and a symbol can be differentiated. In order to stimulate an analogy, a clue is needed, which is to be inevitable or possibly ambiguous impression.

It is a danger of analogy to be a direct association of clue with expression, as it draws one image from another fact. An analogy constructs a mutual communication with spectator, but author suggest his own logical system. A route of experience, prepared by author is shared in more direct manner than in metaphor.

ALLEGORY Allegory is that a series of words is connected to convey a meaning other than the literal rather than rhetoric use of single word. In another words, allegory is mode of representation by gathering others. In architecture, since it is a phenomenon happening in the gap of the objects, it includes the margin of them. It is also necessary to accept the possible deferment of communication in converting an object to an image or an phenomenon, which enables architecture allow the room for communication of the rest.

For a special observation of the objects, it is often useful to vague them, and it would make the facts more factual. Being skeptical of what the truth is, is a beginning of the process of transformation of meanings in various ways. Doubting the objects activates transformation and cognitive filtering. An allegory reaching phenomenal level becomes a corridor from that richer discourse emerges. When borderline is grasped and blurred, architectural space becomes fertile with a comfortable intimacy.

After all, allegory is an attitude with more literature grasp on object, and it manifests itself in certain level. The architects of "Group 4.3" in Korea, names their offices with literature imagination as well their projects. Not surprisingly, this allegorical manner has been common in Joseon Dynasty. Every building in Joseon has meanings behind, therefore, it was an allegory.

알레고리 알레고리란 '한 단어의 은유적 사용이 아니라, 일련의 단어들이 서로 연결됨으로써 겉으로 이야기되는 것이 아닌 어떤 다른 뜻으로 이해되는 이야기 전개'이다.

좀 더 구체적으로 알레고리는 '다른 것과 모아서 이야기하기'이다. 건축으로서는 대상들 '사이'에서 벌어지는 현상이기에, 디자인은 여餘를 포함한다. 알레고리란 기본적으로는 대상을 현상으로 보는 태도 또는 대상을 이미지로 바꾸는데, 소통의 유예가 어느 정도 필요하다. 관자의 의사 개입을 더 풍부하게 할, '나머지 더'의 의사소통을 위해 필요한 여의 마련이다.

대상을 특별하게 보기 위해서 가끔 대상을 보는 눈을 흐릿하게 하는 일이 유효한데, 그것이 사실을 더 사실답게 하는 일이 되기도 한다. 사물을 여러 의미로 변환시켜볼 일은, 우선 무엇이 진실이라고 규정되기를 주저하면서 시작된다. 의심으로 시작되는 사물은 새롭게 작동시킬 변환적 안경, 또는 인식적 여과를 통해 자기 안으로 들어온다. 물상을 넘어 현상에 이르는 알레고리는 좀 더 풍부한 담론을 끄집어낼 통로가 된다. 건축이 보다 더 활발하게 외곽의 경계를 포섭적 구조로 하면, 그 안의 공간은 얼마나 풍부하고 편안한 내밀성이 있을까.

결국 알레고리는 대상의 포섭을 보다 문예적으로 이루는 태도인데, 한국 현대건축에서는 한 특정 층위에서 증변으로 나타난다. 4.3그룹 즈음 사무소 이름을 제제提題하는 사람들이 생겼다. 승효상사무소라 하지 않고 '이로재履露齋'라 하고, 우경국사무소라 하지 않고 '여운헌餘韻軒'이라 한다. 그러한 습성은 설계된 건축에도 곧잘 나타난다.

그러한 건축의 문예적 알레고리는 이미 조선조에는 보편적 사실이었으니 새삼스러운 일은 아니다. 맑고 깨끗해지는 곳 소쇄원瀟灑園, 북송의 황정견黃庭堅이 말하기를 춘릉春陵의 주묵숙은 인품이 몹시 높고 가슴속이 담백솔직하여 광풍제월과 같다고 한다. 그래서 광풍각光風閣, 제월당霽月堂 등이 그렇다. 소쇄원에는 담장 하나에도 일일이 문예적 의미를 두는데 애양단愛陽壇은 빛의 사랑 애일愛日만이 아니라 효심의 의미다. 소쇄원은 중후의 시 소쇄원 48영詠을 낳는다. 여하튼 조선의 모든 현판이 붙은 건물은 알레고리다.

애양단 | 소쇄원 | 담양 AEYANGDAN, SOSOEWON | DAMYANG, KOREA

Itami Jun

ItamiJun Architects + ITM Architects, Japan/Korea

Pinx Museum_Water Art Museum

Seogwipo, Jeju-do, Korea
2004-2005

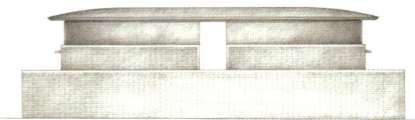

제주도의 토착 소재를 취한 수水미술관은 강한 입방체를 타원형으로 도려내어 하늘의 움직임을 수면에 투영시켰다. 더불어 사람들이 간과하고 있는 자연이나 유년시절 등을 연상시키며, 반짝반짝 빛나는 자갈의 묘한 아름다움, 졸졸 흐르는 물소리를 표현했다. 수변에 놓인 돌 오브제는 직접 만든 작품으로 벤치처럼 그곳에 앉아 무심無心이 되었으면 하는 바람을 담았다.

The Water Art Museum employs the native materials of Jeju Island, with strong cubes cut into oval shapes, reflecting the movement of the sky on the water. The museum reminds one of nature, often overlooked by people, as well as childhood, and features the exquisite beauty of shining pebbles on the bottom of a stream along with the murmuring water flowing in the stream. Stone objects d'art placed on the waterside contain my wishes that with the stonework that I created myself I also want to sit down on a bench and be devoid of worldly desires, like a stone.

WATER
REFLECTION
MOVEMENT OF THE SKY
REMINDS ONE OF NATURE
BEING DEVOID OF WORLDLY DESIRES
STREAM

제주도의 풍경과 미술관 | 한국에서 방方과 원圓은 땅과 하늘이라는 뜻을 갖는다
LANDSCAPE AND MUSEUM IN PICTORIAL NATURE OF JEJU-DO | RECTANGLE AND CIRCLE MEANS EARTH AND SKY IN KOREA

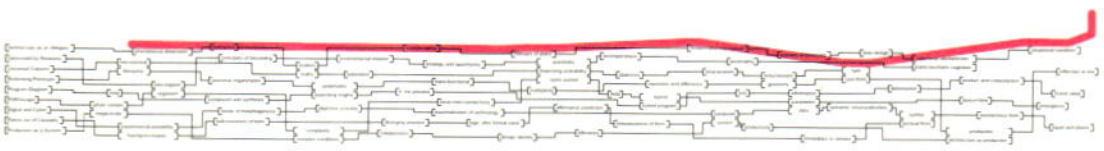

핀크스미술관_수水　　이타미 준
이타미준 건축연구소

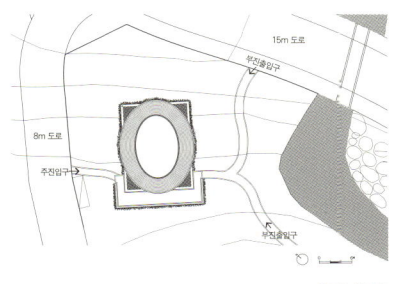

SITE PLAN

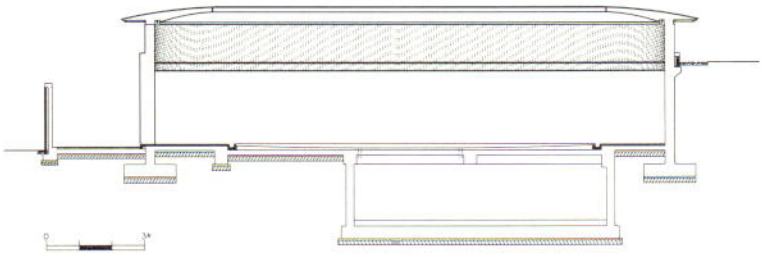

A-A' SECTION 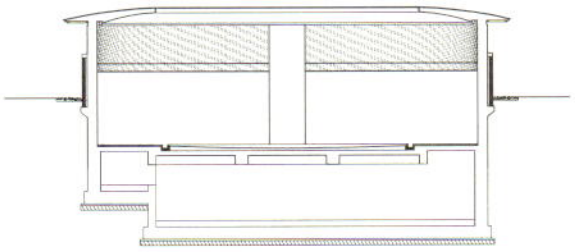　　B-B' SECTION

타원의 하늘과 물이 거울로 동조한다　SKY IN ELLIPSE RESONATES WITH WATER

Itami Jun

ItamiJun Architects + ITM Architects, Japan/Korea

Pinx Museum_Wind Art Museum

Seogwipo, Jeju-do, Korea
2004-2005

바람의 공간도 잃어버린 자연과 기억을 연상시키고자 한 공간이다. 오두막 개념으로 설계한 나무상자는 한쪽 입면이 활처럼 호를 그리고 있다. 나무판의 틈새로 바람이 통과하면 소리를 낸다. 바람이 강한 날, 판과 판 사이에서 마치 현弦을 문지르는 것 같은 소리가 들리는 것은 뜻밖의 놀라움이었다. 그곳에 놓여 있는 돌 오브제는 의자로, 바람 소리를 듣는 명상의 공간이기도 하다.

This is a space intended to remind one of nature and memories that are deprived of the space of wind. The concept is a wooden box designed with a cottage that has a curve like an arrow on one side. Wind makes a sound as it passes through the gaps between the wooden panels. It was an unexpected surprise to hear the sound coming from between the panels on a windy day, as it sounded as if the strings of musical instruments were being rubbed against each other. The stone object d'art placed there acts as a chair, providing a place for meditation while listening to the sound of wind.

WIND

MEMORY

COTTAGE

WOODEN PANELS

PASS

GAP

MEDITATION

비오토피아의 대지와 미술관 MUSEUM AND TOPOGRAPHY OF BIOTOPIA

핀크스미술관_풍風   이타미 준
이타미준 건축연구소

미술관 허리 부분의 출입구   ENTRANCE TO MUSEUM

미술관의 허리에서 내부 공간 | 소리와 빛의 은유
INTERIOR VIEW FROM THE MIDDLE OF THE MUSEUM |
METAPHOR OF SOUND AND LIGHT

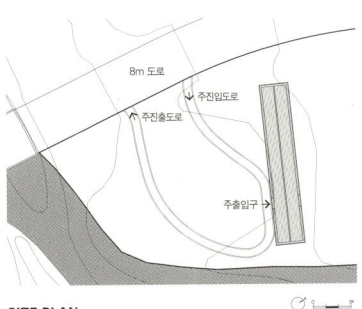

SITE PLAN

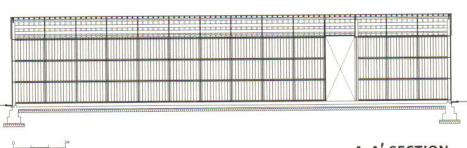

A-A' SECTION

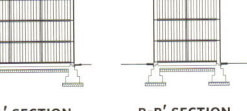

B-B' SECTION

Itami Jun

ItamiJun Architects + ITM Architects, Japan/Korea

Pinx Museum_Stone Art Museum

Seogwipo, Jeju-do, Korea

2004-2005

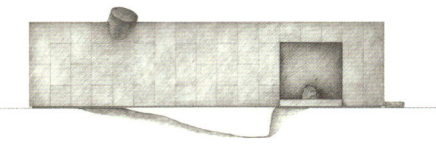

석석미술관은 하나의 사유이자 시적인 환상이다. 돌의 공간은 단단한 상자, 그것도 암흑 속에 의도적으로 구멍을 열어 인공적인 쇠의 꽃으로 삼았다. 그 구멍을 통해 쏟아져 들어와 이동하는 빛을 주역으로 연출한다는 환상. 그리고 보는 사람을 통해 제한없이 무엇인가를 연상시키는 공간이기도 하다.

The Stone Art Museum is a philosophical and poetic fantasy. The space of stone is made up of a hard box with holes that intentionally open into darkness, making it an artificial flower of steel. It is a fantasy that presents the light passing through the holes as the main actor. This is a space that freely reminds viewers of something.

언덕이 흘러내리는 지형에 미술관이 얹힌다 THE MUSEUM SITS ON THE FLOW OF TOPOGRAPHY

핀크스미술관_석石 이타미 준
이타미준 건축연구소

안과 밖의 풍경화 LANDSCAPE BOTH IN AND OUT OF MUSEUM

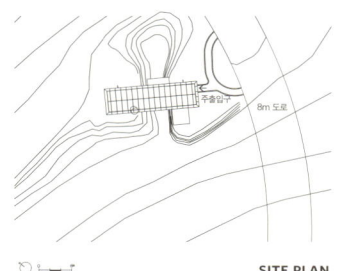
SITE PLAN

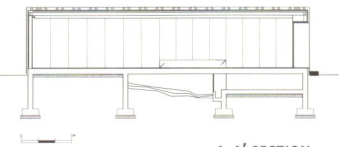
A-A' SECTION

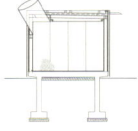
B-B' SECTION

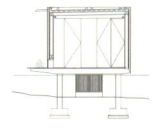
C-C' SECTION

ANOUK LEGENDRE + NICOLAS DESMAZIERES
X-TU ARCHITECTS, FRANCE
JEONGOK PREHISTORY MUSEUM
COMPETITION PROJECT_WINNER
GYEONGGI-DO, KOREA
2006-

이 프로젝트는 굽이치는 강을 따라 생긴 두 언덕의 아름다움과 한반도에 첫 주민이 등장하는 것을 지켜봐온 한탄강변의 경관을 기리고자 했다. 이미 존재하고 있는 땅의 형상과 지층이 돋보일 수 있도록, 건축물이 시각적으로 차지하는 크기를 줄이고자 했다. 이를 위해 언덕을 파내 그 속으로 건물을 배치하고 설비·창고시설을 지하에 조성했다. 건물 중앙부에서 곡선의 변화를 주어 절벽과 하늘, 땅과 연결될 수 있는 지점을 연출했으며, 반사재료를 사용하여 벼랑의 모습을 위에서 비춰낼 수 있게 하는 등 건물 외피에 변화를 주었다. 이렇게 배치된 건물은 멀리 도로에서부터 인지되며 두 언덕 사이에 걸쳐진 다리와 같은 형상으로 나타난다.

절벽과 이루는 자연스러운 건물의 경계와 그것을 통과하면서 오는 감동은 선사 시대로 들어가는 관문을 상징함과 동시에 선사유적공원을 향해 열린 문이 된다. 이와 함께 건물과 언덕의 곡선을 따라 길이 조성되는데, 선사 시대에 자연스럽게 조성된 길들을 떠올리며 그것을 모티브 삼아 산책로를 계획하였다.

We wished to honor the riverside landscape that saw the birth of the first inhabitants of Korea, and acknowledge the beauty of the curves of the two hills echoing the meandering river. How can such a preexistent form and its geological underground chasm be enhanced? By digging the chasm to let the Earth tell its history, and by alleviating the visual hold of the project in order to let the chasm express itself. For this, the building will be encased into the hollowed out hill, and the stock-rooms will be located underground by curving the central part of the building so as to unveil the geological crack by clothing it in a shimmering skin that will reflect the precipice from underneath.

Thus set up, the project appears like a bridge stretched between two cliffs that can be seen from the distant motorway. The precipice as a natural threshold and the emotion it induces will realize a symbolic threshold into the prehistoric era that will also give access to the prehistory park. Then we will create many paths around the curves of the project and of the cliffs, because the paths, which were made by nature, belonged to the landscape of the first human beings.

BRIDGE

LANDSCAPE

GEOGRAPHY

VALLEY

SHIMMERING SKIN

ROAMING

DOUBLE WALL

CIRCULAR PATH

SCENOGRAPHIC SPACES

전곡선사박물관
현상설계 당선작

ANOUK LEGENDRE + NICOLAS DESMAZIERES
X-TU ARCHITECTS

구릉 밑에서의 경관 | 대지의 은유 VIEW FROM THE FOOT OF HILL | METAPHOR FOR LAND

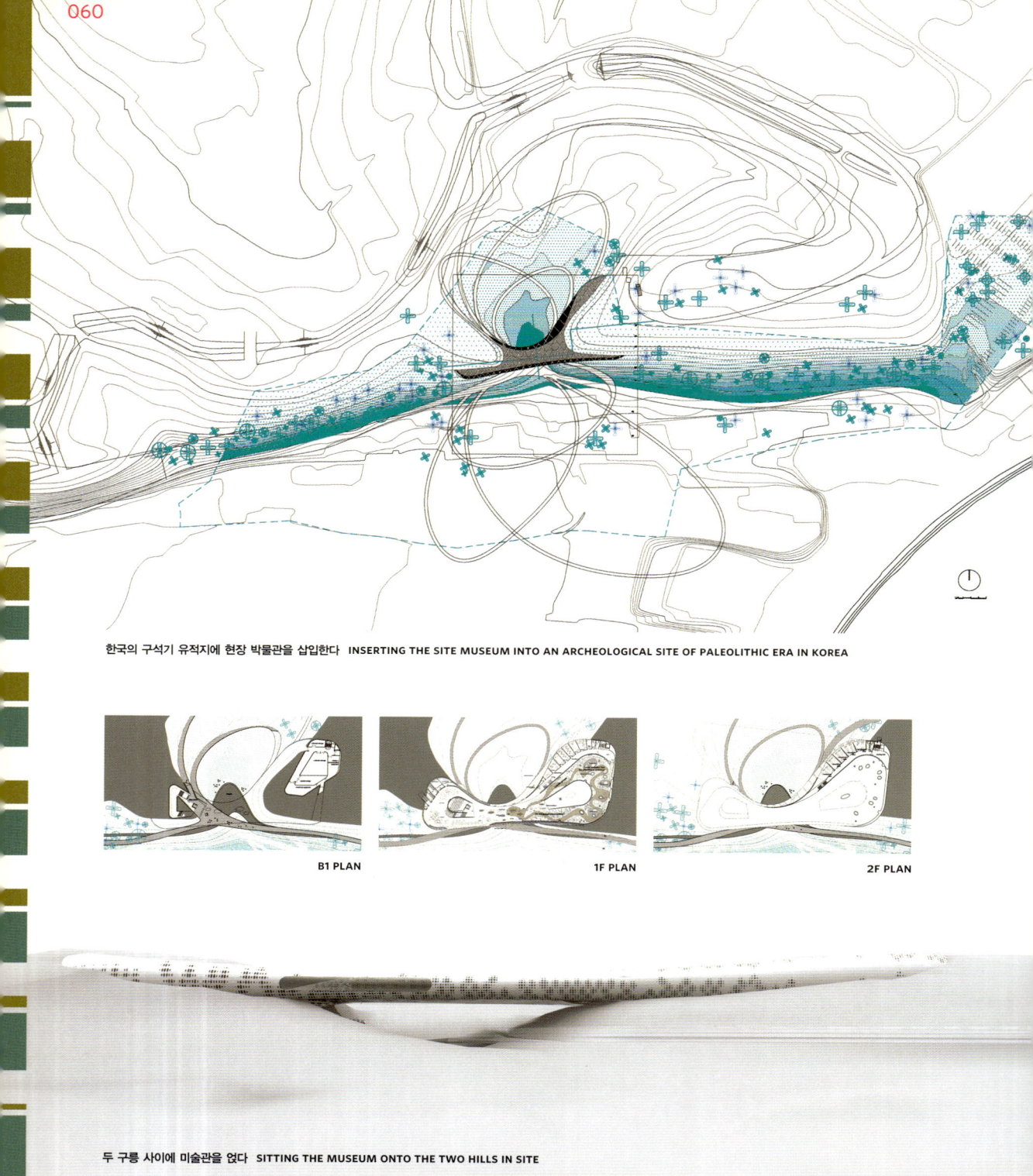

한국의 구석기 유적지에 현장 박물관을 삽입한다 INSERTING THE SITE MUSEUM INTO AN ARCHEOLOGICAL SITE OF PALEOLITHIC ERA IN KOREA

B1 PLAN 1F PLAN 2F PLAN

두 구릉 사이에 미술관을 얹다 SITTING THE MUSEUM ONTO THE TWO HILLS IN SITE

061 전곡선사박물관 현상설계 당선작   ANOUK LEGENDRE + NICOLAS DESMAZIERES
X-TU ARCHITECTS

전시공간 내부 | 유물에의 산책 INTERIOR VIEW OF EXHIBITION | PROMENADE IN TO THE RELICS

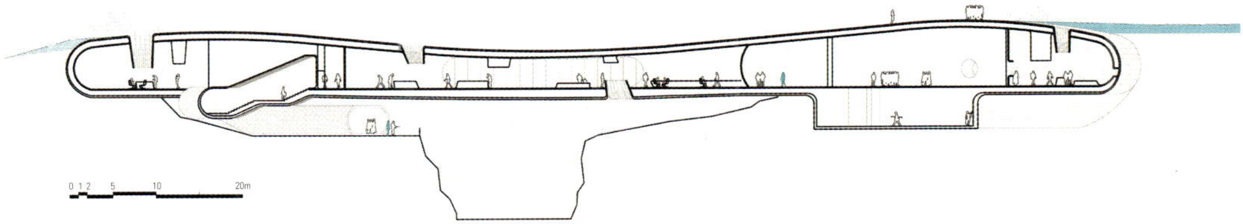

한탄강을 조망에 둔 산책 테라스 브리지 PROMENADE ON THE TERRACE BRIDGE

Anouk Legendre + Nicolas Desmazieres
X-TU Architects, France
National Library of Czech Republic
Prague, Czech Republic

이 프로젝트의 목표는 100개의 교회 첨탑이 특징을 이루는 도시를 배경으로 자연스럽게 어우러져, 구축된 볼륨을 상대화하고 거주지 느낌을 그대로 유지하는 것이다. 이를 통해 공원과 유대관계를 형성하고, 도시와 조화를 이루며, 국립도서관이 도시에서 표방해야 하는 프로그램의 상징적인 측면과 이 모든 것을 융화시킨다.

일반인에게 공개되는 모든 스터디룸은 돔 아래 거대하고 관통된 하나의 공간에서 여러 레벨에 위치한다. 공중에 뜬 통로를 통해 도서관의 대형 홀과 연결되며 모든 방향에서 들어오는 빛이 통로를 따라 흐른다. 대형 열람실 위 메자닌은 십자가형의 플랫폼을 구성하며, 현수교처럼 공간 위로 사지를 뻗는다. 공간을 나누고 실의 경계선을 구분지었기 때문에 벽은 필요없게 된다. 이동은 메자닌의 그늘을 따라 자연스럽게 이루어지고, 건축은 빛으로 한발짝 더 다가서게 된다. 타워, 솔리드 부분, 매설 공 구조물은 콘크리트로 이루어져 내화력을 최적화했으며 돔 구조물도 내화강재로 만들어졌다. 메자닌은 파티오 구조 위에 놓인다.

통풍 더블 레이어의 패리에토다이내믹 시스템을 통해 에너지 손실률 제로에 도전하는 것이 이번 프로젝트의 야심찬 목표였다. 타워 파사드는 이중 레이어로 구성됐다. 바깥쪽 레이어는 방수 세리그래프 유리로, 안쪽 레이어는 필요한 경우 전통 유리가 더해진 프린트 콘크리트로 만들었다.

내외부측 사이에는 궁륭vault 구조를 삽입했다. 바깥쪽은 방수되며, 판유리와 미끄럼 방지 유리 콘크리트로 이루어져 있다. 나무 계단은 공원과 조화를 이룰 것이다. 궁륭 안쪽은 세 장의 세리그래프 EFTE로 이루어져 겨울에는 절연이 가능하고 여름에는 통풍을 가능하게 하는 에어쿠션 역할을 한다.

SYMBOLIZE

FUTURISTIC SKYLINE

HEAP

DOME

VERTICAL CONNECTION

OPEN SPACE

PATTERN

VENTILATED DOUBLE LAYER

RELATIVIZE

ACCESSIBILITY

도시에서 장면 VIEW FROM THE STREET

체코국립도서관　ANOUK LEGENDRE + NICOLAS DESMAZIERES
X-TU ARCHITECTS

레이스 섬유와 같은 투과성 표질로서 낮과 밤의 풍경　SCENES IN DAY AND NIGHT WITH PERFORATED SURFACE MATERIAL LIKE LACE FABRIC

Our aim is to integrate the project into the city typical feature formed by its 100 church steeples to relativize the volumes and to safeguard the residential character of the neighborhood to create the essential link with the park so as to secure a link with the town to reconcile all this with the symbolic side of the national library's programme. All the study rooms open to the public have been located under the dome on several levels in a single of the national library space perforated, spanned by suspended gangways the large hall. The light glides between the gangways. The mezzanines above the large reading room form cross-shaped platforms like suspended bridges. They divide the space into rooms, so that partitioning becomes unnecessary. The flowing circulations run naturally in the shade of the mezzanines. So, architecture leads one's steps towards light.

　　　　The structures of the towers are made of concrete. The structures of the dome are made of fire-protected steel. The mezzanines rest on the framework of the patios. It is the ambition of the project to make a building with "zero loss" energy through a parieto dynamic system made of a ventilated double layer. The tower facades are made of a double layer; The air circulation between the two layers allows the recovery of energy in winter and ventilation in summer.

　　　　The vault structure is included between the inner and the outer sides. The outer side is made of glass panes and non-skid glass concrete. The inner side of the vault is composed of three sheets of serigraphied EFTE making an air cushion which is insulating in winter and ventilated in summer.

도서관 내부와 자연채광　STUDY ROOM AND STACKS AND NATURAL LIGHT

공원에서의 장면　VIEW FROM THE PARK

Hyunsik Min
Korean National University of Arts / Kiohun Architects & Associates, Korea
Pyounghoa Nuri Peace Park + Youth Training Center
Imjingak, Paju, Gyeonggi-do, Korea
2006

임진각은 지난 반세기 동안 금기의 땅이었고, 철저히 외면되어왔다. 이데올로기와 총기로 더럽혀지면서 전쟁의 처절한 상흔들이, 쌓여온 깊은 증오와 갈등의 아픈 기억들이 그리고 그리움과 회한의 고통들이 깊고 넓게 펼쳐진 산과 들에, 하늘과 바다에, 땅 깊숙이 흠집으로 각인되고 무늬같이 새겨져 마치 손에 그려진 손금과도 같이 담겨 있다.

우리가 이 땅을 외면해오는 동안, 이곳을 온통 점용한 끈질긴 생명의 자연이 인위적 상처를 치유해오면서 거대한 잠재력을 길러왔다. 그래서 이곳은 오히려 위대한 가능성의 땅이다. 이제 어두운 기억들을 극복하고, 새로운 삶을 담아 평화의 별명을 가진 생명의 땅으로 화려한 부활을 시도하려 한다.

임진각에 그려진 집은 일반화된 '건축공간'에 반하여 자연과의 관계를 집요하게 추적한, 건축이라기보다는 자연의 한 요소이기를 바란 구축물이다. 재구축된 자연 지형과 이 속의 인공구축물은 건축적 산책로로 조직되고, 건축적 풍경을 이루게 된다.

음악의 언덕, 언덕의 틈새에 자리잡은 촛불 파빌리온, 물 위에 뜬 카페 안녕, 언덕 넘어 거대한 인공의 언덕으로 마련된 연수시설, 호안에 열 지은 숙박동 그리고 이들을 사라지듯 이어가다가 큰 호수를 이루는 물길 등 이곳을 구성하는 모두가 쓰임새에 충실함을 넘어 전체가 임진강을 가로질러가는 들판과 행복한 관계를 맺는 풍경을 만들어나갈 것이다.

평화누리 전경 | 야외무대를 향한 옅은 둔덕이 참여를 모은다
REBUILT TOPOGRAPHY OF THE PARK | SLIGHT MOUNT TOWARD OUTDOOR STAGE INVITES PARTICIPATIONS

평화누리 + 청소년수련원　민현식
한국예술종합학교 | 기오헌

축전 공연 풍경　CONCERT-SCAPE IN THE PARK

Put under taboo during the last half century, this land has completely looked away. Scars of war and memories of hatred are inscribed on every segment of this land like lines of a human palm. Now with the wave of reconciliation, this land is about to be reborn in the name of peace.

Scars made by man on the land have been cured by nature, and through this process of recovering it has become a land of enormous potentials. As a historically and politically important location, the site also has ecological significance.

A hill is rebuilt; waterways and lake are irrigated once more. The former is the Hill of Music, and in the cave that cuts through its section sits the Candle Pavilion. On the other hand, Cafe "Anyoung" floats above the waterway and wet land. Going over the hill, a huge artificial hill is located which contains spaces of training facilities and along the lake bank, the lodging complexes form a row.

This reconstructed geography, along with man-made constructions contains in it, forms an architectural landscape through which an architectural promenade is organized. By continuous dialogue with the existing landscape, it will one day become an important factor of the overall nature.

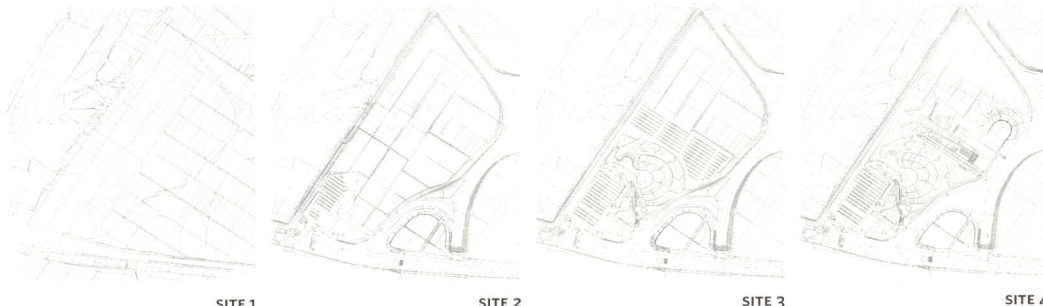

SITE 1　　　SITE 2　　　SITE 3　　　SITE 4

촛불 파빌리온 CANDLE PAVILION

연결 브리지 CONNECTION BRIDGE

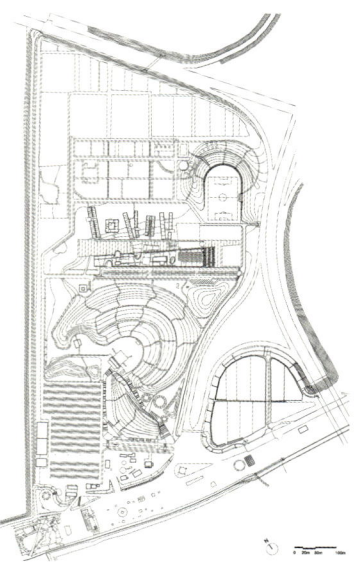

배치도_하부가 평화누리, 상부가 청소년수련원
LAYOUT_YOUTH TRAINING CENTER(UPPER) AND THE PARK(LOWER)
gross floor area: 1,859m²(built)+21,506m²(2nd project)
site area: 288,255m²

평화누리 + 청소년수련원

민현식
한국예술종합학교 | 기오헌

늪과 카페 안녕 POND AND CAFE "ANYOUNG"

위쪽의 공원과 아래쪽의 수련원이 일군을 이룬다 BIRD'S-EYE VIEW OF THE WHOLE SITE

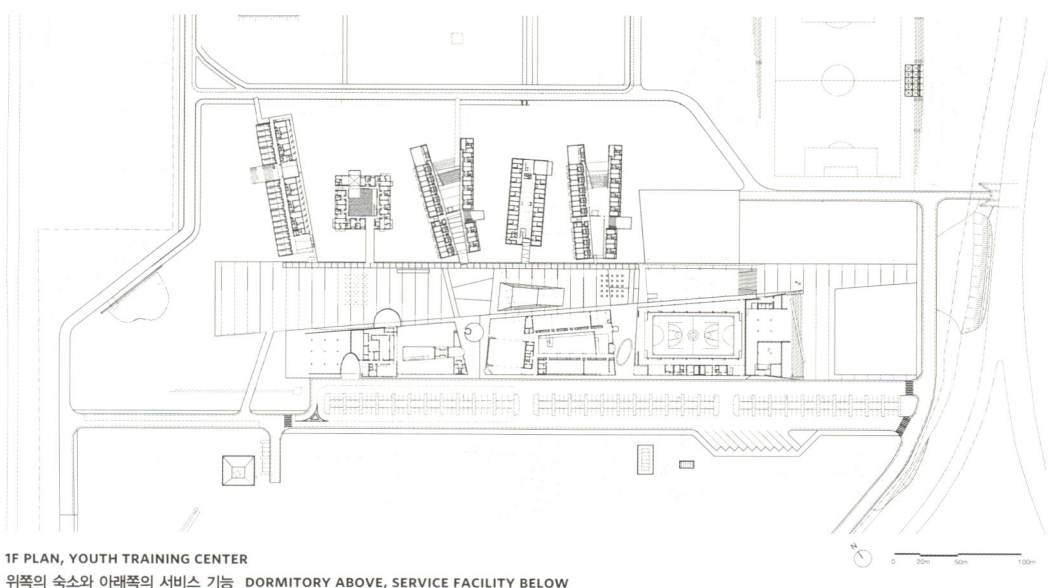

1F PLAN, YOUTH TRAINING CENTER
위쪽의 숙소와 아래쪽의 서비스 기능 DORMITORY ABOVE, SERVICE FACILITY BELOW

평화누리 + 청소년수련원 민현식
한국예술종합학교 | 기오헌

수련원 전경 BIRD'S-EYE VIEW OF YOUTH TRAINING CENTER

Byungyoon Kim

School of Architecture, Hongik University, Korea

Asia Publication Culture and Information Center

Paju, Gyeonggi-do, Korea

새로운 도시를 만들어가는 과정에서 거리, 소공원, 공원 등 비워진 장소는 매우 중요하며 필수적이다. 이런 관점에서 혹자는 불확실한 공간이라 부르는 이 공백과 비움을 통해 아시아출판문화정보센터를 계획하였다. 이 공백들은 대지를 크게 확장하고, 이들을 통해 뒷산과 파주출판도시를 따라 앞으로 평행하게 흐르는 한강이 투명하게 이어진다. 이들의 불확실한 공백으로 인해 우리들은 어디에 있는지 어느 정도 규모인지 어떤 형태의 건물인지 알아차리기 어렵다. 공백의 형태를 결정짓는 일은 개방장소 개념을 통해서 생활환경 형식을 구축함을 의미하기도 한다. 비워진 장소, 공백의 장소들은 서방에서는 공공장소로 인식되어 왔으나 우리 사고에서 이 개념은 단지 공간으로서만이 아닌 대지의 투명과 경관에 대한 사유에서 비롯된다.

출판문화정보센터는 지대한 공간형태학적 사고를 바탕으로 공공의 장소가 이루어지기를 기대하고 있으며 이들의 규모와 비워진 자연성은 도시, 경관과 어우러져 공공 행위들을 자유롭게 한다. 이 건물이 지닌 또 다른 능력과 공백의 장소들이 지닌 성격은 분명 재조명되어야 하나 경관을 담기 위한 공간으로서 다양한 평면으로 이들과 연결된 많은 공공 장소로 사용될 공간들은 다층적 지형을 지닌 새로운 형태의 대지로 인식될 것이다.

HYPERSENSUAL

MEDIA OF TACTILITY

REDUNDANT PLACE

RECORDING OF TIME

SYSTEMATIZE

CONTRACTION AND DIFFUSION

MULTIPLEX MENTALITY

CIRCULATING STRUCTURE

CONNECTION SYSTEM

NATIVE INTELLIGENCE

KOREAN TACTILITY

Initiating a certain size of planes and allocating the distances between, linking them together and finally flowing them without any disturbance to move, might be a work in terms of emerging of new lands. Whether a void or emptied space is called as a street or as a square or as a park, that was not a great problem in the time of the previous growth of new town development. But it is very consequential matter nowadays and must be an essential element in citing the new cities. According to this concerns, the void - some one called "it becomes the uncertain space" - was a specific matter for the planning of the Center in the book city and also the emptied space as well.

They wide opened the sites which became transparent, possible to see the mountain backyard towards the Han River front which flows in parallel with the land of Paju Book City. Certainly we were not able to realize the area of the place and location, form of the building as well, because of the uncertain spaces- the "void". Deciding its form means establishing the form of the inhabited environment which was once a wild and wet land in terms of the opening of the place.

The opened space or the emptied space are often considered to be public space in the western world. But in our term, open and void is not just for the space, it rather concerns about transparency of the land when conceiving the landscape. The core center of the Paju Book City has plenty of spaces of morphological scheme - the public spaces - considering that the major issues of forming this center, multi-layered planes and its empty nature are what have allowed to recognize public activities in both the city and the landscape. Surely the qualities of the building and the nature of those empty spaces need to be defined; the spaces could be fully used for public spaces as space within the landscape which is interconnected with the all of the planes and permits it to imagine new types of land, geographically with natural multi-layers of the land, where the center locates itself.

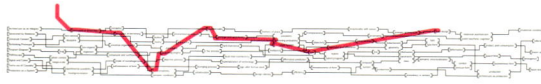

아시아출판문화정보센터

phase-1 전시정보지원동 | phase-2 교육연구지원동 | phase-3 교육연수지원동

김병윤
홍익대학교

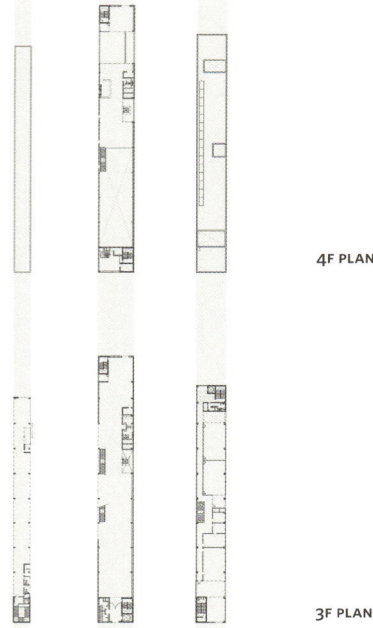
4F PLAN

3F PLAN

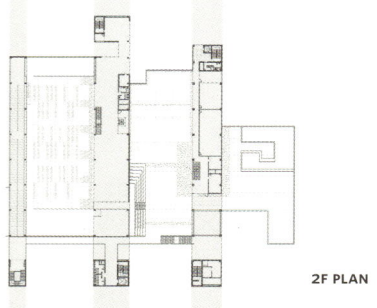
2F PLAN

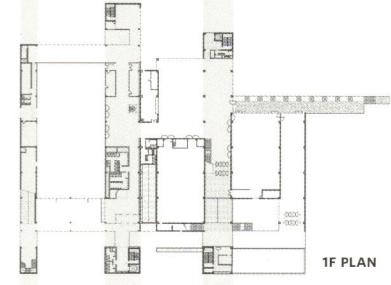
1F PLAN

분절된 공간을 엮는 동선과 시선 CIRCULATION AND GAZE CONNECTING THE ARTICULATED SPACES

아시아출판문화정보센터 김병윤
phase-1 전시정보지원동 | phase-2 교육연구지원동 | phase-3 교육연수지원동 홍익대학교

중간 옥상에서의 새로운 풍경 NEW LANDSCAPE OF THE INNER DECK

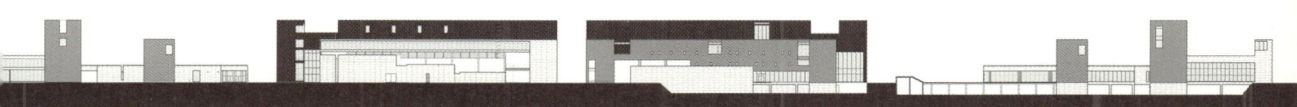

ELEVATION

Byungyoon Kim
School of Architecture, Hongik University, Korea

A-Municipal Library
Gyeonggi-do, Korea

모든 전달체계에 변화가 오는 현대의 초감각적 지각력을 바탕으로 생태학적 지능을 증진하는 대지와 주변의 관계를 현시하며, 이전에 조성된 대지의 기존 습성과 흐름, 감촉성의 관계 등을 연속된 구축 미디어로 실현시킨다. 도시의 여백으로서 사용할 대지의 흔적 읽기와 시간의 기록이자 일련의 사건처럼 미디어를 구축한다.

현재 시간에 대한 기록을 위해 다양한 변화를 지각하는 시간성 개념을 실현한다는 명제 아래 일상의 틀을 이곳을 이용할 하나의 알리바이 같은 틀에서 조직하고 이러한 대전제에 따라 모든 디자인 개념은 단계와 정지, 유동의 원칙을 유지한다. 대지의 인식 흐름을 따라 자연 재현의 재구축적 랜드스케이프가 이루어지고 생태자연의 흐름과 조경공간간 교차, 수축, 확산으로 비고정의 생태지대가 유동한다.

건축이 일으키는 기록들은 생명의 근원인 자연을 감소되게 하는 반면 장소 구축성을 통한 지능의 장소로 채움을 전제하나 다시 그 대지를 비움으로써 본래 창조된 대지가 지니고 있던 기억과 지기는 창조하는 자연의 장소로 전환되어 보다 현실적인 공동의 구축장소로 기록되기를 바란다.

비운 자리는 자연을 바라보는 새로운 틀이 될 것이며, 기억과 존재의 현실성을 담보하고 현존하는 각종 지능적 교류체계를 확보함으로써 미래를 예측하는 지혜의 장이 되기 위한 여백의 공간 등으로 내외적으로 수식과 감성의 영역으로 확대된다.

도시 자연주의적 구축 스케일의 중첩이 이루어지며 이를 바탕으로 공공적 의미, 그 이상이 실현되는 공동성을 위해 공간조직의 단계별 구축과 간격을 두고 이어지는 흐름의 문맥이 생성된다. 삶의 보다 투명함이 이루어지는 실존의 공간으로 여백은 치환되어갈 것이다.

분절과 결합 그리고 영역마다의 개별성 ARTICULATION AND INTEGRATION WITH THE SINGULARITY

RECORDING

MEMORY

INTELLIGENCE FLOW SYSTEM

URBAN NATURALISM

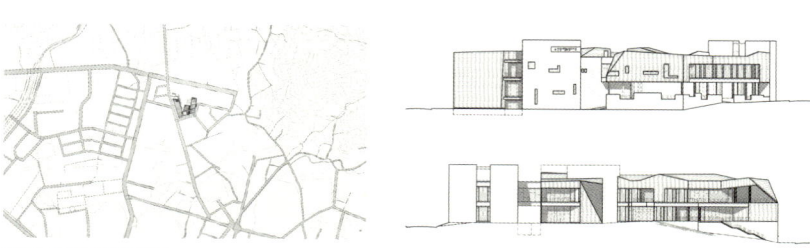

SITE | 도시와 자연의 접경 BORDER OF CITY AND NATURE AT A PROVINCIAL CITY

A-시립도서관 김병윤
홍익대학교

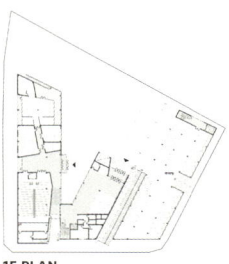
1F PLAN

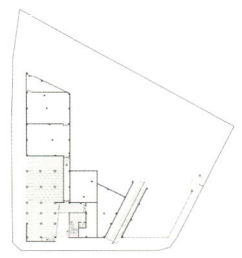
2F PLAN

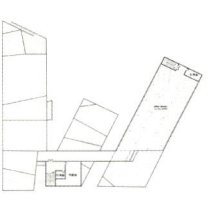

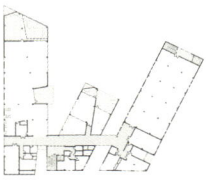

아래쪽의 직선은 도시에 접하고 뒤쪽의 열린 공간은 자연에 접한다 STRAIGHT LINE(BELOW) ENCOUNTERS THE CITY : OPER SPACE(ABOVE) MEETS THE NATURE

Looking through the context and its ecological attitude of the site with the hypersensual deliberation, this project intend to form the media of tactility and potentiality. As a redundant place of the area at the time, but this place will be formed a media towards the recording of time.

For recording against at the present time it comes true the hour sincerity concept which perceives the change which is various under proposition one alibi which will use this place from the same frame to systematize the daily life, it follows on like this large prerequisite and all design concept maintains the principle of phase and standstill and flow. The earth recognition it flows to follow and the re-construction landscape of nature reappearance becomes accomplished and mode of life nature flows with the remark just mode of life zone flows with range, contraction and diffusion of landscape architecture.

Recording which the emergency of architecture it will fill at place of the intelligence which leads the other side place construction characteristic which does to be diminished the nature which is an origin of life prerequisite one that it will rebel again and it empties as origin with the memory which the earth which is created keeps it falls and at the natural place which it creates at construction place of actuality commonness to be converted, compared to it is recorded it wishes.

Emptied place sees a nature and becomes the new frame and reality a memory and existence mortgage and secures the various intelligence flow system which exists as becomes the market of the wisdom which predicts a future and in territory of numerical formula and sensitivity with the space back of the margin for inside and outside it is magnified. The reiteration of urban naturalism tectonic scale becomes accomplished and the joint characteristic where public meaning and above that are come true with the character which will reach phased construction and flows the context becomes lifestyle. The transparency of life and existentialism of space which accomplishes a substitution for the bare land.

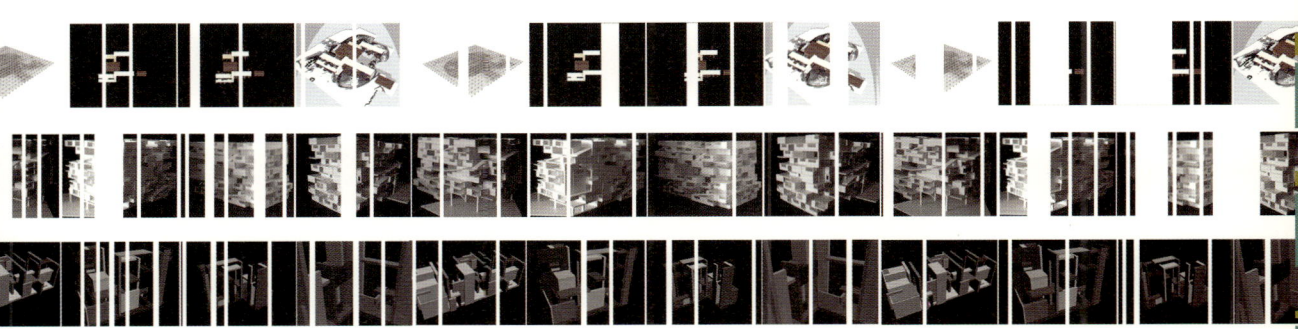
디자인 과정 모형 스터디 MODEL STUDY IN DESIGN PROCESS

T.KUBOTA + A.LUNDSTRÖM + S.MATTHYS
ELEPHANT OFFICE, SWEDEN + GANSAM PARTNERS, KOREA
SEOUL PERFORMING ARTS CENTER
COMPETITION PROJECT_2ND PRIZE, FINAL
SEOUL, KOREA
2005

서울공연예술센터는 서울시민의 문화 수준과 도시의 풍요로움을 상징하는 새로운 건물이 될 것이다. 부지에는 공연예술센터의 특징과 중요성을 부각시킬 만한 특별한 오브제가 필요하며, 이를 통해 노들섬은 그에 어울리는 새로운 정체성을 갖게 될 것이다.

계획 초기 한강에 활짝 피어난 수련을 생각했으며, 그것은 가까이서는 물론 먼발치서 바라보는 사람들에게도 신비롭고 모호한 모습으로 보일 것이다. 밤낮으로 변하는 빛에 따라, 바라보는 각도에 따라 이 거대한 원통형 건물은 다양한 모습으로 변화할 것이다.

공연예술센터 내부는 오페라하우스와 콘서트홀로 구성된다. 오페라하우스와 콘서트홀이 수직으로 적층되고, 중앙 로비에서는 시야를 전혀 방해받지 않고 도시 전경을 바라볼 수 있다. 도시 한가운데에 위치한 노들섬의 지리적 특성이 시각적으로 드러나는 것이다. 건물 내부로 빛과 영상이 투과되어 낮 동안에는 거의 파악되지 않는 건물의 부피감이 밤이 되면 점점 더 명확해진다.

파사드는 반투명 금속 스크린으로 만들어진 섬세한 외피로, 밤에는 표면에서 점멸하는 색색의 빛점들이 만들어내는 이미지로 파악되어 특유의 복합적이고 다양한 모습을 드러낼 것이다.

ECOLOGICAL

SPAC

NATURE

URBAN IMPACT

CULTURAL CONTEXT

UNITIES OF ACTION, PLACE, AND TIME

COMPOSITION

SKIN

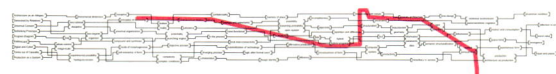

서울공연예술센터
현상설계 최종2등작

ELEPHANT + 간삼파트너스

한강교, 노들섬 그리고 아트센터의 상징적 형태 HANGANG BRIDGE, NODEUL ISLAND AND MONUMENTALITY OF CYLINDRICAL ARTS CENTER

The Seoul Performance and Art Center (SPAC) reveals a mystic blurring complexity to both the passing spectator and the distant viewer. What might this vast cylindrical object be, changing according to the lights of day and night, according to the angle of view?

At night her appearance changes to almost its opposite. The spectator can only guess the exterior volume by the colorful light points on its skin, the pixellized images emerging and disappearing on it. Inside, lights and reflections come up and express a dense interior life. Volumes that have been nearly invisible during daytime become more and more perceivable. The interior is a poetic universe of its own, complex, diversified, attracting, longing to be discovered.

Sited on an island with evocative history the SPAC will become a new symbol of the cultural richness and energy for the city and the people of Seoul. The SPAC consists of two main interior volumes; the Opera and the Concert hall in a vertical arrangement, and a main foyer cutting in between to become visually aware of the prominent position of Nodeul Island in the heart of the City - by its extraordinarily unobstructed panoramic view. The sculptural volumes are inscribed in the contrastive geometry of a cylindrical envelope, consisting of a semi transparent metal screen facade.

OPERA HOUSE

CORES

CONCERT HALL

PUBLIC CIRCULATION

REHEARSAL ROOM

SERVICE

SKY RESTAURANT

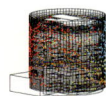
SKIN

COLOR GLASS BLOCKS

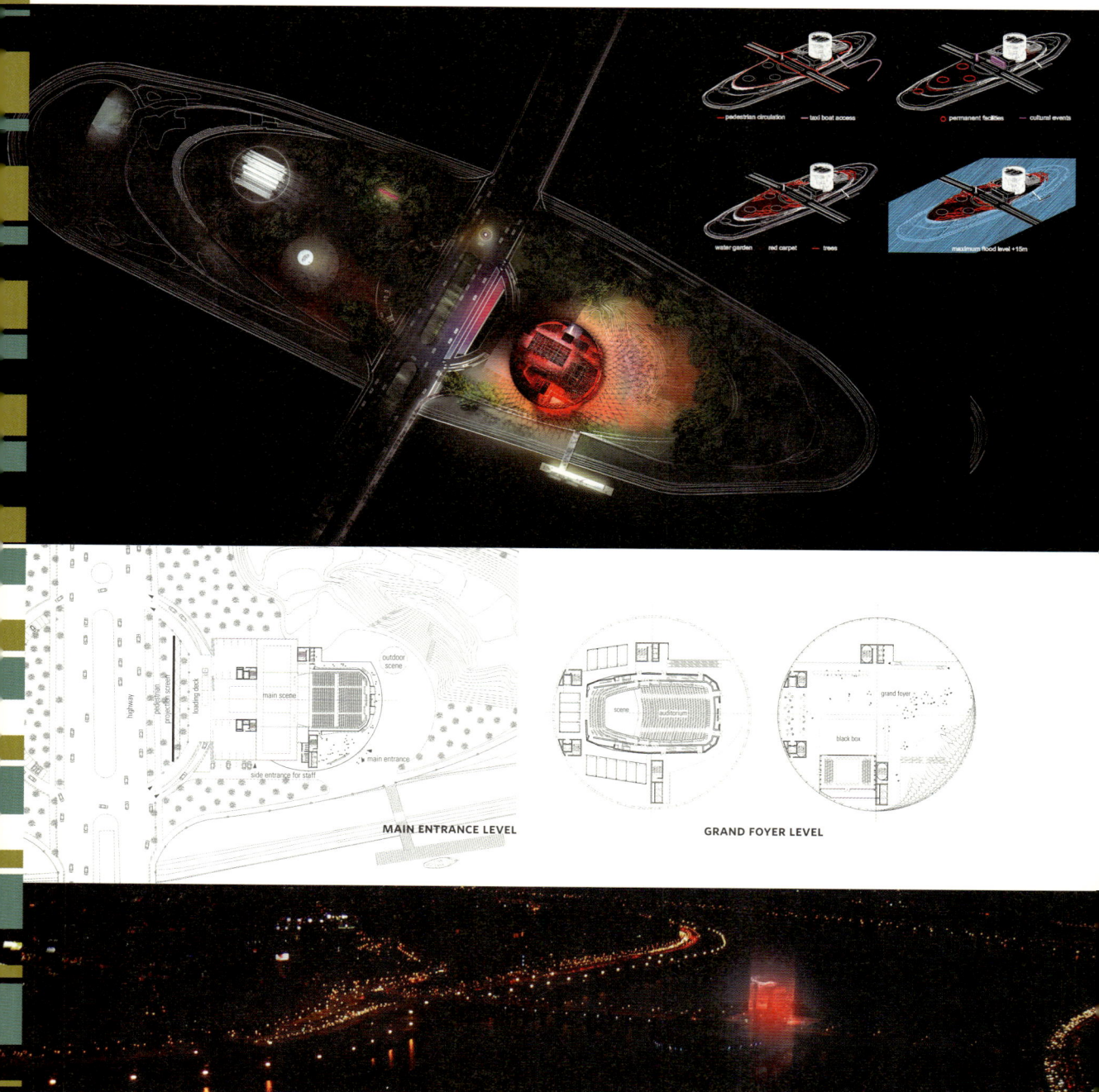

한강 야경의 비스타 NIGHT VISTA OF HAN RIVER

서울공연예술센터
현상설계 최종2등작

ELEPHANT + 간삼파트너스

LEGEND
- VISITORS / PUBLIC
- SERVICE / LOADING DESK
- VIEWS
- BLACK BOX

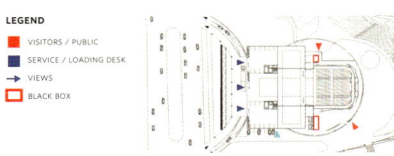

ACCESS · SCENES & WORKSHOPS

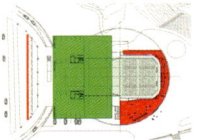

THEATRE

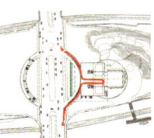

TAXI DROP-OFF

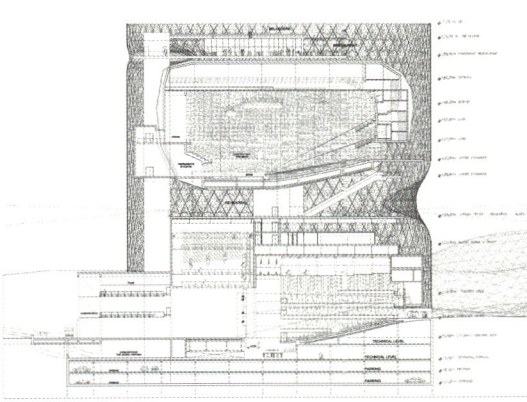

SECTION

오페라하우스 INTERIOR OF OPERA HOUSE

반투명 금속 스크린 외피가 수면과 만드는 풍경 LANDSCAPE OF WATER WITH SEMI TRANSPARENT METAL SCREEN FACADE

JEAN NOUVEL
Ateliers Jean Nouvel, France + Samoo Architects & Engineers, Korea
Seoul Performing Arts Center
Competition Project_Winner, Final
Seoul, Korea
2006

한강은 주변 산세와 함께 서울의 가슴을 흐르며 풍경을 만드는 유산이다. 아주 오래전부터 한강을 주제로 하는 산수화가 즐겨 그려져 왔으며, 서울사람의 심성을 씻어 왔다. 아마 장 누벨이 본 인상이 조선의 전통 산수화인 진경산수일 것이다. 그리고 그는 이를 노들섬에서 재현한다.

우리는 전통적으로 석가산이라는 정원의 조경기법을 가지고 있는데. 그것과 장 누벨의 생각이 닮았다. 다만 여기에서는 media tactility가 함께 구사된다. 현대의 기술과 한국의 산수는 대립적이지만, 그 둘의 결합이 현상을 일구면서 새로운 미적 체험을 이룬다. - 박길룡

The Han River is a heritage that flows through the heart of Seoul and creates a landscape in harmony with the surrounding mountains. Since a very long time ago, landscape painting on the theme of the Han River had often been produced, purifying the minds of the people of Seoul. Jean Nouvel's impression of Seoul would probably have been one of Jingyeongsansuhwa(real landscape painting) which is a traditional landscape painting of the Joseon Dynasty, and he attempts to reenact this on Nodeul Island.

Traditionally, Korea had a landscaping technique for gardens called Seokgasan (artificial hills), and this is similar to the ideas of Jean Nouvel. However, here it is combined with "media tactility". Although modern technology and the mountains and rivers of Korea are conflicting concepts, the combination of the two creates a phenomenon, achieving a new aesthetic experience. - Text by Kilyong Park

LANDSCAPE
SCENERY
BLURRED
POETIC METAPHOR
KOREANESS
URBAN CONTEXT
LINK
IMAGE

한강, 한강교 그리고 새로운 섬의 풍경
HAN RIVER, HANGANG BRIDGE AND THE LANDSCAPE OF NEW ISLAND

081 서울공연예술센터 현상설계 최종당선작 JEAN NOUVEL + 삼우설계

한국 전통 정원 석가산의 확대된 개념 CONCEPTION SIMILAR TO KOREAN TRADITIONAL LANDSCAPE

한국의 진경산수 풍경과 기술 KOREAN TRADITIONAL LANDSCAPE AND TECHNOLOGY

새로운 지형 속에 묻힌 공연 기능 PERFORMING HALLS BURIED INTO THE NEW TOPOGRAPHY

자연과 혼합하는 공적 영역 PUBLIC AREAS MIXED WITH NATURE

서울공연예술센터
현상설계 최종당선작

JEAN NOUVEL + 삼우설계

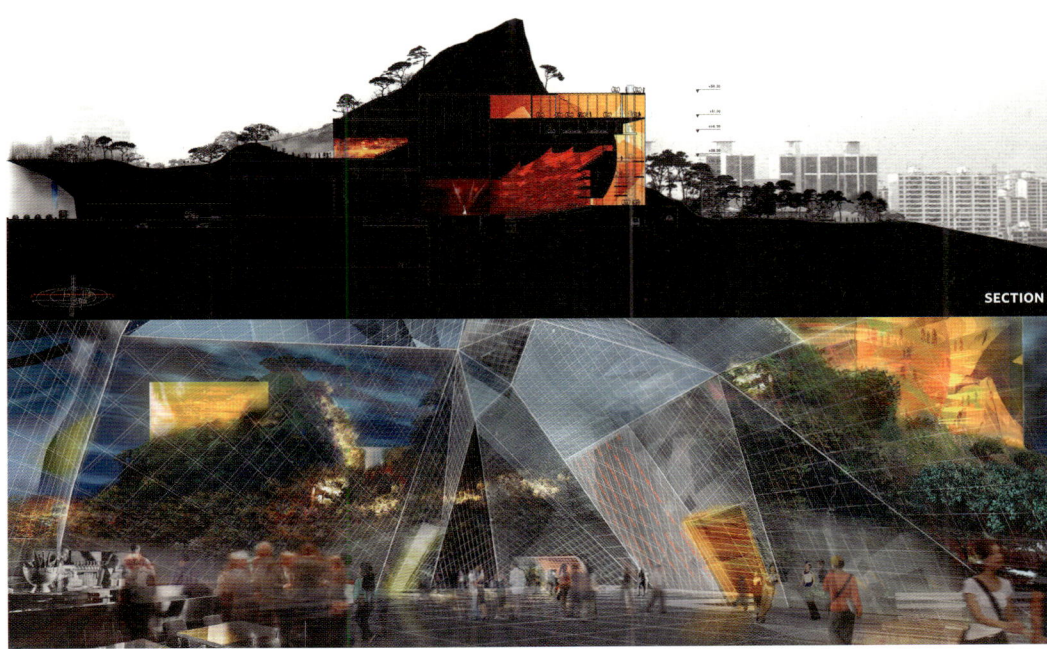

SECTION

내부 투시도 INTERIOR PERSPECTIVE

SECTION

SYMPHONY HALL

OPERA THEATER

오페라하우스 INTERIOR VIEW OF OPERA HOUSE

Youngsun Chung
Seoahn Landscape Architects, Korea
Seonyudo Park Landscape
Seoul, Korea
1999-2002

선유정수장은 한강 물을 직접 끌어올려 정수하던 곳으로, 섬 전체가 물을 가득 품고 있던 물의 섬이었다. 그러나 선유정수장이 선유도에 자리잡은 1970년대 후반 급속한 산업화의 부작용으로 인해 한강의 수질은 날로 악화되고, 한강개발사업으로 강변의 자연생태계가 파괴되어 한강은 오염된 물길로 전락한다. 선유정수장이 그 기능을 다하고 한강이 이전의 푸름을 되찾아가는 즈음 선유도의 공원화는 자연스럽게 한강과 환경 그리고 생태 문제로 그 방향을 잡는다.

공원 설계는 유기적으로 구성된 시설물들이 만들어내는 공간과 땅의 모양을 이용하여 선유도와 선유정수장이 간직한 기억과 환경, 자연 그리고 미래를 어떻게 담아낼 것인가에 초점이 맞춰졌다. 선유도가 도시에서 가지는 지리적·공간적 잠재력을 드러내는 것도 생각의 중심을 이루었다.

선유도는 크게 4개 부분으로 구분된다. 첫째는 선유도를 둘러싼 옹벽 하부의 둔치로 한강의 생태복원을 시도하는 공간이다. 둘째는 옹벽 둘레의 언덕 부분으로 숲과 조망이 있는 놀이와 휴식, 문화의 공간이다. 셋째는 물의 흐름을 따라 전개되는 환경과 생태를 주제로 한 정원들이다. 넷째는 공원이 가지고 있는 환경과 생태교육의 기능을 지원, 강화하는 정보, 전시, 관리의 공간이다.

기능을 벗어나버린 정수장의 구조물들은 공간과 입체의 조형성으로 우리에게 다가온다. 너무 빽빽하게 자란 숲의 나무를 솎아내듯 재활용의 가능성과 공간의 잠재력을 가진 구조물과 건물을 선별하고 나머지는 철거하여 공간을 만들었다. 남겨진 건물들은 정수장 시설의 핵심이던 지하공간과 함께 땅의 입체적인 요철을 체험할 수 있는 극적 공간을 형성하는 동시에 전체 공간에 강약 리듬을 준다. 건물들을 철거한 빈자리를 대신 채우듯 질서정연하게 심겨진 미루나무와 곧게 뻗은 동선을 따라 일직선으로 늘어선 나무, 단정한 모양의 띠를 이루며 모아 심겨진 흰 자작나무는 정수장 구조물이 지닌 강한 직선 형태를 공간 전체에 공명시키는 역할을 한다.

GEOGRAPHICAL

MEMORIES OF HISTORY

ECOLOGICAL RESTORATION

PLAYGROUND

RELAXATION

ECOLOGICAL

CIRCULATION

SYMBOLIZING

GEOGRAPHICAL DESIGN

ENCOMPASSING VIEW

PURIFICATION STRUCTURES

SEDIMENTARY LAYER

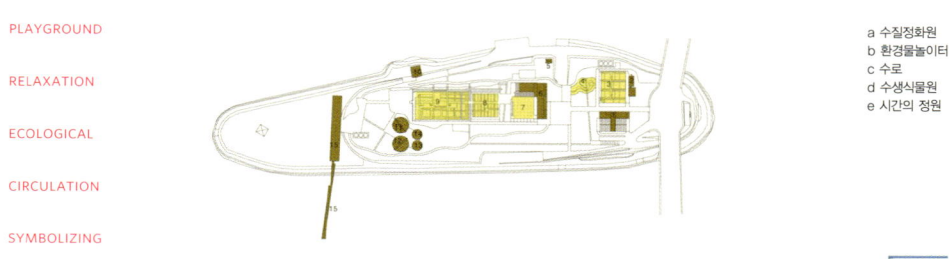

a 수질정화원
b 환경물놀이터
c 수로
d 수생식물원
e 시간의 정원

1 방문자 안내소
2 온실
3 수질정화원
4 환경물놀이터
5 선유정
6 한강전시관
7 녹색기둥의 정원
8 수생식물원
9 시간의 정원
10 카페테리아 '나루'
11 원형극장
12 환경놀이마당
13 환경교실
14 화장실
15 선유교

SITE PLAN

주제정원을 흐르는 물은 수질정화원→환경물놀이터→수생식물원→시간의 정원→물 저장탱크를 거쳐 다시 수질정화원으로 순환한다

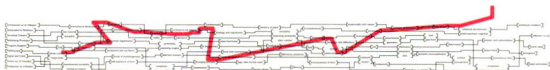

선유도공원 조경   정영선
서안조경

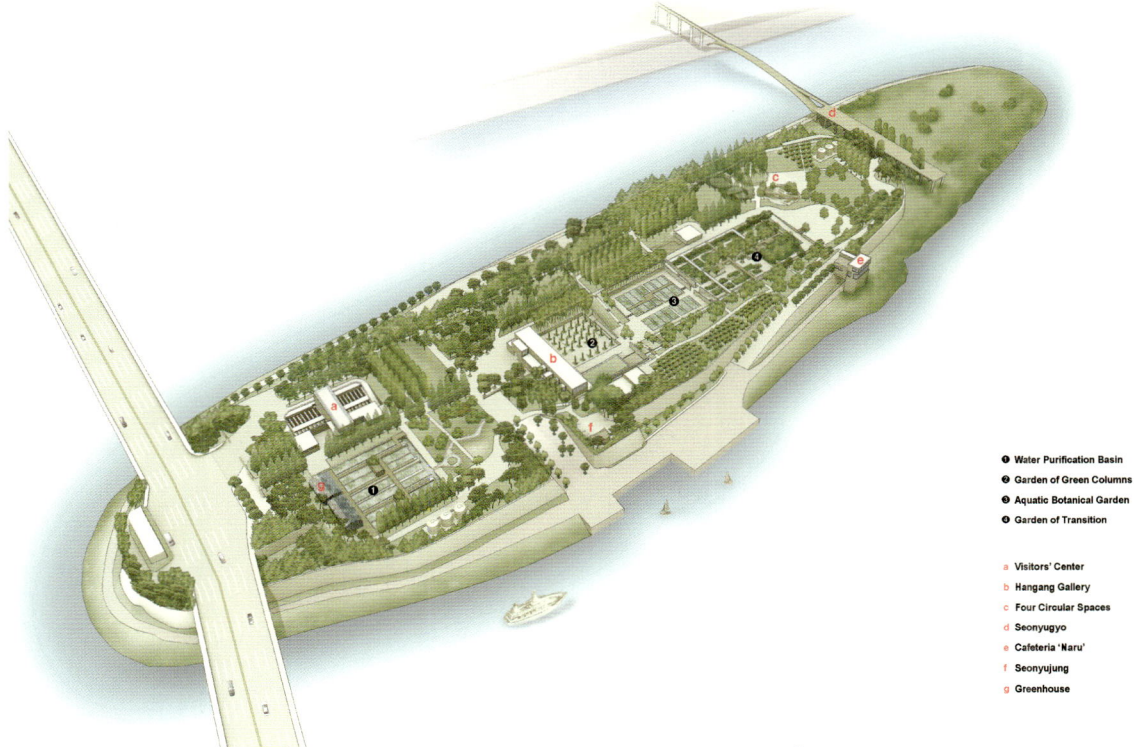

1. Water Purification Basin
2. Garden of Green Columns
3. Aquatic Botanical Garden
4. Garden of Transition

a. Visitors' Center
b. Hangang Gallery
c. Four Circular Spaces
d. Seonyugyo
e. Cafeteria 'Naru'
f. Seonyujung
g. Greenhouse

제2한강교와 선유도공원 2ND HANGANG BRIDGE AND SEONYUDO PARK

Seonyudo once was an islet full of water - since it used to accommodate a water purification plant that drew water directly from the Han River. However, in the late 1970s when the water purification plant was constructed on this island, rapid industrialization, had a negative impact on the environment, including deteriorating water quality in the Han River. Moreover, Han River development projects destroyed the ecosystem of the riverside, reducing the Han River to a polluted waterway. Nowadays, as the plant has ceased to perform its original function, the Han River is regaining its green scenery of the past, the reborn Seonyudo Park is resolving environmental and ecological problems affecting the Han River.

The main design concept for the Seonyudo Park was to reveal the geographical and spatial potential of Seonyudo, situated in the midst of the city of Seoul. At the same time, the design concept focused on expressing memories of Seonyudo's history and the water purification plant, by utilizing the peculiar shapes of space and land created by organically composed facilities, for the environment, nature and the future.

The newly-transformed facilities of the former water purification plant are now viewed as structures with and formative beauty of spaces and solidity. As if to cut away trees in an over-grown forest, only the structures and buildings with potential for recycling are sorted out to create more space. The selected remaining facilities, together with the underground space once key to the water purification plant, create a dramatic space where visitors experience the solid unevenness of the ground, simultaneously giving the whole space rhythmical sense of strength and weakness.

In the empty space where some facilities have been removed, cottonwood trees are planted in smart order, and trees line the straight circulations. Also, the white birches planted together in tidy belt-shape function to resonate throughout the whole space the shape of intense straight lines of water purification structures.

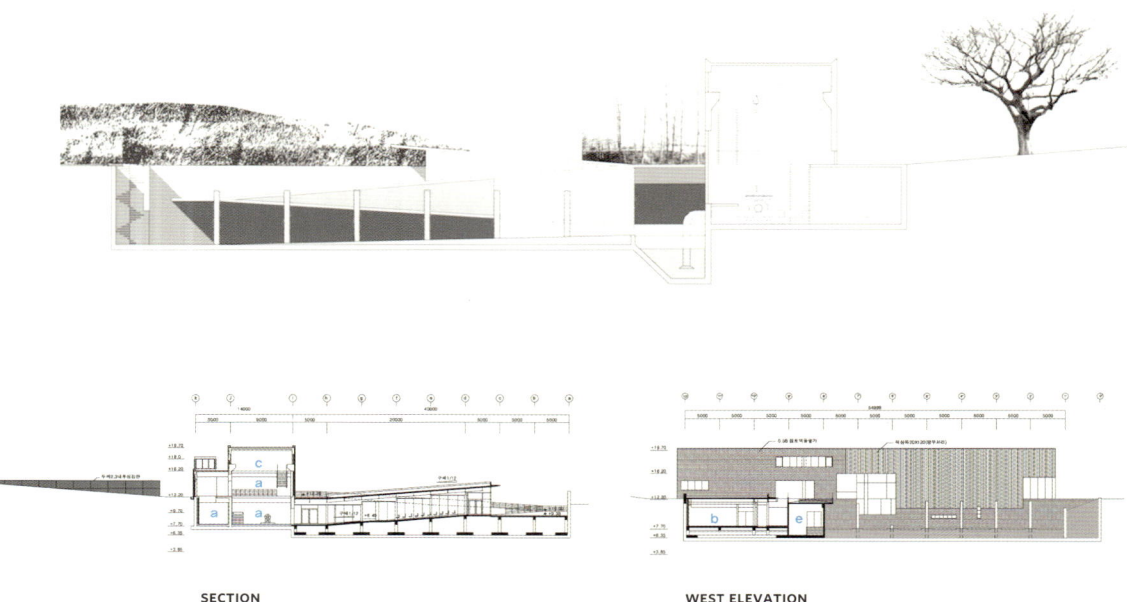

SECTION

WEST ELEVATION

기존 시설물이 전하는 시간의 기억 MEMORY OF REMAINS

선유도공원 조경 정영선
서안조경

기존 시설물들의 텍토닉한 성질
TECTONIC CHARACTER OF EXISTING STRUCTURE

저수조의 정화 기능 PURIFYING FUNCTION OF RESERVOIR

새로 디자인한 인공 연못
NEWLY DESIGNED ARTIFICIAL POND

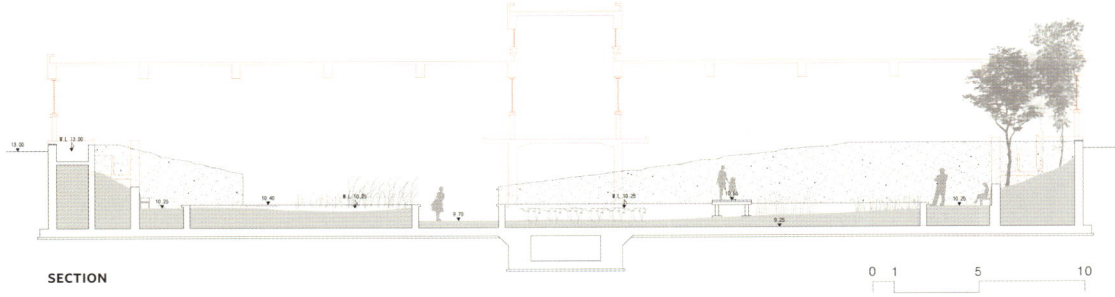

SECTION
기존 구조물과 자연조경의 얼개 FRAMEWORK OF EXISTING STRUCTURE AND NATURAL LANDSCAPE

직선의 수로와 자연의 유기성 RECTILINEAR WATERWAY AND ORGANIC NATURE

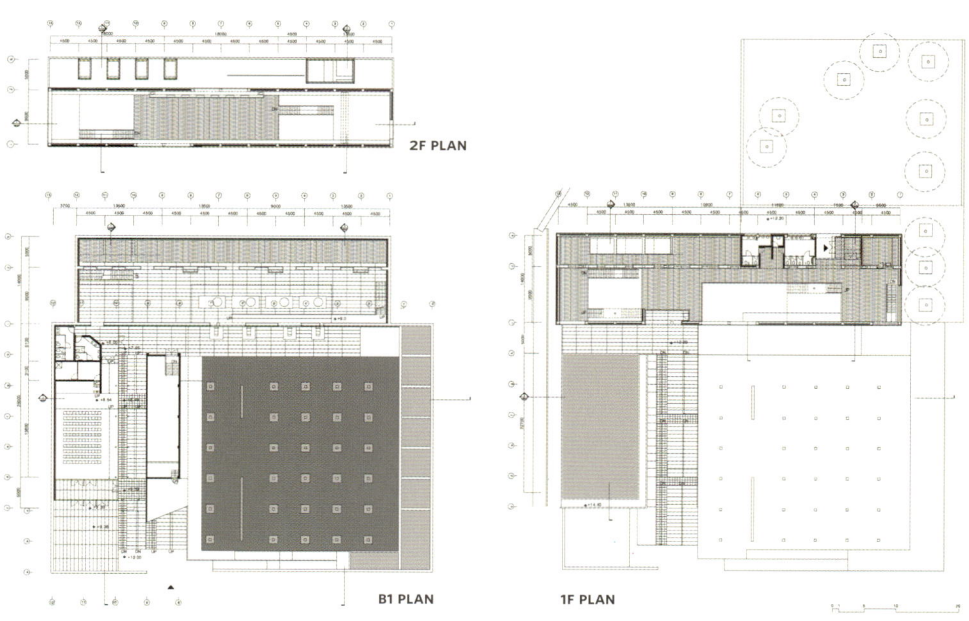

전시관 _ 조성룡 EXHIBITION HALL DESIGNED BY ARCHITECT SUNGYONG JOH

선유도공원 조경　정영선
서안조경

기존 시설의 잔해를 이용한 야외소극장　AMPHITHEATER WITH REMAINING RUINS

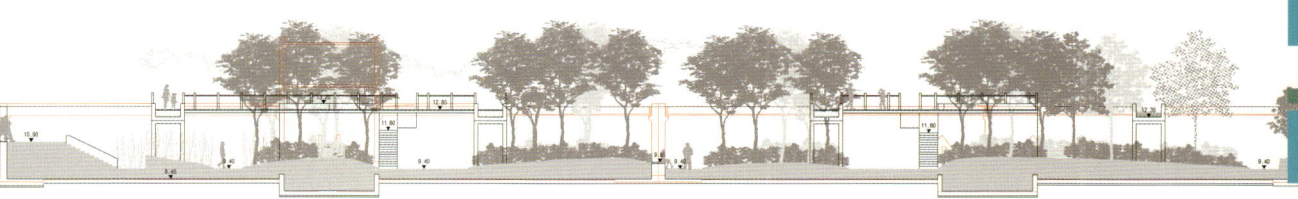

자연환경과 인공구조물의 교직　INTERCONNECTED NATURAL LANDSCAPE AND ARTIFICIALITY

Andrés Perea Ortega
Spain
Seoul Performing Arts Center
Ideas Competition Project_1st Prize
Seoul, Korea
2005

서울공연예술센터는 모두가 꿈꾸는 장소를 만드는 데 의의가 있다. 강물의 흐름, 낮과 밤, 시간 흐름, 계절 변화 등에 따라 행사를 기획할 수 있다. 이 프로젝트는 모든 형태의 커뮤니케이션을 위한 울타리가 된다. 사회·문화·종교적인 것에서부터 개인적이고 친숙한 것에 이르기까지 다양한 프로그램을 수용한다. 이러한 환경을 만들기 위해 직접적인 교류나 활동의 공유와 같은 전통적인 시스템에서 출발하여, 필요한 정보시스템을 제공하기 위해 최신 테크놀로지를 사용할 것이다.

　　　　　이 프로젝트는 인간에너지, 현상학적 에너지, 정보에너지 그리고 기술에너지에 이르기까지 모든 에너지를 수용한다. 에너지 자원의 재활용 측면에서 풍력·수력·태양에너지, 재생에너지 등을 사용해서 건물을 운영할 것이다. 그리고 이 건물의 유지와 변형이야말로 미래의 소프트웨어 기술이 될 것이다. 매개방법, 사적·공적 영역, 보행자, 제품공급, 조용한 환경을 유지하기 위한 내부의 교통, 주차 등의 기능적인 문제를 해결하고 순환의 충격과 범람 관련 문제도 연구했다.

　　　　　서울공연예술센터는 열린 가능성을 보여주는 장소여야 한다. 독특한 정원, 단체 이용이 가능한 온실, 주거, 창조적인 장인들을 위한 워크숍과 스튜디오, 연구센터 등은 건물의 사용과 유지 그리고 생명력을 담보할 것이다. 이 모든 것은 별도 기능을 추가하지 않아도, 여러 활동을 결합하고 중복 공간을 조율함으로써 얻을 수 있다. 이는 상호작용 및 변화와 지속성 측면에서 효율성을 보장하는 간결하지만 복합적인 제안이다.

ENCLOSURE

COMMUNICATION

RECIPIENT OF ENERGY

MAINTENANCE AND TRANSFORMABILITY

LIGHTNESS AND FLIMSINESS

COINCIDENTAL AND CONTRADICTORY

COMPLEX WITH POSSIBILITIES

COMPACT AND COMPLEX

INTERACTIVITY

MUTABILITY

DURABILITY

The project proposes an event stranded in the river, surrounded by the flow of the river, the days, nights, months, autumns, winters, springs, episodes of social, cultural, religious, familiar, personal events. The project is an enclosure for all forms of communication. These include, introspection, interpersonal interactions, familiar or tribal affairs, and metropolitan, cosmopolitan and international relations. And for this environment, we provide systems adapted to information needs, with the latest technologies, starting from the traditional systems of direct relation, or shared activities.

　　　　　The project is a recipient of energy: human energy, phenomenological energy, information energy and technological energy. The project will be a state-of-the-art answer to the demands of a sustainable environment. Architecture will use all natural resources in running the complex, and its maintenance and transformability will be tomorrow's soft technology. The time for lightness and flimsiness has arrived.

　　　　　Capable of traveling through the future events of Seoul. The kit program of the contest is incremented by complementary activities that enlarge the character of the anticipated functions, or that establish new overlaid programs, coincidental and contradictory in some cases.

BUILDING

NOT BUILT

ACCESS POINT

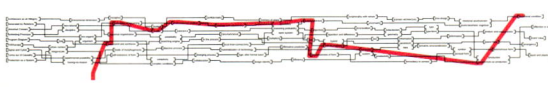

서울공연예술센터
아이디어공모 1등작

ANDRÉS PEREA ORTEGA

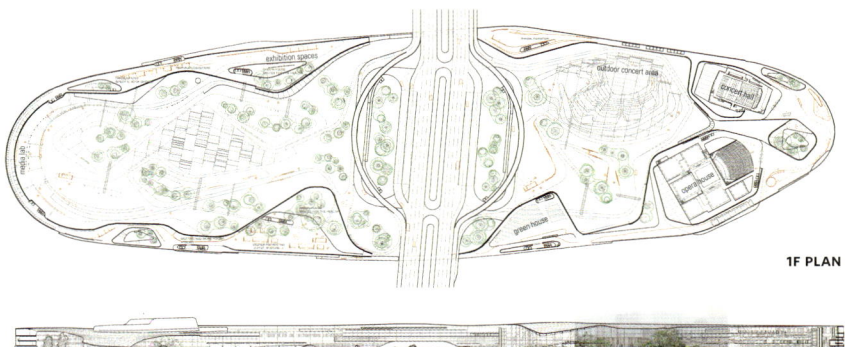

1F PLAN

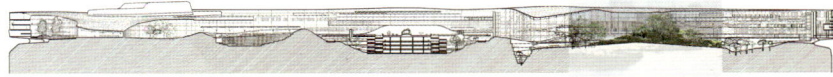

SECTION

WEST ELEVATION

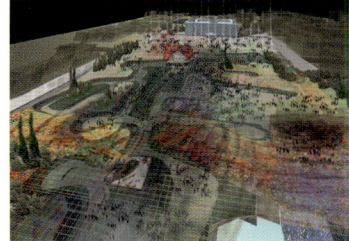

백만 개의 레이어를 갖는 정원
THE GARDEN OF ONE MILLION LAYERS

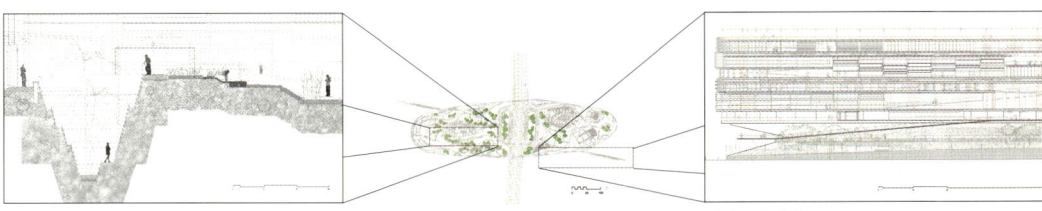

정원의 종단면
THE GARDEN OF ONE MILLION LAYERS_SECTION

입면 부문 상세
DETAILED ELEVATION FRAGMENT

한강, 한강교 그리고 섬 VIEW OF HAN RIVER, HANGANG BRIDGE AND THE ISLAND

Andrés Perea Ortega
Spain
New Multi-functional Administrative City
Competition Project_Winner
Yeongi, Chungcheongnam-do, Korea
2005

특화되지 않고 세대를 초월한 공동의 건설 프로젝트인 이 도시는, 정치가가 아닌 시민들이 적극적으로 일구는 역사로 이루어지는 곳이다. 도시공간은 상징적인 장소가 아니라 활용할 수 있는 도구이다. 사유재산을 공동재산 개념으로 받아들이고, 특히 땅을 사용할 때 거주행위와 자연의 사유화를 구분한다. 또한 이 도시는 도시성과 자연환경이 균형을 이뤄 농촌, 도시, 사이버공간이 중첩되는 복합적이고 효율적인 환경체계가 구현되는 활력 넘치는 곳이다. 우리는 기존 마을과 조경이 미래 도시와 행복하게 공존하는 건강하고 자연치유적 도시를 추구한다. 자연조건이 뛰어난 이 도시에서는 나들이, 산책, 여행 등을 즐길 수 있다.

이 도시는 경계가 없고 지속적인 구조와 형태를 띠며, 2만 명이 거주하는 25개 도시가 응축된 복잡한 곳이다. 도시 기반시설은 중첩되고, 공간은 여유롭고 완충지가 있으며, 다양하게 변모하여 수천 가지의 시각적 자극으로 가득 찬 도시가 될 것이다. 도시는 변이할 수 있도록 배열된 염색체와 같다. 기반시설을 중첩하고 누적해 기타 활동들은 인근 혹은 전국의 시설을 이용할 수 있게 한다. 이는 상호중첩된 활동과 흐름을 통해서 구역화 개념을 거부하는 것이다. 우리가 제안하는 도시관리 및 개발전략은 개방적이므로 시간에 따라 다양한 방향으로 나아갈 수 있다.

TURE AND UNMISTAKABLE CITY

USEFUL CITY

CITY OF USERS

CITY OF SOLIDARITY

DIVERSE AND CHANGING CITY

THE VALLEY AS URBAN UTOPIA

CONFIRM AND INTEGRATE THE EXISTING VILLAGES

CREATION OF A RECREATIONAL LAKE

COMPLEX AND COMPACT CITY

LENGTHWISE STREETS

CROSS STREETS

INTERSTICE

외곽으로 서클을 이루는 주거와 시설 RESIDENCES AND SERVICES FORMING THE OUTER RING

행정중심복합도시
현상설계 당선작

ANDRÉS PEREA ORTEGA

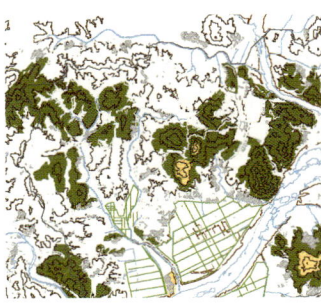
자연 NATURE

기반시설망 INFRASTRUCTURE

구조와 도시 형태 STRUCTURE AND URBAN FORM

This city is the "city of democracy" and "city of people", in which the only representation is the people. Being a very different place, this is an ideal city for people to realize their dreams and desires. This city is an unspecialized and timeless cooperative development project, a city established by people's affirmative public works, and not by politicians. Therefore, the urban space is not a symbolic one, but a practical one.

The concept of private property is replaced by that of common property, and especially when it comes to the management of land, the residence and the privatization of nature is distinguished. This will be a vibrant place where a complex and efficient environmental system for life is established, in which rural, urban and cyber spaces intersect with each other in a harmony of urbanism and the natural environment. A healthy and self-healing city in which preexisting villages and landscapes happily coexist with a future city is pursued. Filled with sunshine and clear air and corresponding with natural conditions, the city will be suitable for physical activities such as outings, promenades, travel, and more.

This city has a continual structure and form without any boundaries. It is a complex and condensed place of 25 cities with 20,000 dwellers. The city's foundation systems intersect with each other, and there are spare spaces and buffer zones. The city, changing diversely, is full of thousands of visual stimuli. The city is like a chromosome arranged for mutations, and the intersecting and accumulating foundation systems make it possible to use national and local facilities for other activities. This intersecting activities and streams defy the concept of localization. The city management and development strategy we proposed is an open strategy that can progress into various directions as time goes on.

빈 중심을 향한 경관 PERSPECTIVE VIEW TOWARD CENTRAL VOID

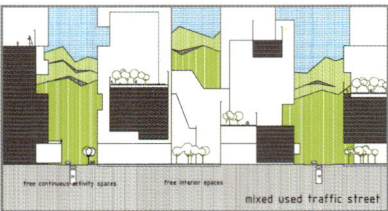 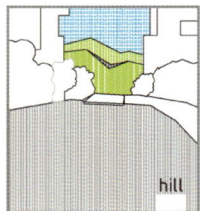 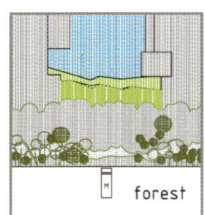 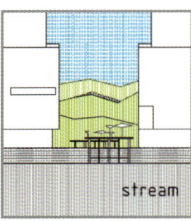

도시 경관의 다양성 DIVERSITY OF URBAN LANDSCAPE

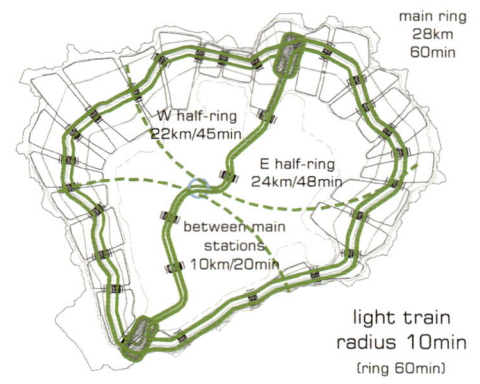 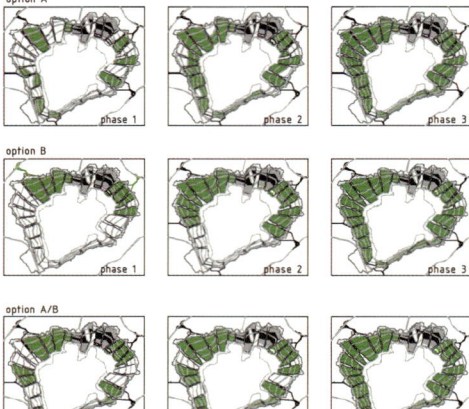

25 MIN FROM ANY STATION TO THE VALLEY

비워진 도시 중심에서 채워진 외곽을 향한 경관 VIEW FROM CENTRAL VOID TO PERIPHERAL SOLIDS

행정중심복합도시
현상설계 당선작

ANDRÉS PEREA ORTEGA

 Natural-sphere
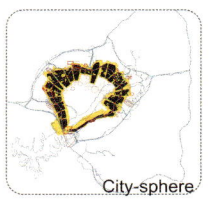 City-sphere
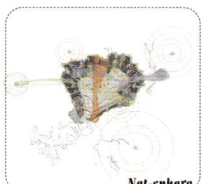 Net-sphere

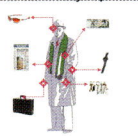

MVRDV + Youngwook Joung
The Netherlands / Korea
Anyang Peak
Anyang Art Park, Gyeonggi-do, Korea
2006

이 공공시설물은 안양유원지에 조성된 예술공원에서 전망대 기능을 하며 공원 전체에서는 시각적 비스타를 이룬다. 전망대는 삼성산의 등고선을 원뿔형으로 연장하여 산의 형태를 확장시킨 것이다. 작은 첨탑으로 안내하는 길은 안양유원지 중심부의 정상에 있어, 공원의 핵심 부분인 언덕을 오르다보면 어느새 나선형을 그리며 언덕 위로 이어지는 타워가 된다. 정상 아래로 두 개의 등고선을 이용해 길의 윤곽을 결정했다. 두 개의 등고선 중 하나는 외부 나선을, 다른 하나는 내부 나선을 형성하며 이 두 곡선이 안쪽에서 만나면 길의 폭에 변화가 생긴다. 1.5m를 최소 폭으로 두 길의 기준을 삼고 있으며, 길의 경사는 10분의 1로 고정시켰다.

네 개의 고리가 이어진 듯한 146m 길이의 길은 14.6m 높이의 타워를 형성하며, 160m² 규모의 지역을 감싸게 된다. 내부의 빈 공간은 소규모 전람회 같은 행사를 유치하는 전시관으로 사용될 수 있으며 공연예술을 위한 무대로도 사용할 수 있다. 이때 관객들은 언덕 정상에서 무대를 내려다볼 수 있다.

A path which is leading to a small peak is located on top of central summit in Anyang resort. One way to revitalize this area is to emphasize the natural wonders that are there already. Nature could be intensified. A viewing tower is proposed, supercharging the hill into Anyang Peak. The path leading up the hill, an essential element of the park, is used as a tool to generate this idea. The spiral path becomes the tower, extending the hill seamlessly. Reshaping the peak. Two contour lines from the top were used to shape the path. One forms the outer spiral and another one forms the inside spiral line. As these two contours offsets inwards, the width of the path varies. The minimal width which is 1.5m was the guiding line for these two contours. And the inclination of the path was fixed as 1/10 slope.

A 146m long path with a four rings forms the 14.6m height peak which covers 160m² area. The internal void acts as a pavilion; it can hold a small exhibition or installation. The space can also be used as a performance space, allowing visitors to look down on the stage on top of the hill. The path encircles the peak turning it into a destination.

SPIRAL PATH

CENTRAL SUMMIT

PARK

HILL TOWER

EXTENDING CONTOURLINE

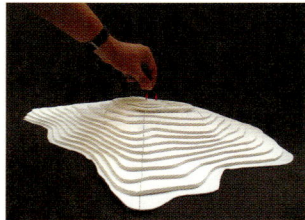 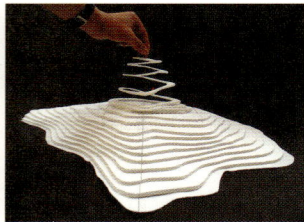

지형의 연장으로서 인공 구축물 ARTIFICIAL CONSTRUCTION AS A TOPOGRAPHICAL EXTENSION

안양 피크   MVRDV + 정영욱

주변과의 교감 SYMPATHY WITH SURROUNDINGS

내부에서 하늘을 봄 INTERIOR VIEW TO THE SKY

접근에서 읽히는 생성 원류
TORRENT OF BECOMING FROM APPROACHING POINT

MVRDV + Youngwook Joung
The Netherlands / Korea
City Sofa
Competition Project_Participation
Busan, Korea

서로 다른 프로그램들은 각각의 독립적인 성격을 수행하는 지구를 형성하며, 이 지구들은 다시 하나의 레이어로 융합되어 거대하게 꺾여 들어올려진 볼륨 형태로 표현된다. 이 캔틸레버 구조로 들어올려진 제스처는 지상층에서 시네마 빌리지로 극적으로 접근하는 동시에 상층부에서는 자연스럽게 형성된 슬로프 위에 야외상영극장이 배치된다. 도시 쪽에서는 꺾여 올려진 이 건물의 제스처가 멀리 있는 타워의 공간적인 볼륨과 연결되며 이 집합체로 접근을 유도한다. 내부 공간에서는 다양한 박스들이 앉혀진 거대한 밸리가 형성되며 이 밸리의 땅과 맞닿은 부분에 콘코스가 있다. 콘코스 서쪽 부분에 위치한 각각의 극장들은 콘코스에서 모두 보이도록 되어 있으며, 그 반대편에 컨벤션 홀과 높이 들어올려진 사무실들이 자리하고 있다.

The needed volume is imagined as a single layer of the demanded program. The programs are sorted in clear functional zones that allow for independent usage throughout the year. Bending the program opens up the area at ground level. It creates a dramatic cantilevered gesture towards the river. It allows for the required slope for the seating area of the open air cinema on top. On the city side the bending act leads to an inviting access gesture. Here the volume raises, towers up, connecting the volume spatially with the towers further away. The interior is conceived as a giant "valley" in which all programmatic "boxes" are situated. At the bottom of this "valley", there where the building touches the earth, the concourse is positioned. The concourse is surrounded with stairs on both sides that acts as giant tribunes; where one watches the visitors. it is positioned in such a way that every cinema can be seen from the concourse. On the other side of the concourse, the convention hall, the rentable spaces and higher up the offices are situated. All bending acts create a perfect open air cinema on top: sloped seating on the western side and a screen on the northern side.

BENDING ACT

PROGRAMMATIC BOXES

CONCOURSE

ZONE

SINGLE LAYER

DRAMATIC CANTILEVERED GESTURE

SLOPE

GIANT VALLEY

REMARKABLE VOLUME

URBAN CHAIR

'접이' 공간 FOLDING TO MAKE SPACE

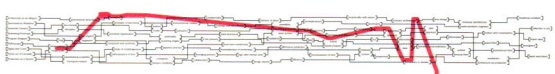

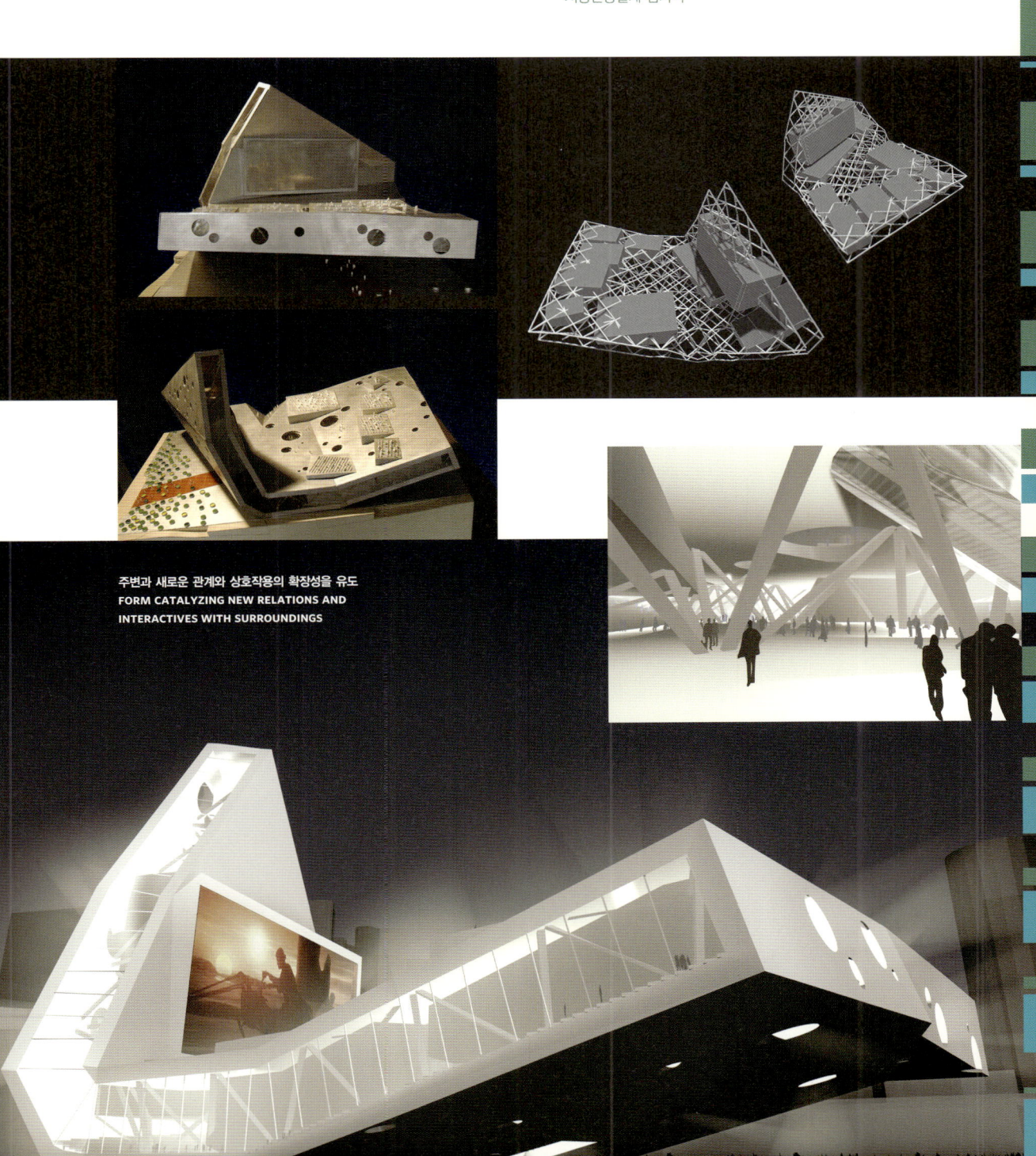

보편적 창의

이공희
국민대학교 건축대학

REALITY

NEO-RATIONAL

REDUCTIONISM

STRUCTURALISM

OBJECTIVITY

REASONING PROBABILITY

TYPE

SUBSTANCE

가치와 향위

건축의 가치가 불변하지 않음을 알면서도 그에 관한 논의의 궁극은 늘 그 가치를 추적하는 일이 된다. 때론 건축이 시대의 거울인 연유로 그 시대에 사유의 가치로 추적하기도 하고, 그 실천의 향위를 통해서 주변성과의 관계를 밝혀내는 에너지에 관심을 두기도 한다. 그 에너지는 전통이나 관습에 의해 운용되기도 하고 아방가르드와 같은 특별한 현재에 간섭을 받기도 한다. 한편 그 실천은 비교적 하나의 가치 속에 보편적 프로세스를 우선하기도 하지만, 사유를 달리하면 과정이 다를 수밖에 없다는 건축의 속성에서 현재적 가치의식을 배려한 약속일 수 있다. 그런 연유로 그 가치에 관한 약속은 시대의 에너지에 관심두고자 할 때 개연적인 탐색 대상이 된다.

논의의 종착은 현재 한국 건축을 이루는 가치에 주목하여, 그 건축적 사유와 실천들의 잠재적 궁극을 추적하고, 그 향위를 그림으로써 건축에 관한 시대정신을 공유하고자 함이다. 따라서 본 논의는 두 가지 목적을 갖는다. 하나는 한국 현재건축의 잠재적 궁극의 향위가 어떤 지형도를 그리는가를 밝히는 일이고, 다른 하나는 그 향위를 우리의 사실로 보여주는 건축가를 드러내 가늠해보는 일이다.

시대의 향위를 밝히는 일에 건축가를 주목하는 습성은 건축의 역사가 인류 문명과 그 시작을 같이하지만, 전통적으로 본격적인 건축가의 등장을 창조적 소수의 건축가가 전체의 담론을 생성했던 르네상스 이후로 보기 때문이다. 한편 주로 사회경제적이고 통계적인 면을 강조하는 도시사 연구에 반하여, 건축사 연구는 근대적 주체로서 건축가의 생각과 방법들을 그 중심에 놓을 수밖에 없다.

보편적 창의와 관계로부터 생성

한국 현대건축의 지형도를 그리자면 여느 서구 현대건축과는 다른 몇 가지 빗겨갈 수 없는 우리 형편이 만든 사실을 기억해야 한다. 서구 건축이 곧 근대건축이라는 막연한 믿음이 갖는 개념적 오류, 변방의 경험과 소위 개발 시대라 하는 1970, 80

Universal Creativity

Gonghee Lee
School of Architecture, Kookmin University

Values and Directions

Even though we know that the value of architecture never changes, we always eventually seek that value through discussions. For the reason that architecture is the mirror of the era, we sometimes trace it as the value of the idea of the era, or we sometimes take interest in the energy that is used to figure out its relationship with what surrounds it. The energy either gets operated by traditions or customs, or gets intervened by an extraordinary present such as avant-garde.

Meanwhile, although the practice puts the universal process first in the relatively identical value, the property of architecture that different thinking only leads to different processes might be a promise that has the awareness of the present value in mind. That is why the promise about the value becomes the possible goal of search when the era's energy emerges as the interest of architects.

The end point of the discussion is reached after focusing on the value that constitutes the contemporary Korean architecture, tracing the architectural thought and the potential eventuality of the practice, and sharing the spirit of the era about the architecture. In this regard, this discussion has two objectives. One is to show part of the topographic map of the potential eventuality of the contemporary Korean architecture, and the other is to reveal the architect who shows us that we are actually going in that direction.

The habit of focusing on architectures in figuring out the direction and the status of the era started with the beginning of the human civilization. But this is also the reason why the traditionalists view that the fully dedicated architects only arrived after the renaissance, where a small number of creative architects dominated the whole discourse.

Meanwhile, compared to the study of the urban history that mostly emphasizes the social, economic and statistical aspect, the study of the history of architect adds ideas that cannot help but putting the ideas and methods at the core.

Universal Creativity and Becoming from Relationship

To draw the topographic map of the Korean modern architecture, we have to

년대를 통해 얻은 한국 건조환경을 지배하고만 계획관습은 건축을 보다 개념 전개의 창조적 자아실현에 무게를 두고 실천하고자 할 때 빗겨가기 힘든 원인 중 하나가 되고 있다.

전자는 조선 건축이 근대적 적응에 그 성질을 추스르기도 전에 당면하게 된 일제강점기라는 지역적 특수성을 접어두더라도, 지속적인 새로운 이념에 도전을 어떤 방법으로든 수용해야 하는 대안 없는 번복을 이름이다. 후자는 근대화를 정의함에 있어 '개발홍수'에 이의없음 이면에는 한국 건축이 창조적 소수가 전체 담론을 형성해나가는 생성에 메커니즘의 부재를 인정해야 하는 형편에 있다.

이는 한국 건축이 그간 서구 건축에 대해 예각을 보여, 과거에 관한 대안이나 내일에 관한 이념으로 이데올로기에 집착하기보다는 이미 성숙된 이념을 수용하고 마는 악순환에 익숙했다는 반증이기도 하다.

이어서 벌어진 1990년대 한국 건축은 일면 다양한 가치와 대안들이 때론 급진적 실험과 실천을 시도하지만 또다시 새로운 건축의 생성메커니즘 체계를 만들기보다는 해방 이후 악순환의 또 다른 한 순배를 보여주는 획일성을 보여주었다. 이 악순환의 메커니즘이 생산한 획일성은 한편 보편적 가치에 익숙한 우리 건축의 현실일 수 있다. 긍정적이진 않지만 여기에는 현실이라는 리얼리티는 지역성과 기술이라는 성질처럼 이분적인 원인이기에 앞서 그저 '현재건축에 리얼리티'로 존재한다는 생각이 본 논의의 관점이고 폭이다. 악순환이건 진보를 위한 진전이건 간에 한국 건축은 이 리얼리티 안에 있고, 잠재적 에너지를 갖고 어디론가 향하거나 순환한다.

한국 현재건축의 항해는 다음 두 가지 덕목을 피해갈 수 없으리란 생각이다. 그 하나는 이념과 기술간에 창조적 진전을 보지만 건축은 여전히 그 가치를 개연성에 두려는 사유로, 보편과 합리에 관한 향수로 '신합리로서 보편적 창의'이다. 다른 하나는 인공적인 것과 자연적인 것 사이에 가정되었던 안정된 관계성 상실에 관한 반대급부로 이른바 기하학적 세계관과 대립하는 전통적 개념에 유기적 건축으로부터 땅을 살아 있는 유기체로 보는 새로운 자연관까지를 포함하는 '신유기로서 관계로부터 생성'이다.

보편적 창의로서 신합리

건축은 사유의 실천이며, 때론 의지와 관계없이도 주변 모든 현상과 교섭하여 만

remember several facts that were created by the situation in Korea, that are different from any Western modern architecture. The architectural environment in Korea in the 1970s and the 1980s had the misunderstood belief that the Western architecture is the modern architecture dominate it, in the so-called development era. It is becoming one of the reasons that are hard to avoid, in practicing architecture while focusing more on the creative realization of self.

While the former refers to the change that does not have an alternative, that challenges to the continuously new idea in the architecture of the Joseon Dynasty had to be accepted in any way, apart from the regional uniqueness from the period of the Japanese occupation, the latter has to admit that the mechanism does not exist in the process where the creative few forms the entire discourse, when development projects are flooding the modernization effort.

This is evidence that the Korean architecture has gotten use to the vicious circle of accepting the already matured ideas, rather than getting obsessed with ideologies as the alternatives to the past or the ideas for the future. The following era of the 1990s witnessed the Korean architecture attempt sometimes radical experiments with various values and alternatives, but rather than creating a new mechanism of architecture, it was consistent with the vicious circle after the liberation from the Japanese occupation.

The consistency produced by the mechanism of this vicious circle might be the reality of our architecture, that is used to the universal values. Though not positive, the perspective and scope of this discussion here are that the reality exists in the contemporary architecture as just the "reality", before it is the cause of the dichotomy of regionalism and technology.

Be it a vicious circle or a step toward advancement, the Korean architecture exists in this reality, has potential energy and gets circulated somewhere. The voyage of the Korean contemporary architecture would not be able to avoid the sea of the following two virtues.

One is the idea that creative advancement has been made between ideas and technologies but architectural values lie in the possibility. This is a nostalgia to universality and rationalism, which is the "universal creativity of Neo-rationalism". The other is the compensation for the loss of the stable relation between the artificial and the natural. This refers to the "creation from the relationship of Neo-organism" includes the new view of nature that considers the ground as a living organism in the traditional concept that confronts the so-called geometric view of the world.

든 물성이면서 감성으로 이해되기를 기대한다. 하지만 여전히 건축의 가치는 사유로부터 실천에 이르는 과정에 개연을 존중하는 전통적 습성을 통하여 건축을 보다 합리적인 가치로 보존되도록 하는 원인이 된다.

건축이 상징적 오브제만으로도 그 목적을 이루거나 감성적 임무에 한하더라도 여전히 물성을 이루는 사연은 합리적 개연이기를 우선하는 태도를 합리주의라 하자. 이는 비합리와 우연적인 것에 반하여 이성적, 논리적, 필연적인 것을 중시하는 이성론적 합리주의라 하며, 데카르트 이후 이성을 근간으로 삼는 근대서구 합리주의 철학의 영향력은 건축에서 과학적 합리란 가치를 만들었다.

과학적 합리란 가치는 과학적 원리의 시각화*라는 명제가 되었으며, 건축물에 작용하고 있는 물리적이며 과학적인 원리를 형태적으로 시각화할 때 이것을 아름답다고 하는 시대의 미학적 신념이 되었다. 이는 근대건축의 기계미학에 기초를 이루어 구조적 진실, 장식의 배제, 투명성과 같은 구체적 실천으로 역사주의에 반하여 극복하는 과정으로 진전을 보였다.

* 황두진+이광로, 근대 및 현대건축에 나타난 과학적 합리주의의 형태적 표현에 대한 연구, 대한건축학회 논문집 8권 1호 95쪽

MIES VAN DER ROHE, LUDWIG | NATIONAL THEATRE, MANNHEIM | AVOIDANCE OF DECORATIONS AND TRANSPARENCY

PAXTON, JOSEPH | CRYSTAL PALACE | DISCOVERY OF TRANSPARENCY

하지만 역사 혹은 전통이라는 가치에 관한 합리의 혼돈은 1920년대 이탈리아 합리와 1960년대 신합리를 거치면서 과거의 전통적인 건축 작품들이 갖고 있는 논리적 유형들의 평면과 기본 요소들의 끊임없는 회귀를 통해 영원불멸의 것으로 이해되고, 전통에 근간을 두는 가치로 전환하기에 이르렀다. 기능주의자들에 의한 합목적이며, 효율적 합리성이 아닌 18세기 계몽주의자들이 다원적 구성요소를 법칙성에 근거하여 파악하고자 할 때 개념을 합리라 하고 이를 신합리주의라 하는 개념의 전환을 보았다.

이 신합리는 환원주의, 구조주의 영향이 뚜렷이 보이는 한편 역사에 관심을 두기도 하였다. 결과적으로 건축이 내포하는 정확한 문맥을 구성하는 근본적이고 논리적인 통일성을 명확하게 하는 태도를 보였다. 신합리주의 사유는 건축의 한 분야를 물리적–기호학적으로, 건축물의 디자인 과정에 역사를 이해 혹은 이상적인

NEO-RATIONALISM AS UNIVERSAL CREATIVITY

Architecture is an act to put the thought into practice. It is sometimes expected to be understood as the emotion that is created by interacting with all the surrounding phenomena against its will. But the value of architecture still is the cause to preserve it as a more reasonable value through the traditional habit of respecting the possibility in the process from thinking to putting it to practice. Let us call it rationalism - the attitude that considers architecture as a reasonable possibility, even if its symbolistic objet serves its objective or it is limited to an emotional mission.

This rationalism is based upon the reasonable thought that emphasizes anything that is rational, logical and inevitable, compared to irrational and coincidental aspects. The influence of the modern Western rationalistic philosophy since Descartes has created the value of scientific reason in architecture.

The value of scientific reason has become the proposition of the visualization of the scientific principles*, and the aesthetic belief of the era that calls the visualization of the physical and scientific principles that are applied to buildings beautiful. Based upon the mechanical aesthetics of modern architecture, this has advanced through the process through specific practices such as the structural truth, avoidance of decorations, and transparency.

But the confusion of reason on the value of history or tradition, going through the 1920s' Italian rationalism and the 1960s' Neo-Rationalism, has been converted to a value that is based upon tradition, and that is understood as something that lasts forever, because the basic factors of the logical types of the traditional architectural works of the past constantly come back. The concept was transformed into Neo-Rationalism, which rationalizes the concept for the 18th century philosophers of enlightenment to understand the multipolar factors based upon principles, unlike the functionalists who based themselves upon efficient rationalism.

This Neo-Rationalism was obviously influenced by structuralism, but meanwhile it took interest in history as well. As a result, it showed the attitude that clearly emphasizes the fundamental and logical unity that constitutes the exact context that is implied in architecture.

This Neo-Rationalism gave birth to typicalism that understands it as an ideal extension of the design process of buildings in a physical-semiotic way, accepted collage, and eventually sought the formal order of the collection of languages rather than the creation of independent languages.

Architects with this thought include Aldo Rossi, who bases the world on a few types, and Rob Krier, who follows already determined types.

*Doojin Hwang+Kwanglo Lee, study on the expression of scientific rationalism in modern and contemporary architecture, Architectural Institute of Korea thesis book 8 no. 1, page 95

연장으로 여기는 유형학을 낳았다. 또한 콜라주를 수용하기도 하고, 독자적인 형태 언어의 창조보다 형태언어간 조합이라는 형식 질서를 추구하기에 이르렀다.

　이 사유의 건축가는 자신이 항상 세계를 소수의 유형을 기본으로 하여 이를 변화하는 건축을 구성하는 알도 로시, 형태 근거를 개연적 합목적에 근거하기보다는 이미 정해진 타입에 따르는 롭 크리에 등이 실천 유형이다.

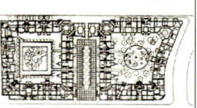

ALDO ROSSI | CEMETERY OF SAN CATALDO　　ROB KRIER | RITTERSTRASSE

　본 논의의 제안이 되는 '보편적 창의'의 개념은 근대가 이르는 과학적 합리로부터 1960년대 이탈리아 신합리로 후근대를 경험하고 현재건축에서도 여전히 합리적 사유를 유지하는 것을 그 기본 생각으로 한다. 거기에 건축이 사유로부터 실천에 이르는 과정에 직관을 최소화하고 객관을 존중하려는 가치를 두었다면 또한 이 보편적 창의의 개념 안에 둔다.

　오늘날까지 한국 현대건축을 설명했던 잣대인 보편과 특수, 서구와 우리 건축과 같은 이분법적인 사고로는 더 이상 우리의 현재인 다가치·다원화 사회를 건축으로 실천하기는 어렵고 불가능하다. 이는 합리주의가 만들었던 방법론과는 다른 새로운 방법과 가치로 실천하는 건축이 있다면 이를 '보편적 창의' 라 하고 이는 다음과 같은 덕목을 존중한다.

- 직관을 최소화하기 위한 객관의 존중
- 합리를 근거로 하는 직관
- 관점을 달리하는 폴드적 사고로서 데이터
- 합리가 프로그램의 기능으로부터 개념 이전에 본질에 관해 물음으로써
　'시설'에 충실하고자 할 때 합리를 '보편적 창의' 라 하자.

The concept of "universal creativity" that is the main issue of this discussion is based upon experiencing the modern scientific rationalism and the 1960s' Italian Neo-Rationalism, and maintaining the rational thinking in the contemporary architecture. If there is the value to minimize intuition and respect the objectivity in the process from thinking to putting it to practice, it would be deemed to be included in the concept of universal creativity.

It is known to be hard and even impossible to create a multi-valued, multi-polar society with the dichotomy of universal and special, Western and Korean, which has actually been the standard that has measured the Korean modern architecture to this day. This is an architecture that is practiced in a totally different way and with a new value system that is different from the methodology of rationalism. This architecture is called "universal creativity" and it respects the following virtues.

- Respect of objectivity to minimize intuition
- Intuition based upon rationalism
- Data as a thought with a different point of view
- Rationalism is called "universal creativity" when it attempts to be loyal to the "facility" as a question on the essence before the concept.

관계로부터 생성, 그 신유기

이공희
국민대학교 건축대학

그간 합리적이라는 명제를 소통하게 한 근거가 되어온 건축언어에 관한 믿음은 합리와 객관이 그린 지구의 위기라는 혼란한 현실 앞에 더 이상 힘을 갖지 못하고 새로운 관계를 희망한다. 이 혼란의 시작은 건축에 관한 사실에 인식 방법까지 의심하게 되었고, 전달수단인 언어에 관한 의심은 단어들이 실재의 물건과 일대일의 대응관계가 성립되는 것이 아니라는 합의가 이루어진다.

사실에 관한 인식 방법의 전환은 언어를 이루는 단어간 차이에서 비롯되는 인식이란 주장을 보면서 플라톤 이후 소위 주류의 서양 철학자들이 유지해온 사고체계는 하나의 체계에 불과한 것이고, 다른 해석이나 사상체계보다 더 옳다고 주장할 근거가 없다는 논리는 건축이 더 이상 전통적 서구문화만으로는 한계가 있다는 것을 인정하는 결과를 보였다. 나아가서는 서구 전통 사고방식으로 만든 서구문화가 창조적 가능성을 배제한다는 우려에 이르기까지 한다.

이성이나 객관적 사실 혹은 진리보다는 수사학적 강압을 통해 자신의 생각을 진리로 가장하여 다른 사고체계 위에 군림하게 된다. 이는 이제껏 미루었던 건축을 이루는 주변성과 새로운 관계를 구축하여 현실화하는 형태 생성 태도를 보인다. 이 새로운 시각의 주를 이루는 자크 데리다의 사고체계에서 우린 흥미로운 사실을 발견하게 된다. 그가 서양철학의 전통적 사고체계를 극복하면서 제시하고 있는 서양철학의 나아갈 길이 동양철학과 많이 유사하다는 점이다. 이성이 지배해온 무미건조하며, 감정이 메마른 서구철학의 비이성적 요소들을 이성의 독재로부터 해방시킴으로써 서양의 사고방식 속에 유희와 생명력을 불어넣으려고 한다.

이를 위해서는 철학, 예술, 문학 등으로 구분하기를 거부하며 절대적인 진리와 절대신은 없고 존재보다는 생성을 이해할 줄 아는 긍정적인 사고체계로의 전환을 주장한다. 이는 전통적으로 정교한 논리보다는 시적이고 심미적인 인식 방법을 강조하여 철학과 예술을 구분하지 않았으며 존재보다는 생성에서 진리를 찾으려 한 동양철학과는 반대 개념에서 접근하고 있어 흥미롭다.

이는 건축이 적어도 관계로부터 만들어지는 생성이라는 불변의 이치라 한

NEO-ORGANIC

POETIC/AESTHETIC COGNITION

PRINCIPLES OF BECOMING

ENVIRONMENTAL RELATION

STRATEGY AND OPPORTUNITY

UNCERTAINTY

POSSIBILITY

INCOMPLETENESS

BECOMING FROM RELATIONSHIP, THE NEO-ORGANISM

GONGHEE LEE
SCHOOL OF ARCHITECTURE, KOOKMIN UNIVERSITY

The belief in the architectural language that has been the basis of communication of the reasonable proposition does not have the power any more in front of the chaotic reality of this planet but hopes for a new relationship. The beginning of this chaos has led to questioning even the method how architecture perceives the facts. The questioning of the language, which is the means to convey the perceived facts, leads to the agreement that language does not establish a one-on-one relationship with objects that exist.

The conversion of the method how architecture perceives the facts is sometimes argued to be stemming from the difference between the words that constitute the language. The logic that the system of thinking that has been maintained by the so-called mainstream Western philosophers since Plato is just a system, and that there is no ground to argue that it is more right than other interpretations or systems of thinking, has resulted in admitting the limitation that architecture faces. Furthermore, it even leads to the concern that the Western culture rules out creative possibilities.

It dominated over other systems of thinking, by masking its thoughts as the truth, through forced rhetorics rather than through reason, objective facts or the truth, which realized the building of new relationships for architecture with what surrounds it. In the system of thinking of Jacques Derrida, which is the mainstream of this new point of view, we discover an interesting fact. It is that his direction for Western philosophy that is supposed to overcome the traditional system of thinking of the Western philosophy has many similarities to Oriental philosophy.

He tries to bring fun and life into the Western way of thinking by liberating the irrational factors of the Western philosophy that is dry and emotionless because it has been dominated by reason. To this end, he refuses the classification of philosophy, art and literature but argues for the conversion into a positive way of thinking that understands that there is no absolute truth and absolute god, only becoming not being. It is interesting that he did not distinguish philosophy from literature like in the Oriental philosophy that traditionally emphasized poetic and aesthetic perceptions rather than sophisticated logics and that he also approaches

다면 동양권에 있는 우리 현재건축을 이루는 대분 중에 하나임이 가능해진다. 그 실천의 실마리는 주변성과 새로운 관계를 모색하여 건축하려는 태도로부터다. 이 실천으로부터 얻는 개념의 확장인 '관계로부터 생성'을 개별 이해로 보자면 기계주의적 세계관에 반하는 전통적인 개념으로서 유기적 건축으로부터 개념을 찾는 것이 가능하다.

현상학적인 관점에 대한 이해는 그렉 린, 로버트 벤투리, 르 꼬르뷔제까지도 개념의 확장을 보아 유기적 관점으로 접근할 수 있으리만큼 광의적이다. 이렇게까지 개념의 폭을 늘리고자 하는 이유는 오히려 전통적으로 유기적 건축으로 해석되는 작업들이 근자에 획기적 전환을 보는 자연과학, 신과학의 관점으로 새롭게 조명되어 건축이 어떻게 자연원리와 부합되는지 수월한 재해석을 보여주고 있기 때문이다.

'유기적'이란 용어는 19세기 프랑스의 병리학자 비샤에 의해 최초로 표현된 이후, 건축에서 유기적 개념은 움직이는 생명체보다는 특정 지역에 뿌리를 둔 생명체의 특성으로 취급되어 비대칭이 유기적 조직체의 특성으로 받아들여졌다. 근대건축의 프랭크 로이드 라이트가 주로 사용한 '유기적'이란 용어는 그 자신의 특정 건물에 대한 용어로 사용되어 통상적으로 비대칭적이고 부지 특성과 결합한 건물을 지칭하는 것으로 정의되었다. 고전 양식과 국제주의 양식의 반대 개념으로 사용되었으며, J. O. 시몬드가 표현한 '유기적 계획'은 부지간의 조화를 극대화시킬 수 있는 프로그램이었다. 즉 배치하고자 하는 각 기능을 다른 기능 및 부지 내 모든 요소들과 최상의 관계를 유지시키는 계획을 말한다. 그로피우스는 대지, 자연, 인간, 예술 등 모든 질서가 있는 우주 상태를 '유기적'이라 규정하였고, N. 뉴튼은 "디자인 과정에서 생물학적인 접근 작업에 의한 것"을 유기적 디자인이라 하여 유기체가 자연 상태에서 서로 결합되어 있는 동적 관계라 지칭하였다.

한편, 경제성장의 말기로 급격하게 환경문제에 대한 관심이 고조되었던 1890년대와 1920년대 그리고 1950년대 말과 1970년대 초에는 모든 사람들이 고도로 발달된 물질적 가치관에 대해 계속 회의를 느끼던 시기이기도 하다. 즉 환경문제는 기존 패러다임의 비판과 함께 새로운 패러다임의 요구가 생태학적 건축, 환경적인 건축과 같이 유기적 건축의 범주와는 다른 새로운 관계로부터 건축으로 이어진다.

이후 이른바 유기적 건축의 향위는 기본적으로 자연을 그 중심에 두면서

the concept of reversion of the Oriental philosophy that attempted to find the truth in becoming rather than being.

If it is an invariable truth that architecture is a becoming that is created from relationships, it can possibly be one of the parts that constitute our contemporary architecture within the scope of Orientalism. The lead of the practice is found in the attitude that seeks a new relationship with what surrounds it. If "becoming from relationship", an extension of concept that is gained through this practice is viewed as an individual understanding, it is possible to find the concept from organic architecture, as the traditional concept that goes against the mechanical view of the world.

From the perspective of phenomenology, even Greg Lynn, Robert Venturi and Le Corbusier can be approached in the organic way. The reason why we attempt to expand the scope of the concept like this is because it shows an easy reinterpretation of how architecture fits natural principles, in the new light of natural science and neo-science that has recently been remarkably transformed in the interpretation of the works that would traditionally understood as organic architecture.

The term "organic" was first used by Bichat, a 19th century French pathologist, and since then it has been accepted in architecture as a concept that rather refers to the characteristics of living organism that take root in certain regions rather than moving creatures. So asymmetry has been considered to be a property of an organic system. The term "organic" mainly used by a modern architect F. L. Wright refers to certain buildings designed by him, usually asymmetrical. It was used as a concept that is opposite to classical type and international type. The "organic plan" expressed by J. O. Simonds was a program that could maximize the harmony between land sites. In other words, each function to be deployed maintains the optimal relationship with other functions and all factors in the site. The definition of "organic" by Gropius refers to the spatial status that has all orders of land, nature, human and art. N. Newton called "design process by biological approach" organic design and a moving relationship where the organism is combined with each other in the status of nature.

Meanwhile, the 1890s, the 1920s, the end of the 1950s and the early 1970s were the periods where interest in environmental problems skyrocketed as they were the latter periods of economic growth. Everyone constantly felt skeptical about the material point of view. In other words, along with criticizing the existing paradigm, the environmental problems called for a new paradigm and eventually led to an architecture from a different relationship from organic architecture such as ecological

그 자연과의 관계 설정을 자연 속에 내재되어 있는 형태 속성 또는 근대적 통념 아래 있는 생태나 환경이니 하는 피상적 이해보다는 근본 원리 속으로 들어가려는 태도에 있다. 이를 '새로운 관계로부터 생성' 이라 하여 건축에 이르는 것을 ' 신유기' 라 하고 다음과 같은 덕목을 우선한다.

– 자연의 형태보다는 새로운 자연관과 그 생성 원리에 충실하려는 태도
– 새로운 생성 원리를 근간으로 건축을 이루는 주변성과 새로운 관계로 구축
– 완결된 작품보다는 전략과 계기를 마련, 불확정이어서 변화의 가능성을 디자인
– 창조보다 생성에 가치를 두는 태도

창의와 생성이 만든 풍경

우리 건축의 현재를 이루는 지도 속에 보편이라는 합리에 대한 향수와 태생의 근거를 새로운 관점에서 발견해내려는 노력을 그리는 일은 생리적으로 개연적이다. 한편 건축이라는 속성에서 새로운 합리적 사고와 유기적 사유는 다양한 가치와 복잡한 프로그램 모두를 수행해야 하는 이 시대에 더욱 가치있는 덕목일지 모른다.

　　　　이미 확정을 본 프로그램일수록 경험적 근거를 중시하여 직관을 줄일 수밖에 없는 연유로 '신합리' 성향을 보이는 것의 향위가 보편적 합리에 있다면, 반면 새로운 프로그램 또는 개념적 진화로부터 진전되는 시설인 경우에는 '신유기' 의 사고로 실천되어 구축은 자연 속에 내재되어 있는 근본 원리와 새로운 관계로부터 생성됨이 우리 환경을 그리는 힘 중 하나일 수 있다.

and environmental architecture.

Since then, the so-called direction of the organic architecture basically put nature in its core and attempted to enter fundamental principles rather than superficially understand the relationship with nature as the property that is implied in nature or as ecology or environment under the modern conventional wisdom. Neo-organism is leading this "becoming from new relationships" to architecture, and it respects the following virtues.

- Attitude that attempts to be loyal to new viewpoints of nature and the principle of its becoming, rather than the appearance of nature
- Building a new relationship between architecture and what surrounds it, based upon a new principle of becoming
- Preparing strategies to designing the possibility of changes that are uncertain, rather than finished works
- Attitude that values becoming more than creation

Scenery Built by Creativity and Becoming

Effort to discover the nostalgia to universality and the ground for becoming in the map of the contemporary Korean architecture is physiologically likely. Meanwhile, the new reasonable thinking and organic thinking might be more valuable virtues in this era where various values and complex programs all need to be performed.

If the direction of "Neo-Rationalism" is toward universal rationalism for the reason that already confirmed programs emphasize empirical grounds more and cannot help but reducing intuition, but if the facility evolves from a new program of a conceptual evolution, it can be put into practice through the "Neo-organic" thinking, and architecture might become one of the powers that enable the fundamental principles that are implied in nature and the becoming from new relationships to duly paint our environment.

Kerl Yoo
iARC Architects, Korea
Asian Culture Complex
Competition Project_Participation
Gwangju, Korea

문화를 생산하는 콤플렉스를 디자인해달라는 요구사항을 문제화하는 데서 이 프로젝트는 출발한다. 아시아문화센터는 문화가 기관을 통해 만들어지기보다는 문화가 발생하는 곳이어야 한다. 이러한 발생은 사회 접촉, 즉 네트워크 복잡성을 최대화해서 달성할 수 있다. 도시 전략으로서 전체 지역을 작은 부분으로 나누는 분화는 기존 이웃의 도시 패브릭을 지속함으로써 실행되고, 프로그램적 해석으로 변형된다. 이러한 부분들은 서브프로그램 사이 특정 관계에 따라 연결되어 중첩 네트워크의 3D 콤플렉스를 형성한다.

그 결과 두 개의 분명히 구분되는 네트워크 조직이 발생한다. 하나는 '프로그램적 네트워크'로 쇼핑, 외식과 음주, 학습, 회의, 전시와 놀이, 근로와 생활 같은 것을 포함한다. 또 하나는 '생태적 네트워크'로 공원, 물과 바람 등을 말한다. 역동적으로 적응하는 융통성 있는 시스템, 일시적 현상과 도시적 도전에 자유롭게 맞추는 창조적 시스템을 만들어내는 도시 역량을 갖추어야 하는 것이다. 각 네트워크의 차별화된 접속은 신생 시스템을 조절하는 데 중요한 역할을 한다. 문화 또는 아시아에 대한 정의는 이 새로운 도시 시스템을 통해 계속 바뀌고 재생산될 것이다.

MAXIMIZATION

NETWORK ORGANIZATION

COMPLEX

LANDSCAPE

URBAN FABRIC

ECOLOGY

INTERCONNECT

PROGRAMMATIC FLEXIBILITY

RHIZOME

FLEXIBLE SYSTEM

SEEMINGLY RANDOM

RECONFIGUABLE SYSTEM

TYPOLOGY

ECONOMICAL SUSTAINABILITY

The competition brief called for designing a complex which manufactures culture. The project started from problematization of this brief. The Asian Culture Complex should be a place where new culture is emerged, rather than manufactured by institutions. Emergence can be achieved by maximizing social contacts, in other words, network complexity. As an urban strategy, differentiation of the whole site into smaller parts is executed by continuing existing and neighboring urban fabric, further being transformed by programmatic interpretations. Then the parts are connected with each other according to specific relationships between sub-programs, forming a 3D complex of nested networks.

Two distinct network organizations get emerged out of it; programmatic network (shopping, eating & drinking, learning, conferencing, showing & playing, working and living) and ecological network (park, water and wind). The interest is in generating urban capability of producing a flexible system that is dynamically adaptable, a creative system that can adjust itself freely to temporal events and urban challenges. The differentiated connectivity of each network plays a vital role in modulating its emergent system. The question of what is culture and what is Asian will be constantly redefined and re?generated by means of this new urban system.

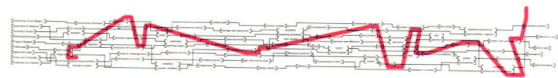

아시아문화전당 현상설계 참가작 유걸 아이아크

PERSPECTIVE VIEW

Kerl Yoo
iARC Architects, Korea
Millennium Community Center
Ilsan, Gyeonggi-do, Korea
2004

밀레니엄 커뮤니티센터는 교회가 주관하는 복합문화 집회시설이다. 많은 사람들을 대상으로 하는 이런 시설이 도심에 고층으로 들어섰을 때 건물 내부 동선과 공간을 어떻게 수직적으로 연결하는가가 과제였다. 8층 높이의 건물이 시각적으로 그리고 동선상으로 모두 연결되도록 도로 면에서부터 5층까지 이르는 경사면을 도입하였다. 이 경사면을 따라 램프가 올라가고 램프 주위는 실내 조경이 된다. 램프와 조경으로 처리되는 부분 이외 남은 부분은 계단식으로 처리되어 노천극장과 같은 다목적 공연장 또는 휴식처로 이용하게 한다. 이 경사면 공간은 노천극장과 램프 사이에 있는 세 기둥을 제외하고는 무주공간이라고 할 수 있다. 그 위에 역시 비스듬히 떠 있는, 건너편 공간으로 연결하는 두 개의 브리지가 매달려 있다.

건물 외곽을 따라 3m 폭으로 공간이 돌아가는데 이것은 3m 폭의 구조물이 되어 60m 정도의 중앙 부분을 가로지르는 구조물을 지지하게 된다. 고층화되어 있는 고밀도의 공간을 여유있는 개방공간으로 만드는 것과 수직적인 층간이라고 할 수 있는 건물 내부를 환기나 조명 그리고 조경에 이르기까지 자연환경과 같이 만드는 것이 이 프로젝트의 설계 목표이다.

The Millennium Community Center is a general cultural and assembly facility run by a church. The project was about creating a linear link between the inner circulation and space, in a cultural facility that is open to the general public and that exists as a tall building in the downtown area. To link the 8-story building both visually and physically, a slope was used from the ground level to the 5th floor. Along the slope is a ramp, around which is the indoor landscape. The areas other than the ramp and the landscape are done as steps, so that the whole area can used as an open-air multi-purpose performance hall or for rest and leisure. The slope is column-free except for the three columns between the open-air concert hall and the ramp. On top of the slope hang two bridges that connect it to the space on the other side.

There is a 3m-space that goes around along the outer wall, forming a 3m-wide structure that supports the central structure that spans 60 meters in diameter. The design concept for the Millennium Community Center is to turn a tall, high-density building into an open space with enough breathing room, and to assimilate the inside of the building (which essentially is a vertical inter-floor space) with nature, from ventilation, lighting to landscape.

Too many buildings in Seoul, including "public" cultural facilities, are not friendly or open to ordinary citizens. The Millennium Community Center, located in the center of the city, was designed to welcome all citizens and pedestrians with open arms, so that anyone can freely come and go anytime. Granted, the process after the design was not free of complexities. But in the end, what is most important is for the building to stay true to its designing principle and become fully open to all citizens.

PUBLICITY

VERTICAL CONNECTION

SPATIAL CONNECTION

LAMP

OPEN SPACE

SECTIONAL CONTINUITY

DENSITY

FLOW

COLUMN FREE SPACE

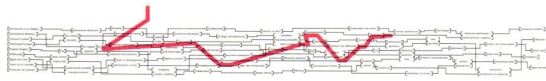

밀레니엄 커뮤니티센터   유걸
아이아크

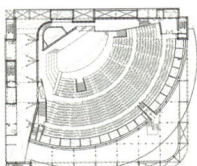
8F PLAN

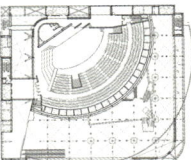
7F PLAN

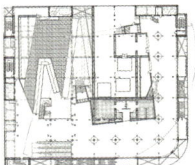
6F PLAN

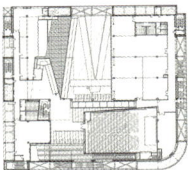
5F PLAN

새로운 체계가 만든 실내 공간 INTERIOR SPACE MADE BY NEW SYSTEM

공간의 상호작용
SPATIAL INTERACTION

교회공간을 담는 외곽구조
STRUCTURAL ENVELOPE CONTAINING THE CHURCH SPACE

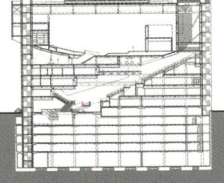
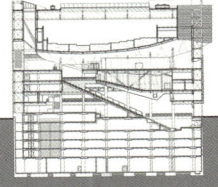
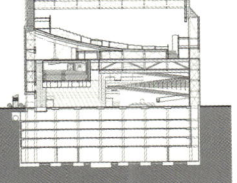
SECTION

Kerl Yoo
iARC Architects, Korea

Graduate School of Architecture Kyunghee University
Yongin, Gyeonggi-do, Korea
2002

건축전문대학원에 필요한 최소한의 공간을 마련하기 위해 대학 측은 공과대학 건물에 붙어 있는 강당을 내놓았다. 강당은 발코니를 포함해 240평 정도의 (바닥) 면적이지만 천장고가 높기 때문에 이것을 최대한 활용해 4개 레벨을 만들어 500평 가량의 가용 면적을 확보했다. 기존 구조와 2층 복도를 보존해야 했으므로 메자닌 층은 최소 단면 구조를 사용했지만 2.1m 정도의 낮은 천장고는 불가피했다. 기본적으로 모든 층의 바닥과 천장은 구조가 노출되어 있다. 따라서 배관과 배선, 전기 설비기구도 노출되어 있다. 이는 단순히 미적 쾌감뿐만 아니라 스튜디오에서 작업하는 학생들에게 좋은 실물 교재가 된다. 이곳은 학생들이 설계사무실을 경험할 수 있도록 유사한 공간으로 구성되었다.

To secure the minimum space necessary to create a graduate school for architecture, Kyunghee University freed up the auditorium annexed to the engineering college. The auditorium had approximately 800m^2 of floor space, including balconies. But thanks to the high ceiling, we were able to create 4 levels to make use of total 1,700m^2. To preserve the existing structure and the 2nd floor corridor, a mezzanine had to be constructed using minimum space possible, but the ceiling still had to be kept low (2.1m). The floor and ceiling on all the floors are basically bare (i.e., the inner structures are revealed), and thus the piping, lining and electric installations are also exposed. The materials used are industrial, such as flooring and steel structure exposing concrete, and archi light.

Exposing the building system was not for aesthetic purposes, but to provide an immediate and actual studying material for the students working at its studios. The Graduate School of Architecture Kyunghee University is akin to one big architectural office. Open studios, archive, conference rooms, private offices for senior architects, and print rooms together create something that looks and feels like an architectural office. This is an intended result, because we thought of the school to be similar as an architect's office.

It is an environment that encourages students to do away with stereotypes, look at reality with an open mind, collect information, and approach issues with creativity, all the while keeping the spirit of cooperation alive with fellow students and/or professors.

RENOVATION

ECONOMICAL EFFICIENCY

CONSTRUCTION

OPEN SPACE

REVEALING FRAME

EXTENSION

CIRCULATION

PROGRAMMATIC ALTERATION

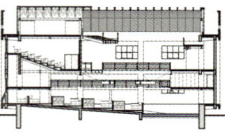 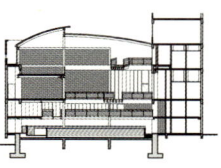 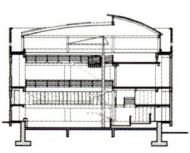

SECTION

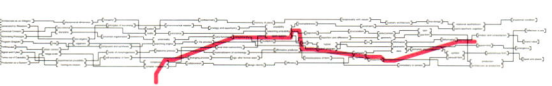

경희대학교 건축전문대학원　　유걸
아이아크

입체화된 창의공간　3 DIMENSIONAL CREATIVE SPACE

역동적 상호교섭　DYNAMIC FLOW OF SPACES

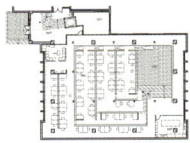

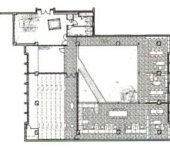
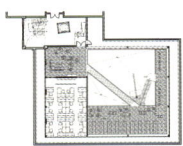
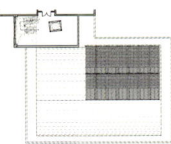

1F PLAN　　MEZ. 1F PLAN　　2F PLAN　　3F PLAN　　ROOF PLAN

David-Pierre Jalicon
D.P.J. & Partners, Korea
Aqua-Art Bridge
Seoul, Korea
2004

우면산에 직접 연결된 원형 구조물은 우면산과 예술의전당으로부터 나오는 에너지를 도심으로 전달하는 '기' 지점 역할을 한다. 이 원형 구조물에 설치된 플렉시글래스로 된 날개 부분의 각도 조절과 우면산의 근본적인 특징이 되는 '물' 이라는 요소를 끌어들여 주간에는 폭포를, 야간에는 여러 이미지를 투영할 수 있는 물 스크린을 만들어낸다.

As a "Ki(Spirit)" point (the ring structure is directly anchored onto cliff), the bridge transfers the energy (Fungsu) of Woomyun mountain and Seoul Art Center to the city. Flexi glass wings create the water fall during a day, the projection water screen at night. Both of them express a fundamental characteristic of this mountain: that is called water.

STRUCTURE
TRANSFER
FUNGSU
WATER
ORIENTAL ENERGY

새로운 자연 창출을 위한 기술미학 TECHNOLOGICAL AESTHETICS FOR A CREATED NATURE

아쿠아 아트 브리지

David-Pierre Jalicon
D.P.J. & Partners

새로운 수경 요소 NEW TYPE OF AQUA-SCAPE

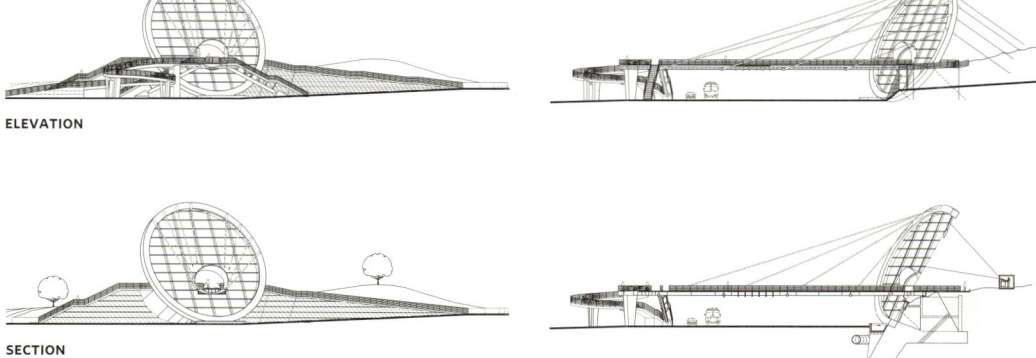

ELEVATION

SECTION

골 원반으로부터 서스펜션된 교량 BRIDGE SUSPENDING FROM STEEL DISK

David-Pierre Jalicon
D.P.J. & Partners, Korea
Nungpyung-ri House
Gyeonggi-do, Korea
2002

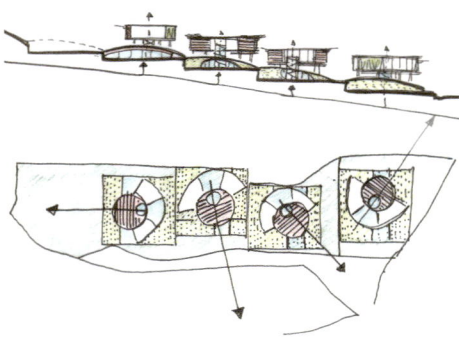

한국 전통주택은 몇 개의 별채로 구성된다. 이 프로젝트는 부지 면적이 부족하여 부모를 위한 공간과 자녀를 위한 공간이 아래위층으로 겹치도록 계획했다. 전통적인 건축계획은 산꼭대기에 위치한 집은 처마가 하늘을 향하도록 하고, 산 중턱이나 지상 가까운 곳에 위치한 집은 처마가 땅을 향하도록 한다. 능평리 주택은 후자에 해당된다. 이 주택의 기념 개념은 한국 전통의 정자를 역전환시킨 것이라 할 수 있다.

Traditional house includes several pavilions. Because of land size, here the pavilions are superimposed: parents are living in bottom, children in top. Traditionally, the design of house on hill top must be orientated to sky, that of house on the mid- slope or ground must be oriented to earth (Nungpyoung-ri house's case). At this time, the house can be regarded as an inverse Korean traditional pavilion.

INVERSE
SUPERIMPOSE
TRADITION
SEPARATION
SENSE OF PLACE

1층의 가족공간 위에 뜬 침실공간 **BEDROOMS FLOATING OVER THE LIVING SPACE**

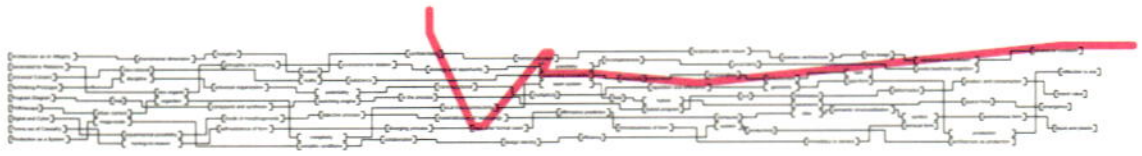

능평리 주택 DAVID-PIERRE JALICON
D.P.J. & PARTNERS

수평적 아크와 수직적 아크 HORIZONTAL AND VERTICAL ARC

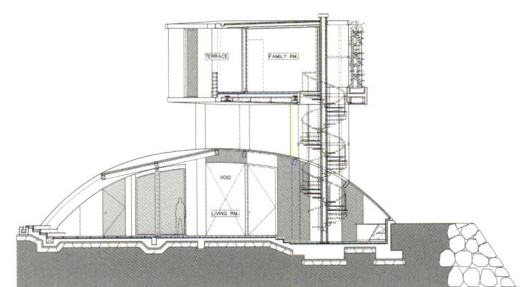

SECTION

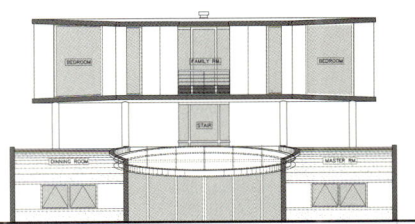

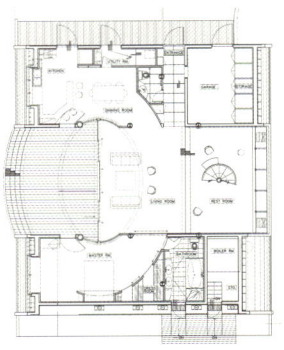

1F PLAN   2F PLAN

실내 수직동선 VERTICAL INTERIOR CIRCULATION

Woongwon Yoon + Jeongjoo Kim
Jegong Architects, Korea
Seoul Performing Arts Center
Ideas Competition Project_3rd Prize
Seoul, Korea
2005

한강 노들섬은 서울 강남과 강북에서 다리로 연결돼 있다. 다리는 남북 도시를 오가는 차들로 채워지지만 차를 세워 섬에 발을 디디는 사람은 아무도 없었다.

오페라하우스는 관객이 일상이라는 현실에서 오페라하우스라는 공인된 공간으로 들어감으로써 인간의 기쁨과 분노, 슬픔이 만들어낸 이야기와 음악이 만나는 곳이다. 오페라하우스는 대부분의 경우 외부 형태와 내부 인테리어 디자인이 계획의 중심이 된다. 우리는 섬이라는 대지의 성격과 오페라라는 음악이 도심의 숨겨진 장소에서 현실이 아닌 다른 세계를 만들어내기를 원했다. 벽으로 둘러싸인 내부 공간은 숨겨진 섬이고 그 안에서 오페라와 음악만의 공간이 만들어진다. 이러한 내부세계는 이따금 보이는 굴곡진 벽 틈으로만 자신의 존재를 도시 사람들에게 열어 보인다.

Nodel Island in Han River is connected by a bridge to the northern and southern part of city, Seoul. The bridge is filled with cars traveling to the south and north of the city, but nobody stops its car and sets foot on the island.

The opera house is a place where the audience can encounter unknown world of stories and music made of delight, sorrow and anger. The opera house, which is completed in its development of form and complicated in its functions, does not allow radical changes in its design. In most cases, the exterior forms and interior design become the center of the project. We are hoping to create a world of reverie with music, the opera using the characteristics of the island in this secretive place, hidden inside the center of the city. The inner space surrounded by walls is the hidden island, and only the opera and music is created in this place. This inner world opens to show its existence to the people of the city only through the small opening of the winding wall.

STORIES AND MUSIC MADE OF DELIGHT, SORROW AND ANGER

DEVELOPMENT OF FORM

COMPLICATED IN ITS FUNCTIONS

RADICAL CHANGES

CREATE A WORLD OF REVERIE WITH MUSIC

SECRETIVE PLACE

HIDDEN ISLAND

INNER WORLD

WINDING WALL

부유하는 예술시설

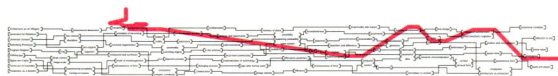

서울공연예술센터 | 윤웅원 + 김정주
아이디어공모 3등작 | 제공건축

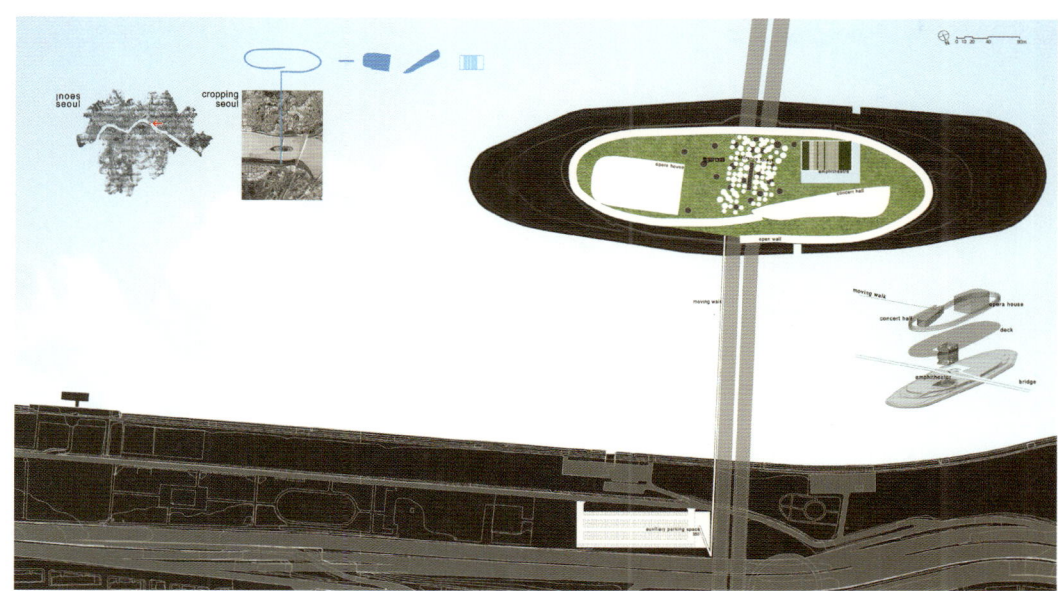

LAYOUT | 섬의 외관을 두르는 순환

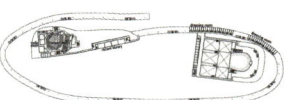
FLOOR PLAN +35.50M

FLOOR PLAN +25.50M

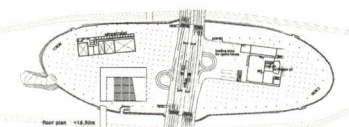
FLOOR PLAN +18.50M

FLOOR PLAN +14.50M

틈의 생성 GENERATION OF INTERSTICE

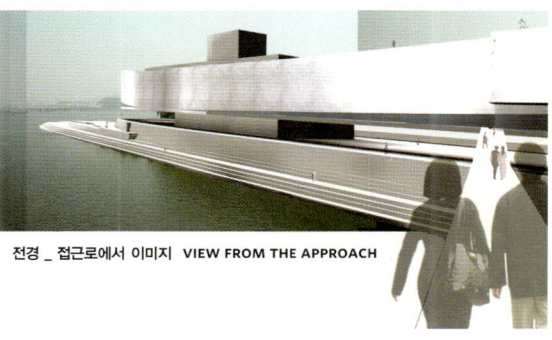
전경 _ 접근로에서 이미지 VIEW FROM THE APPROACH

Woongwon Yoon + Jeongjoo Kim
Jegong Architects, Korea
Myung Film Company Building
Seoul, Korea
2002

혜화동 주택가 막다른 골목 끝에 위치한 1970년대 2층 주택을 개조하고 증축해서 영화사 사옥을 만들었다. 이 프로젝트는 주택의 작은 방들을 영화사에 맞게 적절하게 연결하는 것과 조용한 주택가에 위치한 건물의 단아함을 유지하는 것, 영화사라는 정체성을 드러내야 하는 일견 서로 모순될 것 같은 프로그램을 해결하는 것이 과제였다. 기본 개념은 영화사와 최초의 카메라 '카메라 옵스큐라'를 적용하는 것이 적절해 보였다. 박스 형태인 건물 전면은 조리개 역할을 하는 유리창, 컬러 아크릴 판, 불투명 판 세 겹으로 이루어져 있다. 불투명 판은 빛의 양을 조절할 수 있도록 움직일 수 있고, 원색의 컬러 아크릴 판은 다양한 색깔을 만들어 영화사의 정체성을 표현한다. 영화가 필름에 각인된 이미지가 빛이 통과하면서 화면에 모습을 드러내는 것처럼 색색의 아크릴 판은 낮에는 빛이 투과되어 내부를 변화시켜주고 밤에는 내부의 빛이 외부를 비춤으로써 영화사의 역할을 은유한다.

BACK AN OLD MEMORY

EXPERIENCE OF SEEING

THROUGH A KEYHOLE

SECRET MOVIEMAKING

CAMERA OBSCURA

DARK ROOM

SMALL HOLE

GLASS WINDOW

COLORED ACRYLIC PANEL

빛의 현상을 위한 전면 구성 FACADE DESIGN FOR THE PHENOMENA OF RAYS

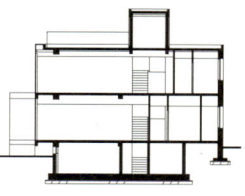

ELEVATION   SECTION

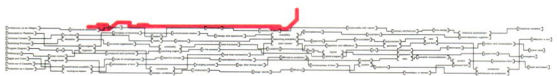

명필름 사옥   윤웅원 + 김정주
제공건축

A film company proposed a project for the building of the company which involved renovation and expansion of a 2 story residence built in 1970s located at the end of an alley of the residential area in Haehwa-dong in Seoul. The most important tasks were to make the residence with the small rooms appropriate to the function of the film company, and the building should display the identity of the film company while maintaining the refined style in the quiet residential neighborhood.

The film company and the first camera, Camera Obscura, became the basic concept of this project. The facade of this box shaped building comprise of three layers--glass window, colored acrylic panel, and opaque panel-- which function like the iris of a camera lens. The opaque panels are moveable to control the amount of light, and the acrylic panels of primary colors create a variety of colors and express the identity of the film company. Like the image on the film that appears on the screen as light penetrates, the daylight entering through the colorful panels during the day changes the interior and the light from the interior shins on the outside at night, the metaphor of the film company.

  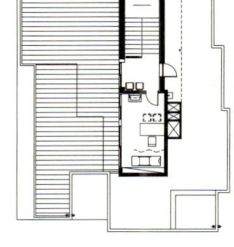

1F PLAN   2F PLAN   3F PLAN

기존 주택을 개조한 업무공간 OFFICE SPACE REMODELED FROM LIVING SPACE

Dominique Perrault
Dominique Perrault Architecture, France
Ewha Campus Center
Competition Project_Winner
Seoul, Korea
2004-

도심부 중앙에 위치하여 복잡한 주변 환경을 가진 이화여대 캠퍼스는 도심의 조직과 결합할 수 있는 총체적인 조망책이 필요했다. 그것은 문화지대로 대두될 수 있는 '캠퍼스 밸리'로 주변 환경을 수용할 수 있는 새로운 지형의 창출을 의미한다. 이화 캠퍼스의 새로운 관문이 될 설계부지는 연중 내내 각종 스포츠 활동과 축제가 펼쳐질 수 있는 공간인 동시에 학교 내로 도시가 자연스럽게 유입되는 매개체가 된다.

정문에서 이어진 광장형 도로를 지나면 보이드로 형성된 캠퍼스 밸리를 마주하게 된다. 캠퍼스 밸리를 횡단하는 다리들과 지하캠퍼스의 내부 동선을 통해 기존 건물들과 연결되어 캠퍼스센터는 학교 전체를 아우르는 통합시스템 역할을 하게 된다. 지하캠퍼스로 들어가는 길은 완만한 경사로 이뤄진 기념비적인 계단으로 조성된다. 이곳은 마치 프랑스의 샹젤리제나 로마의 캄피돌리를 연상할 수 있는 공간으로 포럼과 카페가 있는 광장, 옥외극장, 조각공원 등 다양한 기능공간으로 접근할 수 있는 중심점이 된다. 이러한 캠퍼스센터는 고풍스런 분위기의 이화여대에 어울리는 전원적인 정원으로 둘러싸이게 된다. 건물과 풍경의 경계짓기를 흐리게 하여 건물만이 아닌 캠퍼스를 새롭게 구성하는 유연성을 담고자 했다.

RELATIONSHIP
URBAN RESPONSE
SPOTS STRIP
NEW TOPOGRAPHY
BRIDGES
CONNECTION
PASTORAL NATURE
HYBRID PLACE
LANDSCAPE

캠퍼스 입구의 새로운 조직
NEW TISSUE INSERTED TO CAMPUS ENTRANCE

이화여대 캠퍼스센터
지명현상설계 당선작

DOMINIQUE PERRAULT

The complexity of the immediate site through its relationship to the greater campus and the city of Shinchon to the south demands a "lager than site" response, an urban response, a global landscaped solution which weaves together the tissue of the EWHA campus with that of the city. This gesture, the "Campus Valley", in combination with the "sports strip", creates a new topography which impacts the surrounding landscape in a number of ways. The Sports Strip, like the Valley, is many things at once. It is a new gateway to the EWHA campus, a place for daily sports activities, a grounds for the special yearly festivals and celebrations, and an area which truly brings together the university and the city. It is most importantly a place for all, animated all year long. Like a horizontal billboardm the sports strip presents the life of the university to the inhabitants of Shinchon, and vice-versa.

Once through the sports strip, pedestrian movement and flow through the site is celebrated. A new "Champs Elysees" invites the public into the site carrying students and visitors alike through the campus center northwards, bring together the different levels of the site. The paths connecting the existing buildings are maintained, with new "bridges" crossing the valley creating new east west connections that were previously limited by the EWHA stadium. Underground connections are also suggested, connecting the campus center to Clara-hall, Case hall, Pfeiffer Hall, Thomas Hall, Gibeon Hall and the Gymnasium. A three-dimensional system of connections is therefore possible, truly interesting the campus center with the campus. The Pastoral nature of the campus is perhaps its most remarkable quality.

It should be permitted to grow outwards, or inwards in this case, covering the campus center with trees, flowers, and grass. The Park is re-drawn. An Idyllic garden is the result, creating a special place for gathering, conducting informal classes, and simply relaxing. The notion of weaving together the campus is again evident, blurring the distinction between old and new, building and landscape, present and past.

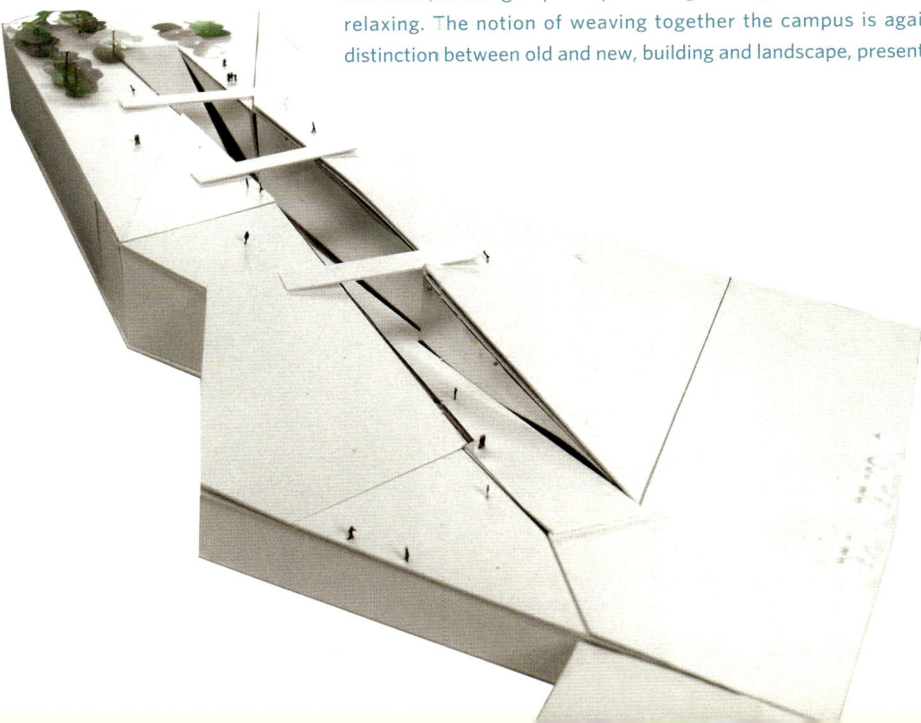

특화된 밸리 접근 SPECIALIZED APPROACH TO THE VALLEY

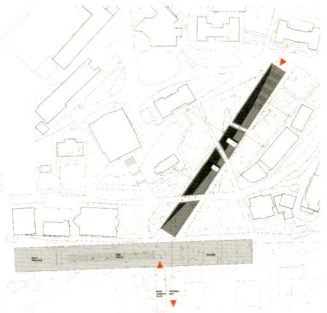

SITE PLAN

도시 풍경으로서 밸리의 야경 NIGHT SCENE OF THE VALLEY AS CITY-SCAPE

이화여대 캠퍼스센터
지명현상설계 당선작

DOMINIQUE PERRAULT

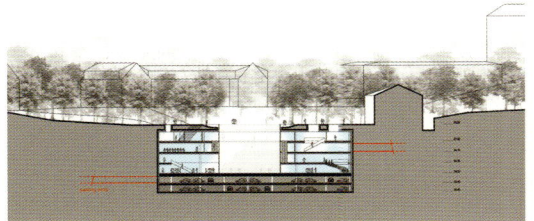
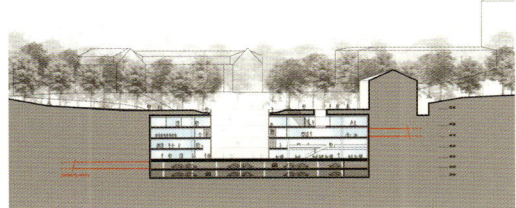

SECTION | 새로운 지하공간의 획득 GAIN OF NEW UNDERGROUND SPACE

아래 층위에서 접속되는 좌우의 서비스 기능 SERVICE AREAS CONNECTED OF THE BOTTOM OF VALLEY

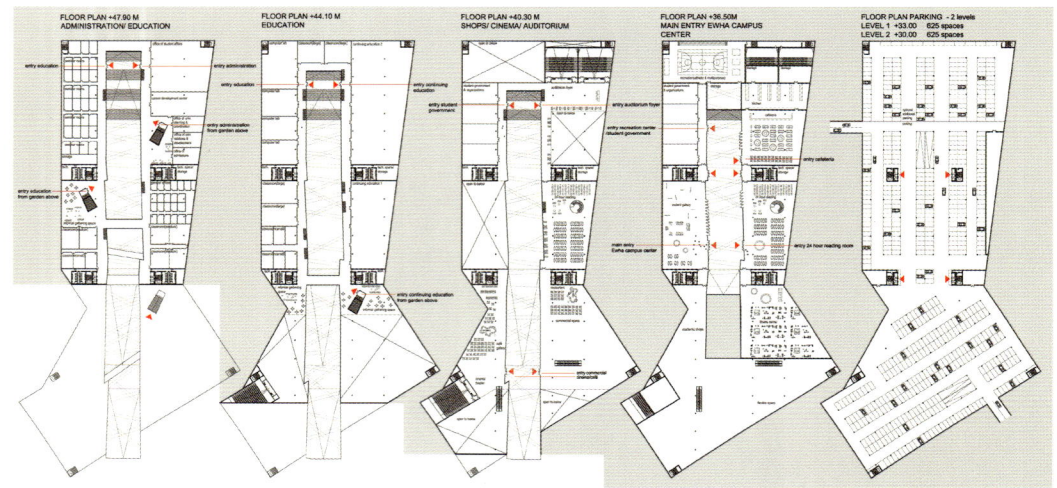

PLAN | 캠퍼스 진입과 서비스 기능 APPROACH TO CAMPUS AND SERVICE AREAS

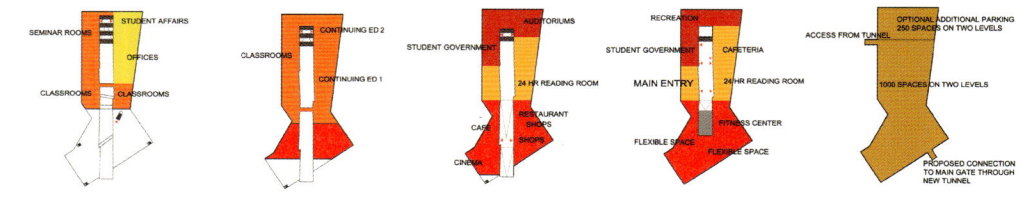

CONCEPT

새로운 행위 생성 GENERATION OF NEW ACTIVITIES

이화여대 캠퍼스센터
지명현상설계 당선작

DOMINIQUE PERRAULT

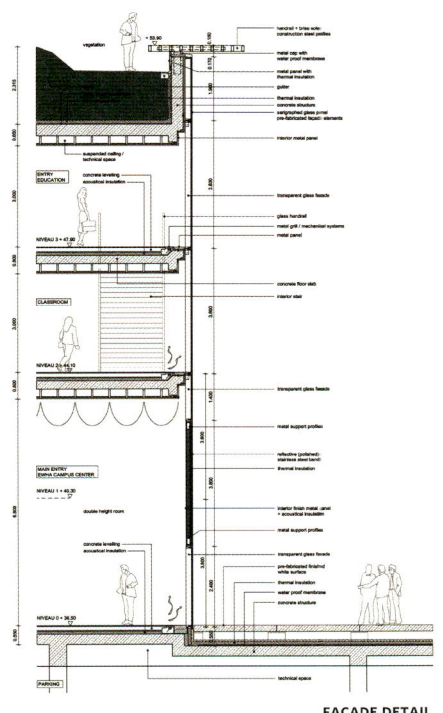

FACADE DETAIL

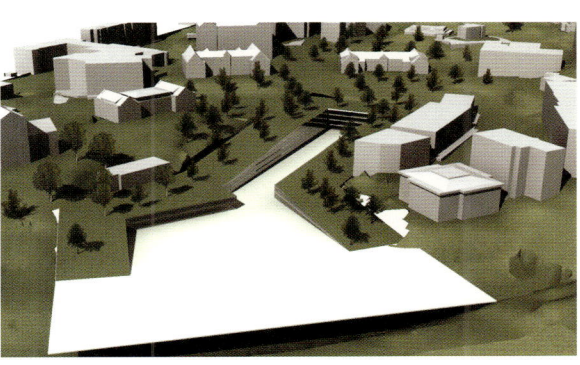

캠퍼스 전면의 새로운 풍경 NEW LANDSCAPE FROM THE CAMPUS FRONT

Jongkyu Kim
Korean National University of Arts / M.A.R.U. Architects, Korea
Design Manual for Heyri Art Valley
Collaboration with Junsung Kim
Paju, Gyeonggi-do, Korea
2004-

헤이리아트밸리의 대지는 경기도 파주시 탄현면 교하리 일원에 '통일동산 개발 촉진지구 조성사업'으로 개발된 지구 내 한 블록이다. 헤이리의 도로체계, 녹지체계, 개별 필지 분할 등 단지설계 마스터플랜은 연세대학교 도시교통과학연구소 김홍규 박사팀이 계획하였다.

이 단지는 기존 도시에서처럼 오랜 시간에 걸쳐 성장하여 늘어나는 도시의 한 자락이 아니라 새롭게 탄생되는 도시의 일부로, 이후의 도시 성장을 유도하는 중요한 계기가 된다. 그러므로 추후에 진행될 불확정적 상황을 유추하여 그 기초가 될 수 있는 도시적 맥락에서 건축의 틀을 마련해야 한다. 헤이리의 독특한 자연 형상

The site reserved for the construction of Heyri Art Village is a district within the city of Paju in the province of Gyeonggi-do. The land has been developed as a part of the "Unification Land Development Project".

The masterplan for the village has been designed by others. As a town to be newly born, Heyri Village carries much significance as an experiment that may serve to guide future developments in other communities as well. Therefore, in order to secure itself as a foundation to serve as an exemplar in the future, Heyri Village

URBAN CONTEXTUAL

UNPREDICTABLE

ENVIRONMENT

ARTIFICIALITY IN NATURE

FLEXIBILITY

SPECIFIC INDETERMINATED SPACE

PATCH

PLATE

LANDSCAPE

INFRASTRUCTURAL URBANISM

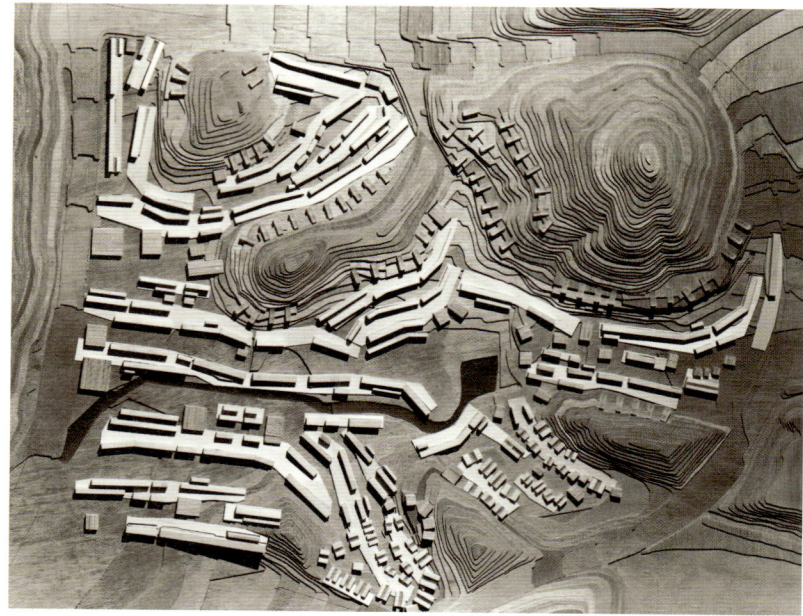

헤이리아트밸리의 공간 조직 _ 지형과 건축의 얼개
SPATIAL FRAMEWORK WITH ARCHITECTURE AND NATURE OF HEYRI ART VALLEY

헤이리아트밸리 건축설계 지침

김준성 협업 | 한국예술종합학교 | M.A.R.U.

김종규

D-12 식물감각_김종규
GREEN SENSE_JONGKYU KIM

에 부합하는 전체 맥락에서 조직을 찾는 작업이 선행되어야 하고, 그 조직의 테두리 안에서 실질적인 프로그램 실현을 위한 세부 작업이 수반되어야 비로소 자연보존의 기본 개념이 성립될 수 있다.

다양한 규모와 형태를 가진 380여 필지가 문화예술마을로서 여러 프로그램을 수행할 수 있도록 자유롭게 조성되는 동시에 단지가 전체 균형을 잃지 않도록 하는 실질적인 건축에 대한 논의가 대두되었다. 이에 따라 개별 필지의 건물 설계가 시행될 때 야기될 수 있는 디자인 난립을 조정하는 동시에 무엇보다도 중요한, 헤이리에서 꿈꾸는 프로그램이 실현될 수 있도록 그 바탕에 놓여서 단지 전체가 일련의 맥락을 유지하면서 구체적으로 체계화될 수 있는 건축 개념 및 설계지침 작업이 필요하였다.

설계지침은 마스터플랜을 근거로 작성하였으며 헤이리아트밸리 건설위원회가 주관하는 현실적인 프로그램 도입과 대지 조성을 위한 토목설계 단계에서 일부 작업이 조정되었다.

지형 해석과 건축 전략

'건축에서의 랜드스케이프', '건축적 랜드스케이프', '지형적 공간', '인프라스트럭처 어바니즘' 과 같은 이론의 배경은 헤이리가 지닌 환경의 틀과 함께 이 단지에서 전개될 다음과 같은 중요한 몇 가지 건축 개념을 풀어내는 데 적용된다.
– 주어진 자연과 인위적인 개발의 조화
– 다양한 프로그램의 융통성 있는 수용
– 시간에 따른 마을의 불확정적인 구성 가능성

한국토지공사에 의해 조성된 통일동산 지구 내 한 블록인 헤이리의 대지는 개발 계획에 의해 잘려진 6개의 작은 산맥들과 그 계곡들로 이루어져 있다. 평지는 마치 손가락을 펼친 형상이다. 서쪽 주진입도로에서 볼 때 계곡은 대지 각 방향의 경계로 방사선처럼 완만한 경사를 갖고 뻗어나가며 이는 대지가 갖고 있는 지형의 방향성을 결정해주고 있다. 단지 내에서 가장 낮은

requires an urban contextual framework that can accommodate and guide its unpredictable expansion and growth.

Heyri's singularly unspoiled setting requires an overall system of organization in order minimize its intrusion into its environment. Such an organization must precede all specific design work. Much discussion has taken place addressing the problem of maximizing the diversity and health of art-related programs on some 380 properties without losing the overall compositional unity of the entire village. It has been found that a concept and design guideline was necessary for the purpose of guaranteeing the realization of the ideals that Heyri symbolizes, on the one hand, while, on the other, helping the architects to control the unnecessary variables that are bound to crop up during the actual design phase. The guideline has been drafted following the masterplan, and addresses the some of the issues discovered during the construction of the infrastructure.

APPLICATION OF THEORIES AND TOPOGRAPHY OF HEYRI

The theoretical backgrounds such as "Landscape vi-à-vis Architecture", "Architectural Landscape", "Landscape" and "Infrastructural Urbanism" are applied at Heyri for the purpose of providing three basic concepts for the actual design. They are:
1. Artificiality in Nature
2. Flexibility
3. Specific Indeterminated Space

The site of Heyri Art Village is comprised of six hillocks and valleys. The plan view is similar to an open hand. Seen from the west, where the main road lies, each of the valleys spreads out in every

B-43 갤러리아고라_김종규
AGORA MUSEUM_JONGKYU KIM

서남쪽 진입부 표고는 14m이고 점차 동북쪽으로 높아지는데 가장 높은 산 정상 부분의 표고는 108m이다. 대부분의 필지는 기존의 지형지세를 이용하여 다소 평탄한 계곡 지역과 그와 연계되는 산자락에 위치한다.

둥근 형상의 단지를 감싸고 있는 6개의 산들이 수려하고 다양한 경관을 만들어내고 있으며 기존 식재들은 보존림으로서 가치가 충분하다. 단지 내에서 수계는 동쪽으로부터 두 갈래로 나뉘어 들어와 한 곳으로 합쳐진 후 가장 낮은 서남쪽 단지 진입부로 향하여 물길이 형성되어 있다. 지형에서 알 수 있듯이 단지 외곽의 동북쪽 고지대로부터 흘러내리는 물은 임진강과 합류하여 한강으로 유입된다. 또한 이 수로와 연계하여 대지 중심부에 넓은 늪지가 형성되어 있는데 환경친화적이며 생태계를 보존할 수 있는 중심축으로서 기틀을 마련할 수 있는 가능성을 제시한다. 단지 내에 있는 6개의 산 중 가장 큰 동북쪽 산 하부를 휘감은 형상으로 농수로가 계획되어 있는데 총 길이의 3분의 1 정도가 지표상에 노출되어 있다.

헤이리가 갖고 있는 자연환경을 보전하는 방법은 필요한 인공 요소들을 한데 모으는 것이다. 그리하여 상대적으로 보존될 수 있는 자연의 영역 역시 넓어지는 효과를 볼 수 있다. 인공 요소는 새로 만들어지는 길과 지어질 건물들 그리고 여러 활동이 다양하게 행해지는 외부 공간들이다. 이러한 인공 요소가 일정한 장소에 집적되면 전체 밀도 측면에서 남겨진 부분이 훨씬 넓은 공간으로 확보되며 인공적인 개발로부터 자유로울 수 있다. 이렇게 남겨진 부분은 마을 전체의 녹지로 공원인 동시에 개인의 넓은 정원이 된다.

헤이리는 생태마을을 지향하도록 환경의 질을 보존하는 기본 시스템이 단지의 전체 맥락을 주도하고 있어야 한다. 생태도시 구성 측면에서 수립되어야 하는 몇 가지 방안을 제안한다. 가장 중요한 원칙은 원 대지의 현황을 살린 수로와 단지 중앙의 늪지를 주축으로 보존될 수 있는

direction, sloping gently. The direction of each valley's orientation indicates clearly the topography of the site as a whole. The lowest point lies to the west, measuring 14 meters high. The highest point lies in the east at 108 meters high. Most of the properties are located in the flatter area of the valleys or in the foothills.

The method of preserving the natural setting is in gathering in one place all the artificial elements. This allows maximum area of the natural setting to remain undisturbed. The artificial elements consist of the roads, buildings, and outdoor spaces where various programs will be held. By placing these elements tightly, the remaining areas of the site are left to serve as individual gardens and a grand public park for the entire community. One of the principles of the preservation method is to leave as much of the stream and the wetlands untouched as possible. Surface covering must be reduced to a minimum, and wherever possible natural elements found at the site must be used to give order to the grounds.

Architectural Landscape of Heyri
Elements of Architectural Landscape
The primary element is the pavement, placed strictly following the flow of the land. Artificial paving is three in kind.
1. "Patch" in the flatter terrain
2. "Stepped patch" in the sloped terrain
3. "Plate" to be placed between patches, or on patches

As a continuous thread throughout the site, patches serve to connect separate properties. The areas between and behind the patches are to be left in their natural state. By having the patches placed strictly in accordance with the slope of the

헤이리아트밸리 건축설계 지침

김종규
김준성 협업   한국예술종합학교 | M.A.R.U.

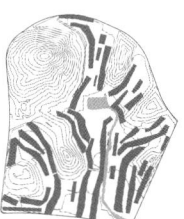
PATCH

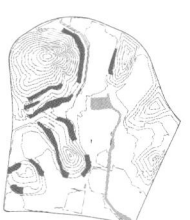
STEPPED PATCH

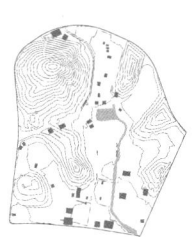
PLATE

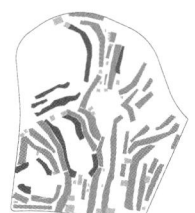
ARCHITECTURAL INFRASTRUCTURE
PATCH_WALLS_PLATES

모든 부분을 최대한 자연 상태대로 남겨두는 것이다. 동시에 이러한 바탕 위에서 적절하게 자연환경과 부합되는 건축 작업이 이루어져야 한다. 가능한 한 재생 가능한 자원을 효율적으로 활용하는 에너지와 설비시스템 도입을 권장한다. 열, 폐기물, 폐수 등의 방출관리도 철저히 이루어져야 한다. 또한 바닥에 토양포장을 최소화하고 인공성이 배제되는 부분은 다양한 야생식물 서식을 위해 대지의 바닥을 자연 상태로 완벽하게 보존해야 한다.

헤이리 지형에 대한 이해를 바탕으로 공간의 성격과 영역을 규정하는 요소를 만들고 이를 조직하여 전체 단지의 체계적 구성을 가능케 한다. 동시에 점진적이고 다양한 변화를 수용하도록 다음과 같은 건축 전략을 갖는다.
– 건축적 하부구조 형성: 건물이 놓일 위치 선정
– 건축 범위 규정: 건물의 전체 규모를 규정
– 불확정성에 대한 고려: 단계별 시행 전략

헤이리의 건축적 랜드스케이프

헤이리에서 구축하려는 건축적 랜드스케이프의 구성요소는 자연과 대비되는 인공의 바닥판이다. 이러한 인공 판은 지형의 흐름에 순응하며 철저히 땅이 지닌 성격을 강화하는 방향으로 놓이게 된다. 인공의 판은 지형 및 필지 상황에 따라 세 가지로 분류된다.
– 다소 평평한 계곡 지역에 놓이는 인공화된 바닥판으로서 패치
– 산자락에 지형에 따라 형성되는 벽에 의한 경사지 패치
– 패치와 패치 사이 혹은 패치에 걸쳐지는 플레이트

인공 바닥판으로서 패치는 앞으로 세워질 개별 건축물들이 구성되는 도시적 장소로서 기본 틀이 되며 이 마을에서 일어날 다양한 행위의 터가 된다. 이렇게 만들어진 판은 여러 필지를 지나면서 연속적으로 펼쳐지고 그외 지역, 즉 패치와 패치 사이의 공간 혹은 패치의 뒷

land, the entire topography of the site becomes clearer as a form. This clarity architecturalizes the natural contours and opens it up to the possibility of architectural manipulation.

ARCHITECTURAL INFRASTRUCTURE

By "architectural infrastructure" we designate a pre-established architectural framework whose function would be to provide a material foundation for all buildings that are to be built in Heyri. Such an infrastructure must satisfy two conditions simultaneously. They are: leaving the natural setting untouched and upgrading the quality of life to be determined by an artificial intervention. The manner of satisfying these two conditions needs to be further studied so as to provide a situation in which the natural and the artificial can maintain a balance within a comfortable range of tension and contrast.

The construction of an architectural infrastructure that can function as required will have to come before any design can be done on the site. This framework will be able to set an example for other towns in the future.

Regardless of the unpredictability of what will be built in the future, the architectural infrastructure can be laid out in advance, with a clearly marked form and boundary. The platform will aid in reading the shape of each of the property as demarcated according to the master plan. Moreover, the infrastructure allows a closer look when laying the patch in areas that overlap, requiring a more site-specific attention.

At Heyri, the patch, walls, and plates function as an architectural infrastructure by pre-determining how a building may be placed on any given property.

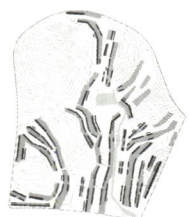

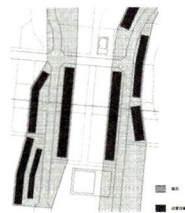

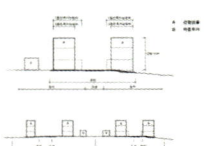

PATCH _ BAR-TYPE

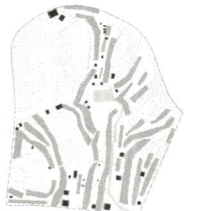

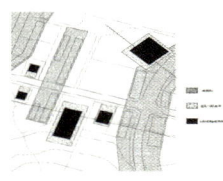

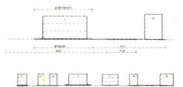

PLATE _ OBJECT-TYPE

부분은 자연 상태로 보전된다. 따라서 산과 산 사이 계곡으로부터 단지 중심부를 지나 단지 입구인 서측의 가장 낮은 지대에 이르기까지 지형을 따라 흐르는 선형 패치들 사이에는 공용의 자연경관들이 그대로 남게 된다. 인공 바닥이 철저하게 지형의 흐름에 따라 놓임으로써 이를 통해 땅이 지닌 성격이 더욱 분명해진다. 그리하여 자연이 가지고 있는 본질적인 구조를 바탕으로 하는 건축적 가능성이 무한히 확장된다. 이러한 인공의 판과 그 판의 조직 및 구성과 맥락을 같이 하여 그 위에 구축되는 띠로서 건축물들, 남겨진 자연들의 관계가 바로 헤이리의 건축적 랜드스케이프를 형성하게 된다.

　　구체적인 사항들이 결정되지 않았더라도 추후에 진행될 모든 건축적 가능성들을 수용할 수 있도록 미리 계획하여 만드는 건축적인 틀을 건축적 하부구조라 한다. 자연과 인위적인 상황의 관계를 잘 파악하여 이를 바탕으로 계획된 건축적 하부구조가 먼저 수반되어야 한다. 헤이리에서 패치와 벽, 플레이트는 건축물이 놓일 자리를 결정해주는 건축적 하부구조 역할을 하게 된다. 건축적 하부구조 틀에서 바닥판과 건물의 관계는 필지가 계획된 원 지형의 여건과 주변 필지 혹은 마스터플랜 상에 분할된 다른 요소와의 관계에 의해 다음과 같이 네 가지 유형의 방식으로 구성된다.

패치 / 선형 건물 유형

패치는 계곡 부분에 자연 형상에 부합되는 전체적인 맥락에서의 조직과 지형에 순응해서 다소 기다란 판의 형상으로 놓이게 된다. 중앙에 길이 방향으로 도로를 포함하고 있어 패치 위 건물들은 도로를 사이에 두고 서로 마주보게 되며 이 건물들은 다양한 각도의 선형 유형이 된다.

플레이트 / 오브젝트 유형

플레이트는 필지 위치 상 패치에 속할 수 없는 필지에서 독립적으로 자연에 놓이게 되는데 이 플레이트 위에 자유로운 볼륨의 건물이 오브젝트로 놓이게 된다. 그러나 이러한 오브젝트는 그

COMPOSITION

The buildings that are to be placed on top of the architectural infrastructure belong to one of the four types

Patch / Bar-type

The patch is designed to follow the topographical flow of the site. The patch is necessarily linear in form as it is located on both sides of the road. The buildings that face each other with a road between them are linear as well.

Plate / Object-type

The plate is provided where it is not suitable for the patch to define the area in question. The building that would go on top of a plate takes on the character of a detached object. Despite its separation, the building must engage itself to the rest of the site by acknowledging the flow of the patch that affects the property.

Stepped Patch / Podium-type

Stepped patch determines the siting of the building on the downslope of a foothill. In such a condition, the building must be elevated on a podium or a plinth so as to maintain a certain horizontal continuity of the overall composition of the block. With a podium-type, a retaining wall is unavoidable, and thus the wall must be incorporated in the design of the building's elevation.

Gate House

Design of the gate house area include the residential-block concept, and must be differentiated from the commercial district. The gate house straddles either two small scale linear buildings that face each other, or it may be situated between a building and a piece of nature. It is free to respond to the topographical flow of the property in terms of its angle of placement.

헤이리아트밸리 건축설계 지침

김종규
김준성 협업 | 한국예술종합학교 | M.A.R.U.

G41 갤러리희원_김종규
HUIWON GALLERY_JONGKYU KIM

자체가 아니라 해당 필지와 접한 패치, 혹은 패치와 패치 사이의 흐름 안에서 서로 관계하면서 건축적 구성의 가능성을 확장하고 있다.

경사지 패치 / 포디움 유형

경사지 패치는 산자락에 위치하는 필지의 건물 구축 가능 범위를 일컫는다. 건물은 포디움 형식으로 지형 레벨에 맞춰 수평적으로 연속성을 갖게 된다. 이 포디움 형식의 구조물을 구축하는 벽이 생겨나게 되는데 이 벽들은 지형이 갖고 있는 경사도에 순응하며 그 자체로 건물의 입면을 형성한다.

게이트 하우스

주거 중심의 블록 개념을 도입하여 별도로 계획된 게이트 하우스 지역은 비즈니스 지구와는 다른 건물 구성방식을 갖는다. 선형 패치 내에는 여전히 도로가 포함되어 있는데 패치 위에 건물이 놓이는 방식은 두 개의 소규모 선형 건물이 한 필지 내에서 마주보고 서 있게 되며, 패치 흐름과는 직교하는 방향을 기준으로 필지 상황에 따라 자유로운 각도로 자연과 인공물에 걸쳐지는 상황이 된다.

불확정성에 대한 고려 - 단계별 전략

프로그램의 불확정성은 필연적으로 단계별 전략을 필요로 한다. 단계별 전략을 위해서는 무엇보다 전체에 대한 이해가 중요하다. 설계지침 작업은 단지 전체에 대한 이해를 가능하게 하며 건물의 설계 방향에 대한 기본 틀을 구성한다. 추후 진행될 개별 필지 작업에서도 단지 전체의 구성을 이해할 수 있도록 하는 단계별 전략이 필요하다. 이를 위해 필지별로 주어진 최대의 볼륨을 전략적으로 이용할 필요가 있다. 이러한 작업은 전체를 구성하여 전체 건물 형태를 몇 개 조각으로 분리한 후 역으로 그것을 단계별로 채워넣음으로써 초기 건축 상황에서부터 전체 건물의 윤곽을 보여주려는 전략이다. 개별 필지의 건물 설계에서 이 전략을 이용함으로써 단지는 설계지침에서 정한 전체 틀 속에서 계속 성장, 진화하는 장소가 된다.

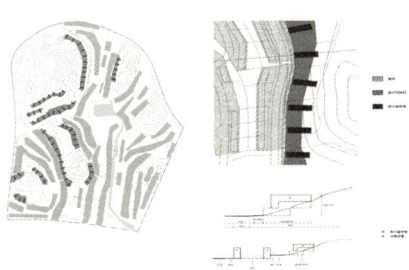

STEPPED PATCH _ PODIUM-TYPE

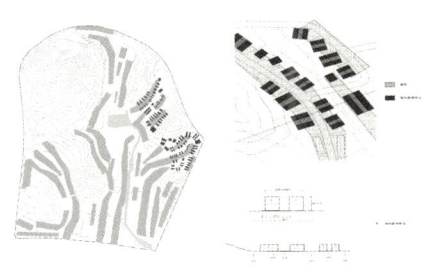

GATE HOUSE

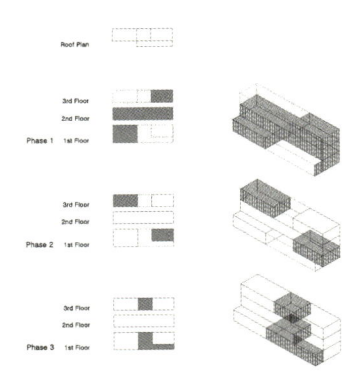

단계별 전략
STRATEGY IN THREE PHASES

프로토타입 다시 생각하기

봉일범
국민대학교 건축대학

프로토타입은 프로토-타입/선(先)-유형, 달리 말하면 유형에 앞선 유형, 즉 유형의 유형이다. 그것은 구체적인 사례로서 반복되는 것이 아니라 일종의 원칙 또는 따라야 할 전형으로서 되풀이된다.

프로토타입은 구조화된 정신의 산물이다. 현상적인 복수성을 묵과하고 있다는 점에서 그것은 추상적일 뿐더러 구조적이다.

일반적인 프로토타입은 대상과 목적, 또는 실제적인 적용에 앞서 예견되는 문제점을 명료하게 드러내기 위해 현실의 모호함을 의도적으로 묵과하거나 배제하지만, 건축에서 프로토타입은 그 자체를 현실 속으로 투사한다.

건축에서 언급되는 프로토타입은 제품 디자인 분야에서 사용되는 프로토타입과는 다르다. 대개의 경우 전자가 역사적인 것이든 그렇지 않든 의미로 가득 차 있다면 후자는 주로 문제 해결을 위한 실용적인 기술이기 때문이다.

테스트를 위한 목적으로 작용하는 프로토타입은, 따라서 지켜야 할 엄격한 규칙이 아니다. 그것은 본질적으로 전략 또한 아니다. 그것은 오히려 전략을 투사하기 위한 원본이다. 건축에서 프로토타입은 이와 같은 방식으로 효용을 발한다.

건축에서 프로토타입은 반복 가능하며, 개별적인 디자인 프로세스마다 그에 앞서 그리고 그 외부에 존재하는 참조항이지만, 결코 그 과정 속에서 소진되는 법이 없다.

프로토타입의 주된 목적이 개연성을 추론하는 것이라는 점에서, 건축에서 프로토타입을 사고한다는 것은 곧 건축을 디자인하는 것과 다르지 않다. 건축의 디자인 또한 아직 존재하지 않는 것에 대한 개연성의 주장이기 때문이다.

건축에서 프로토타입을 사고하는 것은 사물의 본질에 관해 사고하는 것인 동시에, 그 효과의 원활한 작용을 위한 조건들을 규정하는 작업이다. 그러므로 프로토타입은 피상성을 전복시키기 위해 본질을 참조하는 한편 본질을 들추어 표면적인 작용을 만들어낸다.

건축 작업에서 프로토타입은 언제나 아직 미완성이다. 건축가들은 바로 그 미완의 상태를 이유로 그것을 무시할 수도 있고, 여전히 충분히 발전되지 않았다는 점으로 인해

TYPE-FORM

GEOMETRY

UNIVERSAL ORGANIZATION

REPETITION AND DIFFERENCE

POTENTIALITY

TRANS-FUNCTIONAL

REFLECTION TO ERA

MULTIPLICITY

PRODUCTION

Rethinking Prototype

Ilburm Bong
School of Architecture, Kookmin University

Prototype is proto-type, in other words type preceding a type, i.e. type of type. it repeats not as a concrete example, but as a principle or model to follow.

Prototype is a product of structural spirit. It is not only abstract but also structural in the sense that it disregards the phenomenological multiplicities.

While prototype in general intentionally neglects or excludes the vague cloud of reality in order to elucidate the object, purpose, or the problems anticipated before its real application, prototype in architecture plunges itself into the real.

Prototype in architecture is different from prototype in product design because the former is usually charged with meaning historically or not; the latter is mainly practical technique of problem solving.

Operating as a test, prototype is not a rule to obey. It is not a strategy either as it is. It is rather an origin to cast the strategy; so works prototype in architecture.

Prototype in architecture is repeatable, case by case as a reference from outside and before the individual design process, but it never exhausts itself in it.

Prototyping in architecture is not different from the designing architecture in that the main purpose of prototype is reasoning the probability. Architectural design is also insistence on the probability of what it does not exist yet.

To think prototype in architecture is not only to think about substance of a thing but rather to define conditions to make its operation work well. For this reason, while prototype refers to the substance to subvert the superficiality, it emanates surface effects by instigating the substance.

Prototype in architecture has never been completed. Architects may ignore it for its incompleteness, or accept and apply it for its still underdeveloped condition simply as he wants in any case.

원하는 대로 간단히 그것을 받아들이거나 인용할 수도 있다.

유일자로서가 아니라 집합적인 것들의 대표로서, 건축 분야의 프로토타입은 개별 건축가들의 주관에 매여 있지 않다. 그것은 집단적인 동시에 순수하게 객관적이다.

건축에서 프로토타입은 오늘날의 건축이 보여주는 신합리주의적인 분위기와도 상통한다. 그것은 미학적으로 감상해야 할 골동품이 아니라 모사하고 수정하며 합리적으로 재정의해야 할 대상으로서 미리 주어지는 것이다.

건축에서 프로토타입은 분기, 다양화, 또는 눈에 띄지 않게 숨어 있던 것들의 현실화를 있게 하는 근원이 된다는 점에서 '잠재적' 이다.

건축에서 프로토타입은 순수한 의미에서의 환원주의가 갖는 한계를 넘어선다. 하나의 프로토타입은 수많은 그것의 사본들로부터 증류된 결과가 아니라 반대로 수많은 것들의 근원으로서, 실현되어야 할 다양한 가능성들을 내포하고 있기 때문이다.

프로토타입의 패러다임이 결과에서 그것의 원인으로, 물리적인 형태에서 그것을 있게 한 변수들로 옮겨감에 따라, 유일한 전형으로서의 프로토타입이라는 개념은 구체적인 사본들을 생성하기 위한 근저에 놓인 시스템으로서의 그것으로 전이된다.

건축에서 프로토타입은 시간으로부터 벗어나 있다. 그것은 역사적이지 않으며 과거에 속해 있는 것도 아니다. 그것은 언제나 현재에 속하며 미래를 투사한다.

건축을 디자인하는 과정에 투입되는 프로토타입의 기능은 그것이 재현적이기보다는 구축적이라는 점에서 추상기계로서의 다이어그램 개념과도 부합한다.

'원형으로의 회귀' 라는 구호는 모순어법에 다름 아니다. 건축에서 프로토타입이 잠재적이며, 비역사적이고, 다이어그램적인 이상, 그것은 우리가 그곳으로 회귀해야 할 지점이 아니라 그곳으로부터 출발해야 할 지점이다.

건축 디자인 과정의 모든 단계들은, '약식 프로토타입 만들기' 과정의 목적이 온전한 디자인을 완성하는 것이 아니라 문제점들을 명료화하고 그것을 해결하기 위한 대안들을 점검하는 데 있다는 점에서, 하나의 '약식 프로토타입 만들기' 와도 같다.

건축에서 프로토타입은 '다품종 소량생산' 의 개념과 기법들을 만남으로써 단순 반복에서 오는 무미건조함이라는 그것의 고전적인 한계를 극복한다.

무엇보다도 프로토타입은 그 자체로 중성적이다. 그것은 특성을 갖기에 앞서 변형, 치환, 활용, 개조, 조합 그리고 종국에는 이 모든 이유들로 인해 재사고의 대상이 되어야 한다.

Not a singular object but a representative of a category, prototype in architecture is not bound to the subject of an individual architect. It is collective and at the same time purely objective.

Prototype in architecture is not far from the neo-rational atmosphere of contemporary architecture. It is not an antique article to appreciate aesthetically but what is given a priori to trace, to adjust, or to re-defined rationally.

Prototype in architecture is "virtual" in that it is the root of ramification, diversification, or the actualization of what lies latent, or potential.

Prototype in architecture exceeds the limit of pure reductionism because a prototype is not distilled from a large number of instances but in the other way -as a origin of the many- contains so many possibilities to be realized.

As the paradigm of prototype shifts recently from the result to its causes, from the physical form to its parameters, the idea of prototype as singular model shifts to that of a system in the root to generate instances.

Prototype in architecture is out of time. It is non-historic, nor is of the past. It is always in the present and projects the future.

Considering that it is rather constructive than representative, the function of prototype inputted in an architectural design process is consistent with that of diagram as abstract machine.

The slogan "Return to Prototype" is not far from oxymoron. Since prototype in architecture is virtual, non-historic, and diagrammatic, it is not what we return to but what we start from.

Every stage in architectural design process is an act of "rapid prototyping" in the sense that the purpose of "rapid prototyping" is not to build a full design but to make the problems clear and to check the alternatives of resolutions.

Prototype in architecture overcomes its traditional limit of inertness from tedious repetition by the notion and techniques of 'mass customization'.

Most of all, prototype is neutral in itself. Before tinted with qualities, it should be subject to transformation, transposition, conjugation, renovation, grafting. . . and finally for those reasons, rethinking.

SATOSHI MATSUOKA + YUKI TAMURA
MATSUOKASATOSHITAMURAYUKI, JAPAN
JEONGOK PREHISTORY MUSEUM
COMPETITION PROJECT_3RD PRIZE
GYEONGGI-DO, KOREA
2006

박물관은 선사유적지와 야외전시공간, 자연공원, 역사문화마을을 포함하는 광대한 대지의 중심에 위치한다. 이곳을 공원으로 조성하여 방문객을 위한 만남의 장소를 만들고자 했다. 먼저 기존 지형을 따라가면서 주변 환경을 최대한 고려한 단층 건물을 구상해 부지의 변형을 최소화하고자 했다. 밝고 반투명한 지붕을 덮은 플로어가 이어지면서 건물과 부지는 하나로 통합되어, 방문객들은 뚜렷한 경계를 인식하지 못하고 야외에서 숲을 거니는 듯한 느낌을 받을 것이다.

내부는 커다란 공간으로 이루어진다. 적당한 간격으로 작고 완만하게 펼쳐진 경사면이 분할된 공간에서 방문객들은 다양한 활동을 경험하는 한편 전체 분위기를 느낄 수 있다. 건물 내부 어디에서나 주변 경관을 한눈에 볼 수 있다. 공공 편의시설과 교육시설은 3면이 언덕으로 둘러싸인 밸리 지역에 배치시켜 사람들이 이용할 수 있다. 방문객들은 언덕을 오르는 기분으로 전시공간을 돌아다니게 된다. 단 위치에 따라 전시공간의 부분 변형이 가능하여 다양한 전시 규모와 프로그램을 수용할 수 있다. 지붕에는 테플론 코팅이 된 유리섬유막을 사용함으로써 내부 공간을 최대한 자연환경처럼 유지하려고 했다.

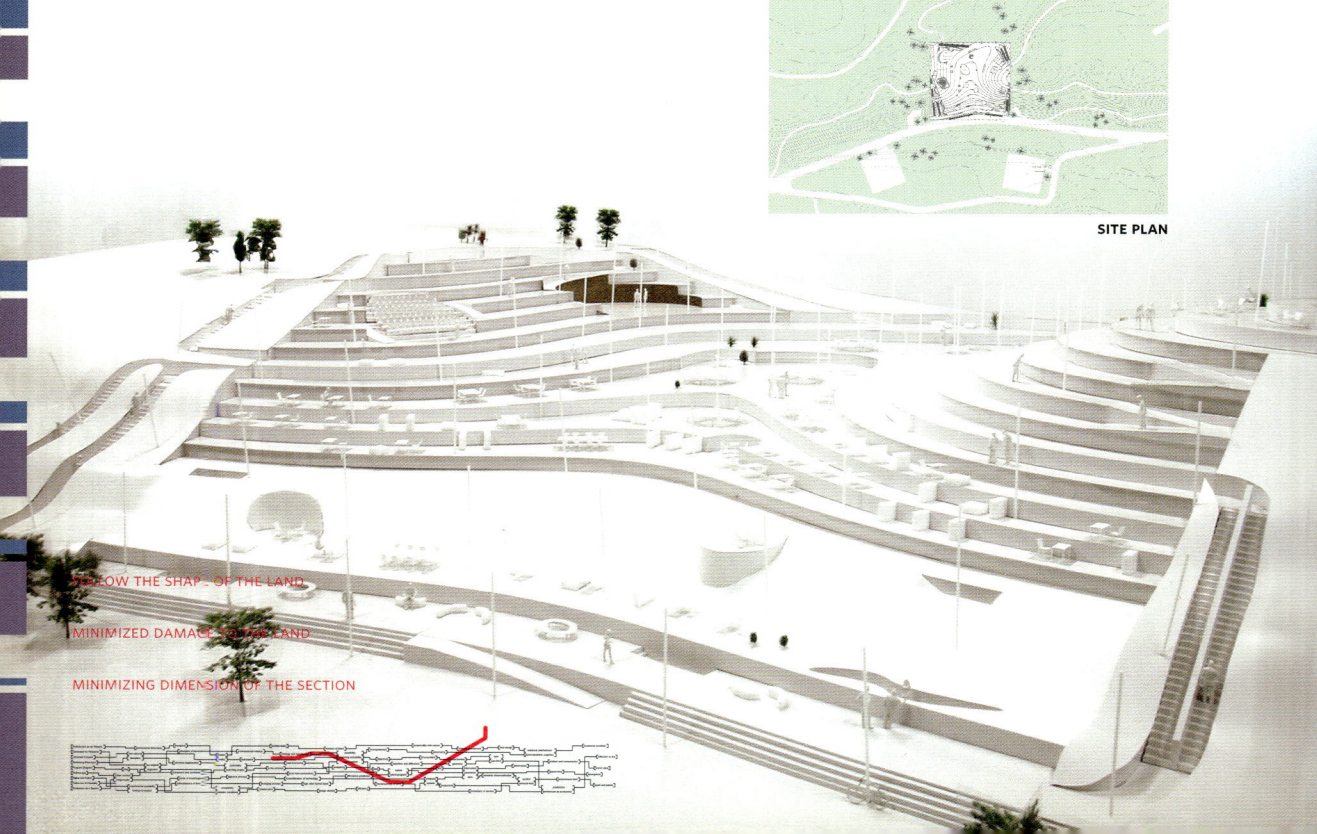

SITE PLAN

FOLLOW THE SHAPE OF THE LAND
MINIMIZED DAMAGE TO THE LAND
MINIMIZING DIMENSION OF THE SECTION

전곡선사박물관
현상설계 3등작

SATOSHI MATSUOKA + YUKI TAMURA

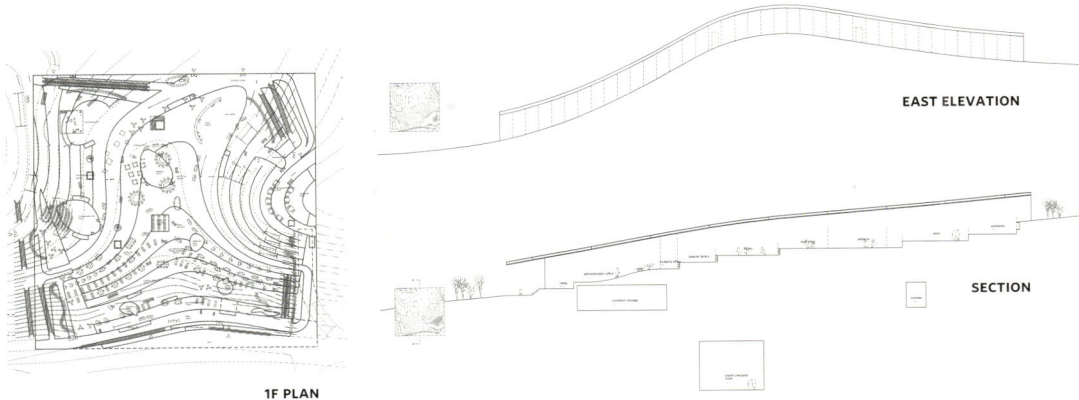

EAST ELEVATION

SECTION

1F PLAN

The site is located in the center of a wide-area master plan that includes a prehistoric site, outdoor exhibition, amusement park, and history culture village. The proposed single-story building follows the shape of the land to respect the local landscape, where the damage to the site is kept to the minimum. The building and site are unified by a continuous floor covered with light and a translucent roof, so that the visitors feel as if they are wandering around the outside forest without a strong boundary.

The interior consists of a big room. Visitors experience various activities in the spaces, which are softly divided by small vertical intervals and gentle slopes in the open landscape while feeling the atmosphere of the whole space. The visitors get a wide view of the surrounding landscape from everywhere in the building. The public amenity zone and educational zone are located in the valley area, surrounded by hills on three sides, so that people feel comfortable while studying or taking a break.

People walk around the exhibition zone as if climbing a small hill, seeing the whole picture of the exhibition. Sectional transformations along the steps accommodate the various sizes and materials of the exhibits. We intend to keep the interior a natural environment as much as possible by using a Teflon-coated glass fiber membrane on the roof.

미술관 내부 _ 단의 축조로서 전시 전개
MUSEUM SPACE_EXHIBITION THROUGH THE TERRACED SPACE

Satoshi Matsuoka + Yuki Tamura
MatsuokaSatoshiTamuraYuki, Japan
Seoul Performing Arts Center
Ideas Competition Project_2nd Prize
Seoul, Korea
2005

우리는 땅을 새로 만들고 그것을 뒤덮음으로써 섬을 음악으로 가득 채우는 것을 목표로 하였다. 첫째로, 한강대교의 서쪽은 남북 방향으로 150m, 동서 방향으로 200m, 최대 깊이 12m 규모로 분화구와 같은 형태로 굴착될 것이다. 그런 다음 거대한 지붕이 다리를 가로지르며 동쪽 끝에 이르도록 전체 섬을 뒤덮게 될 것이다. 이 거대한 지붕은 빛을 받아들여 분화구를 부드러운 빛으로 가득 차게 만들 수 있는 반투명의 막구조로 이루어진다. 대지의 모양을 단순하게 변경하고 뒤덮음으로써 섬은 음악으로 가득 찬 반외부적인 공원으로 변경된다.

교통이나 여타의 소음이 대지를 교란시키지 못하도록 하기 위해, 분화구의 주변부는 천장고가 단지 3m에 불과하도록 바닥판을 들어올림으로써 단면상으로 얇게 눌러 있다. 이 주변부는 또한 갑작스런 홍수 범람으로부터 분화구를 보호하도록 작용한다. 한강대교와 주변부가 동일한 높이에 놓임으로써 대교로부터 대지에 진입할 때 사람들은 한눈에 분화구 전체를 굽어볼 수 있으며 반대편 끝까지 시선을 가로막는 것은 없다. 달리 말하면 이 계획은 섬의 주어진 환경을 있는 그대로 활용하면서도 그것에 새로운 기능을 부가하여 스펙터클과도 같은 새로운 인상을 만들어내고 있다.

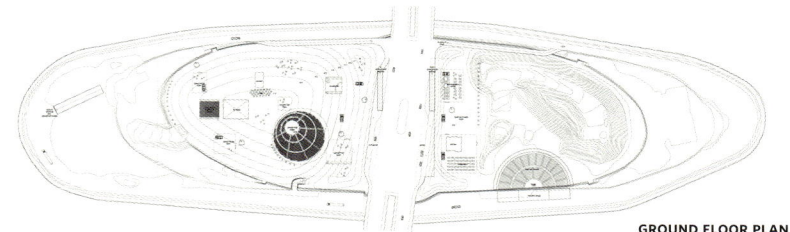

GROUND FLOOR PLAN

RESHAPE

COVER

EXCAVATE

CRATER

TRANSLUCENT MEMBRANE

SPECTACULAR

노들섬을 위한 미니멀한 조형 MINIMAL GEOMETRY FOR ISLAND-SCAPE

Satoshi Matsuoka + Yuki Tamura

서울공연예술센터
아이디어공모 2등작

The site is an artificial island located in the Han River cutting across Seoul. They requested 1,500 capacity opera house, 1,500 capacity concert hall and outdoor concert hall. Our aim is to fill the island with music by reshaping the earth and covering it. First, the western side of the Hangang Bridge is excavated like a crater with a dimension of 150m in the north-south direction, 200m in the east-west direction, and a depth of 12m maximum. Next, a large roof is placed to cover the entire island, spanning over the bridge, reaching all the way to the eastern edge. The large roof is a translucent membrane structure allowing light to pass through, filling the crater with soft light. By simply reshaping and covering the site, the island is transformed into a semi-outdoor park full of music.

In order to prevent traffic and other noise from disturbing the site, the perimeter of the crater is squeezed in section, by raising the floor to a ceiling height of only 3m. This perimeter also acts as a protection against sudden floodwater from entering the crater. The Hangang Bridge and this perimeter are located on the same level so that when entering the site from the bridge, you can look down into the crater in its entirety at a glance, and your vision is uninterrupted through to the other side. In other words, while our plan is to use the environment of the island as it is, but by accommodating new functions, the new impression is spectacular.

분화구 극장 전경 CRATER THEATER

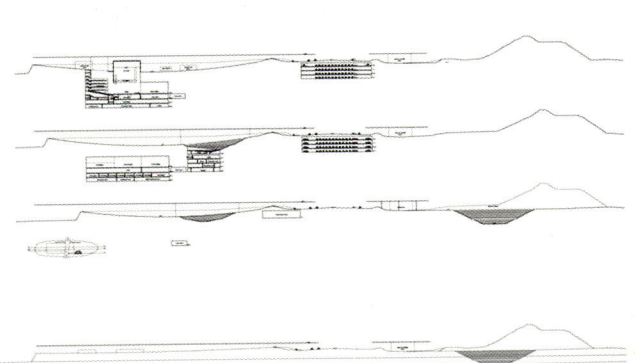

SECTION

지면 아래 공연공간 EXCAVATED PERFORMANCE HALL

Daniel Valle, Spain + Kiohun Architects & Associates, Korea
Imjingak Memorial
Gyeonggi-do, Korea

앞 시대의 기념물이라는 것은 대부분 영웅적 이미지를 위한 '하나의 강한 대상물'로 만들어졌다. 이에 참여하거나 경험하는 방법 역시 직접 전달받는 메시지이거나 일방적인 설득의 경향이었다.

이 새로운 열린 민주주의 시대의 기념물에 대한 우리의 제안은 기념하려는 역사 또는 사건이 관련된 특별한 '자연 요소들(자연물 및 풍경)'과 관계 맺기로부터 시작한다. '개인적 기억'의 숨겨진 의미를 드러내어 여러 사람이 공유할 수 있는 '집합적 기억'으로 전이되어 보편적 가치로 승화되기를 바라는 것이다. 참여하고 경험하는 방법도 교환과 접촉을 통하여, 개별적 기념물들이 관련을 맺어 하나의 총체적 의미에 이르기를 기대한다. 즉 '자연의 요소들과 풍경'이 기념물들과 맺는 깊은 관계이다. 인위적인 것과 자연적인 것의 행복한 건축적 결합으로 여기 임진각에서 평화와 상생의 가치가 고양되는 '새로운 풍경과 의미'가 생산되기를 기대한다.

MONUMENTAL

ONE STRONG MONO FORM

LABYRINTH

INDIVIDUAL MEMORY

COLLECTIVE MEMORY

PARTICIPATION & EXPERIENCE

VARIETY

NATURE

MEANING UNIT

MYTHICAL

TRANSFORMATION OF ORIGINAL TOPOGRAPHY

임진각 기념관   DANIEL VALLE + 기오헌

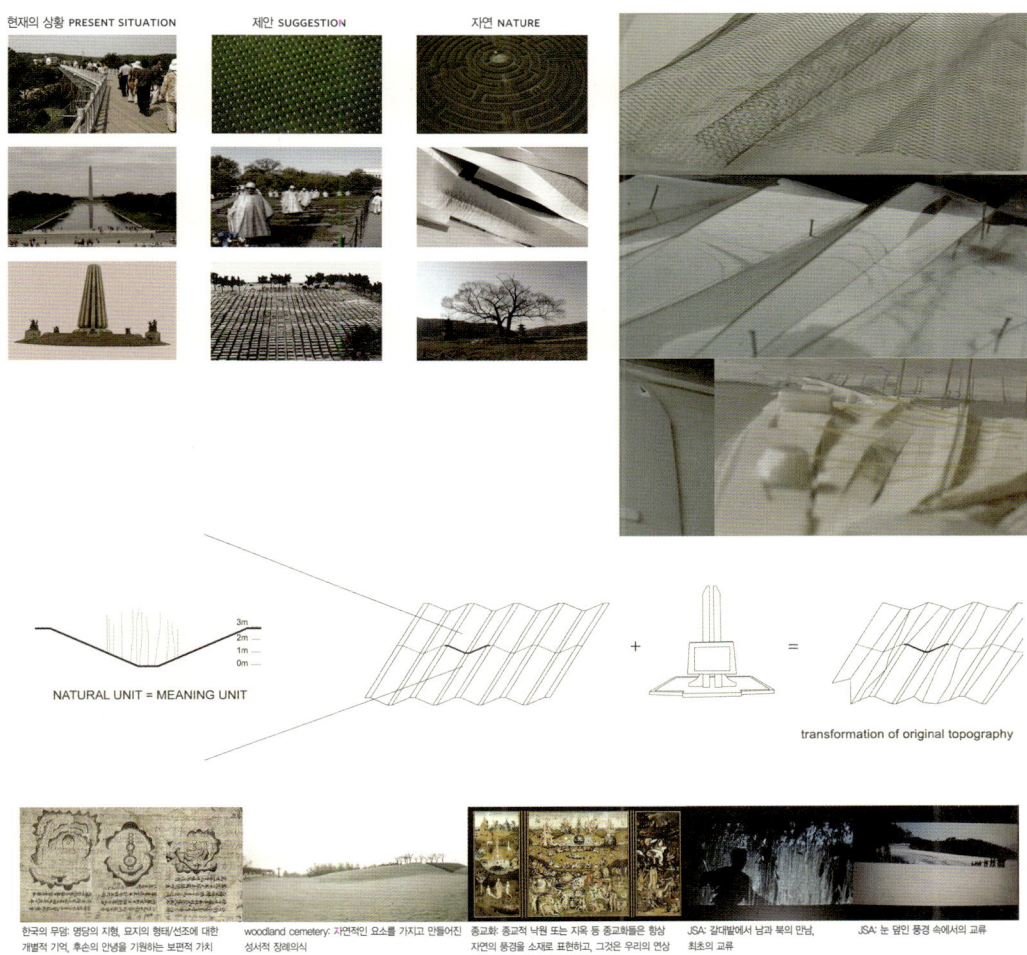

Memorials built in the past era are conceived as "one strong mono form" for the heroic image. The mode of experience or participation is also through the message, or the one-sided persuasion.

Our proposal for new memorial in this new era of democracy is, to begin with, making relation with special "natural components(natural object or landscape)" which have close relation to the history or event to memory. Then the hidden meaning of "individual memory" is transposed to "collective memory" in order to be sublimed to the universal value. The mode of experience or participation is also expected to become the meaning as a whole relating to individual memorials through the communication and contact. This is the deep relationship between the "natural components and landscape" and memorials. "New landscape and meaning" inspiring the value of peace and coexistence will be produced in this area of Imjingak by the architectural integration with the artificial and the natural.

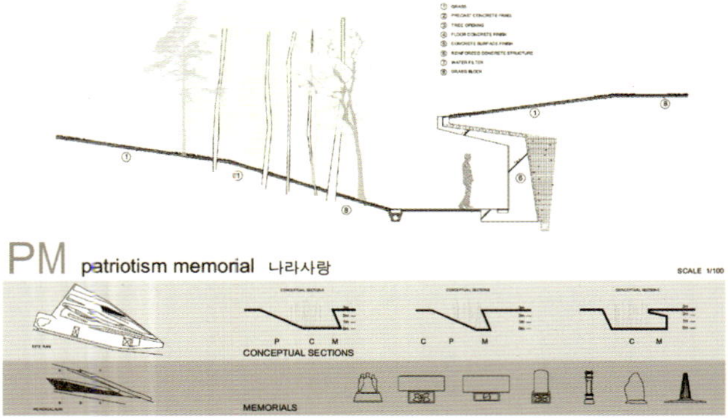

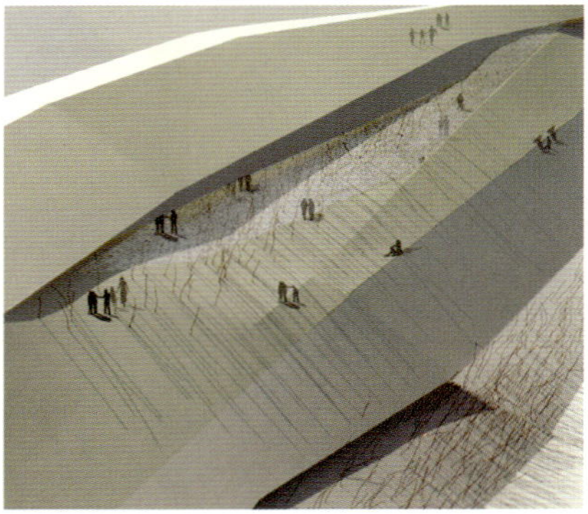

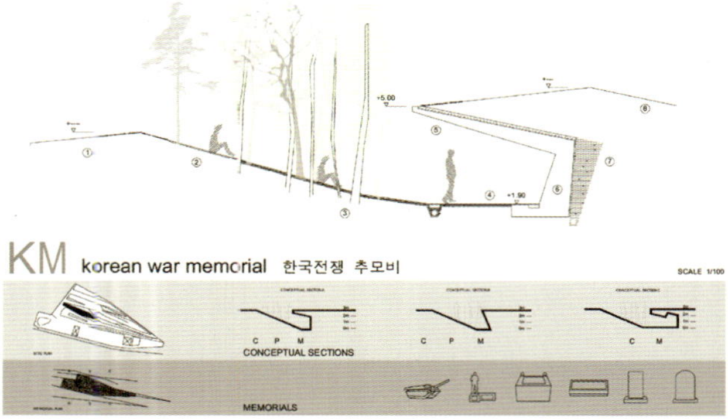

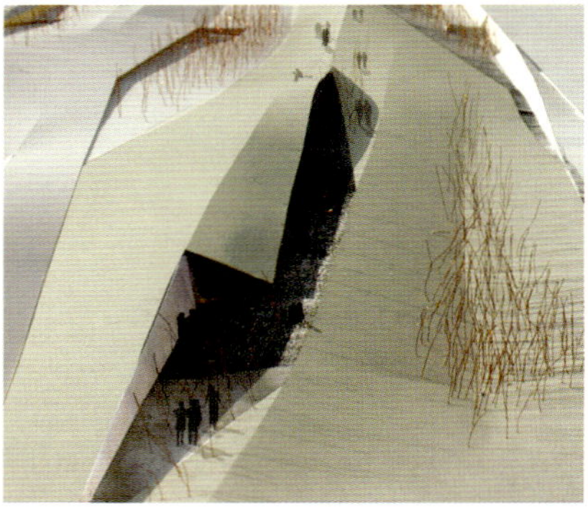

PM patriotism memorial 나라사랑

KM korean war memorial 한국전쟁 추모비

기념과의 접촉 방법 WAYS TO RELATE WITH THE MEMORIALS

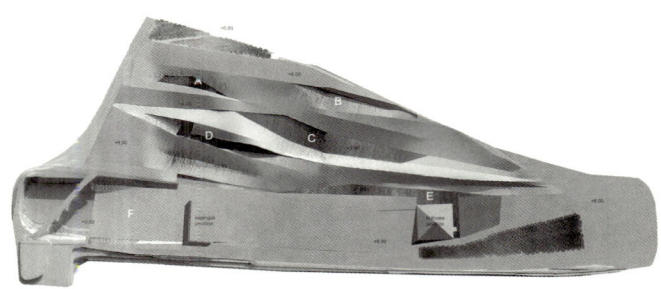

기념물의 그룹핑과 의미의 스토리 GROUP OF MEMORIALS AND THEIR STORY OF MEANING

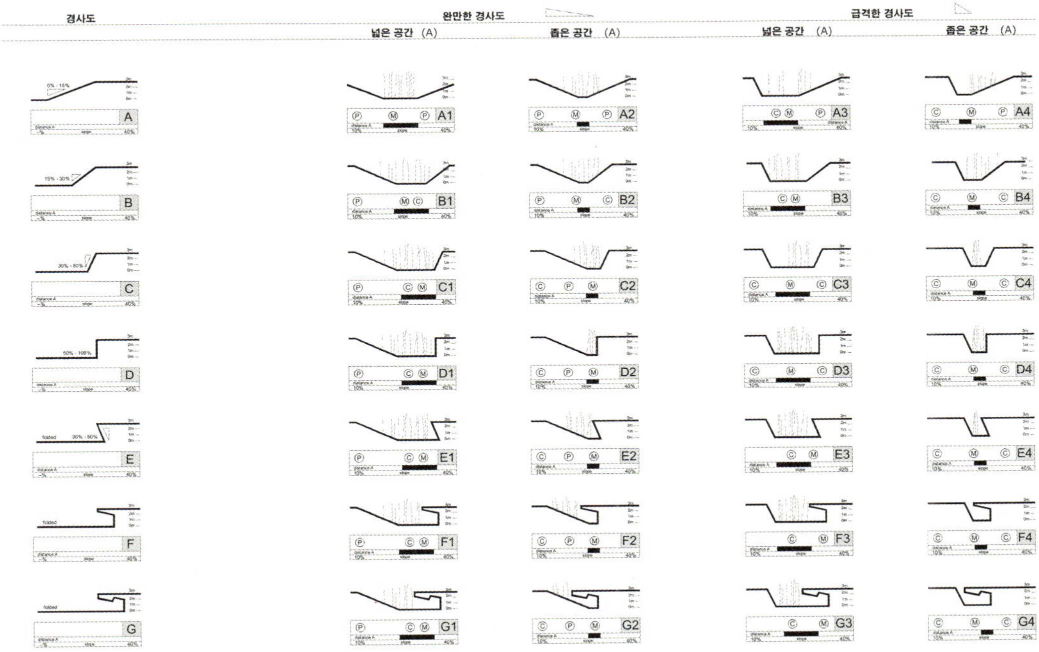

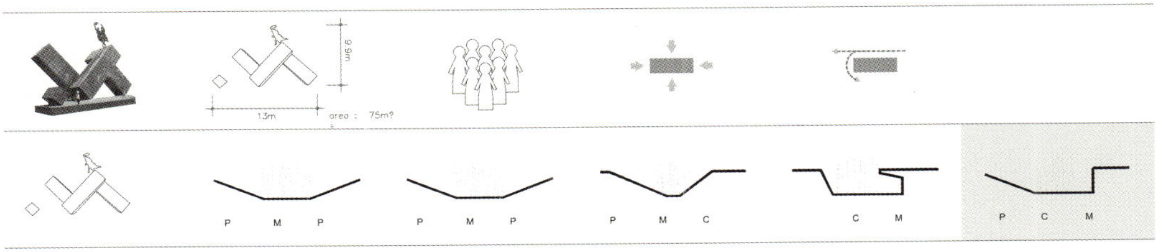

임진각 기념관

Daniel Valle + 기오헌

각도에 대하여 ANGLE STUDIES

MODEL VIEW

Junglim Architecture, Korea
Jungchul Kim + Seonghong Park
Cheonggyecheon Culture Center
Seoul, Korea
2004-2005

청계천의 가장 중요한 이미지는 물의 흐름이다. 이 문화관의 가장 중요한 요소는 사람, 즉 관람객의 움직임이다. 전시관람 동선을 물 흐르듯 구성하고 그 동선 자체를 전시공간화하며 그 공간을 건축화하는 것이 바로 이 프로젝트의 목적이다. 문화관은 박물관처럼 정적이거나 소위 우아한 분위기보다는 동적이고 다분히 대중적인 성격의 전시공간으로 해석된다. 그래서 경사로를 이용한 흐르는 전시공간이 더 적절할 것이다.

청계천 위에 떠 흘러가는 듯한 매스와 외부 전시벽을 보며 접근한 관람객은 펌프로 물을 끌어올리듯 단숨에 옥외 에스컬레이터와 함께 전망대까지 올라간다. 그곳은 콘크리트 건물만 볼 수 있던 과거 고가도로 레벨이지만 이제는 푸른 하늘과 맑은 물 그리고 멀리 남산을 볼 수 있다. 건물 내부로 들어오면 상설전시가 시작된다. 좁고 긴 전시공간에 요구되는 경사로를 적극적으로 전시에 도입하여 구성한 이곳은 청계천의 역사와 미래를 읽으며 흘러 내려간다. 그 속엔 세미나실과 강당의 가변성 및 창을 통해 보이는 또 다른 전시물인 현재 청계천 풍경 등이 포함된다.

SITE

FLOW OF WATER
FLOW OF PEOPLE
MOVEMENT
CHEONGGYECHEON
SLOPEWAY
FLOATING
NATURE
HARMONIZED
PUNCTUATION MARK

청계천과 문화센터
VIEW OF CULTURE CENTER AND CHEONGGYECHEON

청계천문화센터   정림건축
김정철 + 박승홍

열린 진입마당   OPEN ENTRANCE YARD

The cultural center project is aimed at producing movement lines, from which visitors view the exhibition as if they were flowing like water, thus creating an exhibition space and architectualizing the space. In addition, the cultural center is interpreted as an exhibition space with a dynamic and very popular nature, rather than conveying the static and elegant atmosphere of a museum. For this reason, the flowing exhibition space utilizing a slopeway would be more appropriate.

Visitors who have access to the museum while watching the display, which is as if floating and flowing on the Cheonggyecheon, and external exhibition walls, instantly go up to the observatory on outdoor escalators, like pumps drawing up water. The observatory is on the same level as the overpasses that existed there in the past, where only concrete buildings were to be seen; yet now one can see the blue sky and the clear water with Namsan(Mt.) in the distance.

Once inside the building, the permanent exhibition begins. The cultural center, which has imaginatively introduced the slopeway required for a narrow and long exhibition space, flows visitors downward, showing the history and future of the Cheonggyecheon. It includes the versatility of seminar rooms and an auditorium against the backdrop of the current landscape of Cheonggyecheon, seen through windows, which can be viewed as another exhibition item.

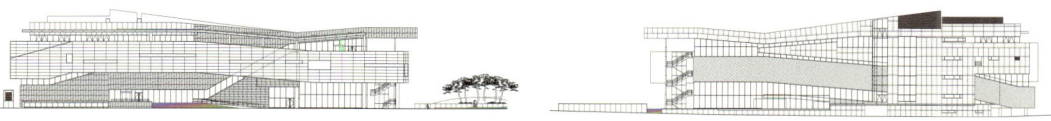

NORTH ELEVATION   SOUTH ELEVATION

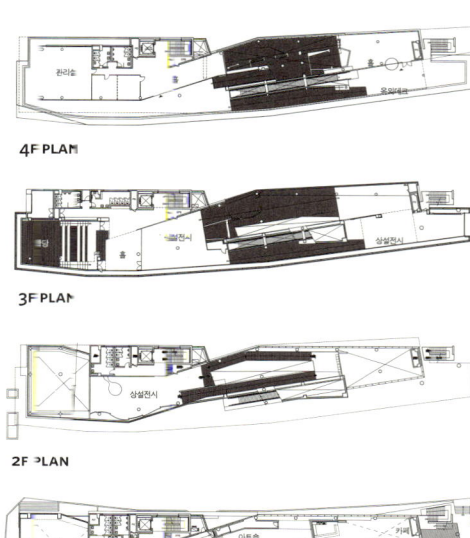

4F PLAN

3F PLAN

2F PLAN

1F PLAN

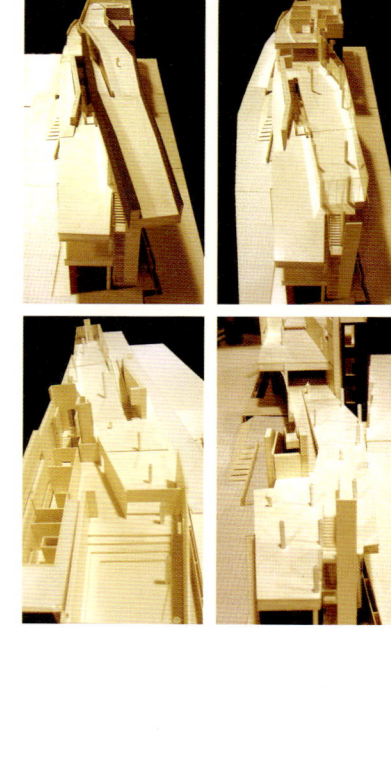

내부 공간 전경 INTERIOR VIEW

청계천문화센터　정림건축
김정철 + 박승홍

조망을 길게 끄는 공중의 길　AERIAL PATH FOR SCENIC DURATION

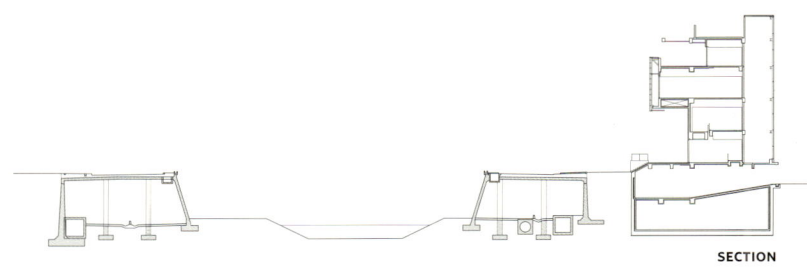

SECTION

Heerim Architects and Planners, Korea
Younghee Lee
Iran Oil Industry Headquarters
Teheran, Iran
2002

이 프로젝트는 지역을 활성화시키는 동시에 이곳을 업무와 교류를 위한 공간으로 자리매김하는 데 그 목적이 있다. 또한 도시의 랜드마크가 될 만한 세계적 수준의 건축물을 통해 국가의 정체성과 첨단산업의 중요성을 더욱 부각시킬 수 있는 계기를 마련코자 한다. 이 업무공간은 친환경 빌딩, 공간의 극대화, 에너지 절감, 건식 건축공법 등 환경친화적 개념을 목표로 설계하였다.

액체의 물리적 성질을 어떻게 디자인 아이디어로 연결시키느냐가 큰 주제였으며 액체의 고유 특성인 유동성, 다양성, 역동성, 내구성, 결집력 등을 이용하여 디자인을 풀어나갔다. 점, 선, 평면, 입체의 결집력으로 표현되는 액체 방울의 기하학적인 면은 설계 발전과정에서 채택한 것으로, 자연에서 볼 수 있는 가장 유동적인 물리적 형태의 하나이며 부피를 가지고 있는 덩어리의 주된 특성을 보여주고 있다.

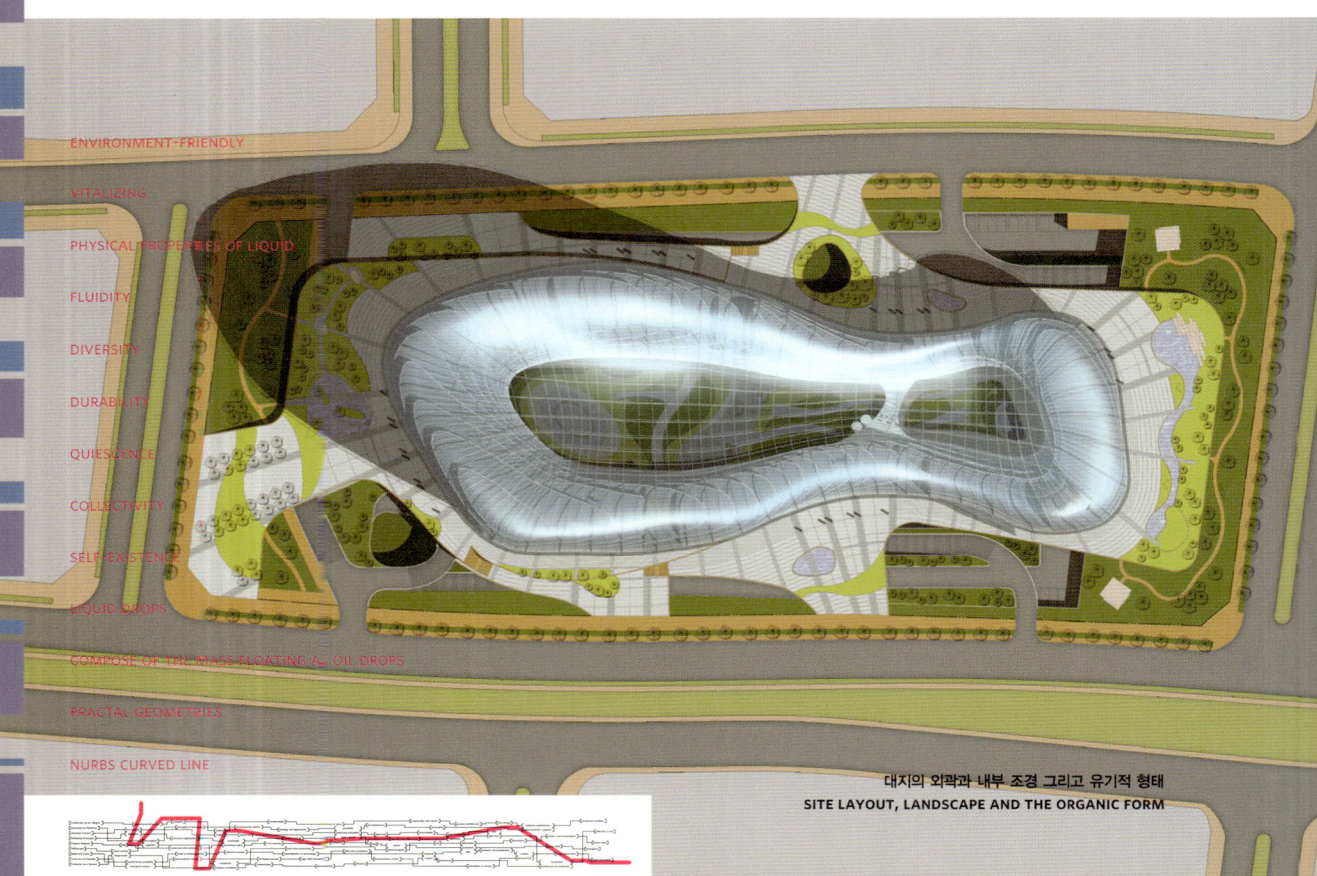

대지의 외곽과 내부 조경 그리고 유기적 형태
SITE LAYOUT, LANDSCAPE AND THE ORGANIC FORM

이란석유성   희림건축
이영희

이중의 표피 DOUBLE SKIN

액상 형태 LIQUID FORM

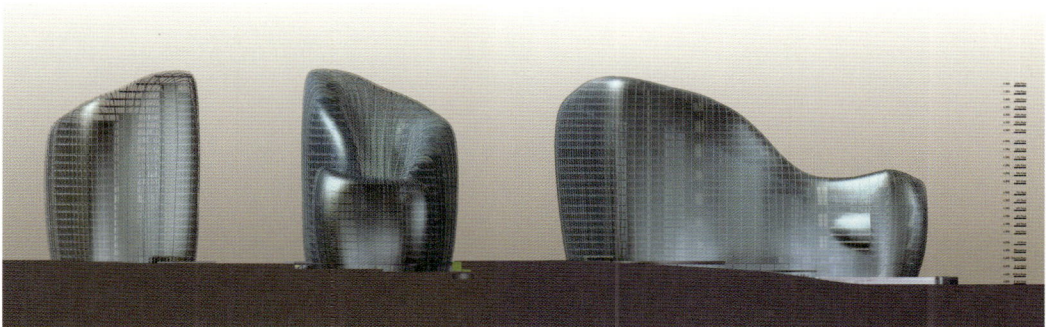
기름의 형태적 은유 FORMAL METAPHOR OF OIL

This project aims for not only vitalizing an area but also making the place firmly for business and exchange. Furthermore, we provide an opportunity of emphasizing the identity of nation and the importance of cutting-edge industry more through this world-class architectural building that will symbolize the city and be a representative icon. The newly coming business workplace is designed with environment-friendly concepts such as an environment-friendly building, maximization of space, energy save, dry construction method, etc.

    In designing this building, the big subject was how to connect the physical properties of liquid with the symbolic design idea. The characteristic natures of liquid like water or oil, such as fluidity, diversity, durability, quiescence, collectivity, self-existence, etc, are used as a method of unraveling design ideas. The geometrical aspects of liquid drops expressed as collectivity of points, lines, planes and three-dimensional objects are adopted during the process of design development, and are one of the physical shapes being the most fluid and show the main feature of mass having a volume.

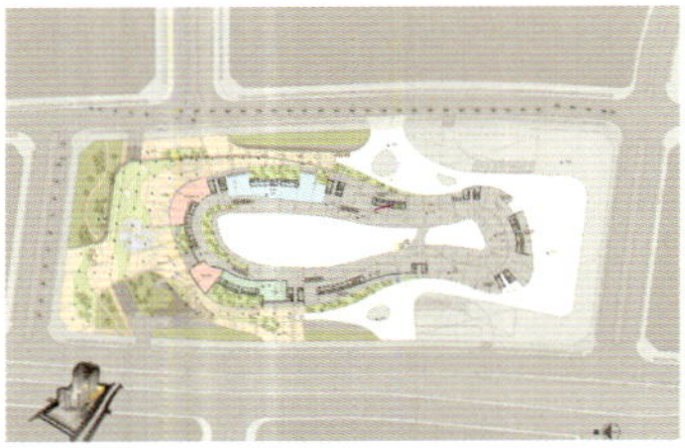

B1 PLAN 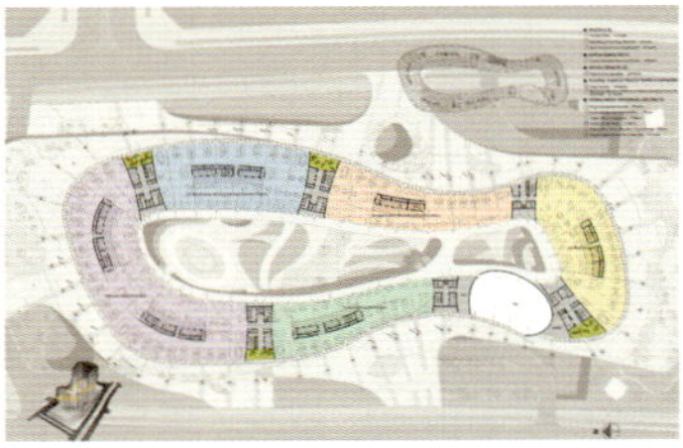　　5F PLAN 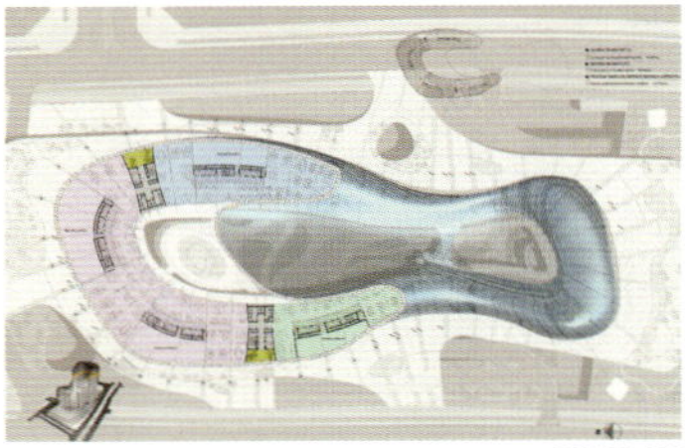　　13F PLAN

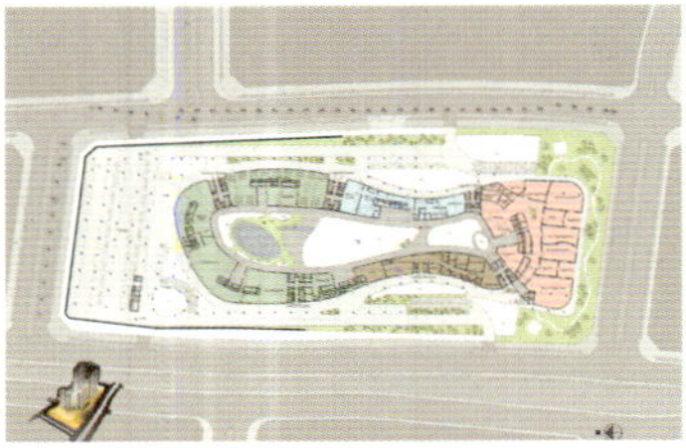

B3 PLAN 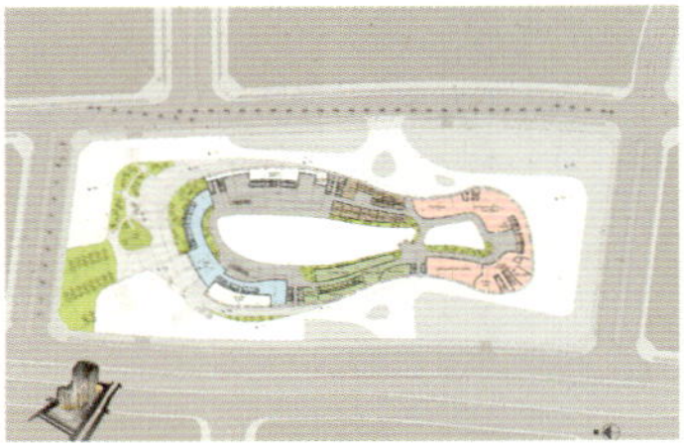　　GROUND PLAN 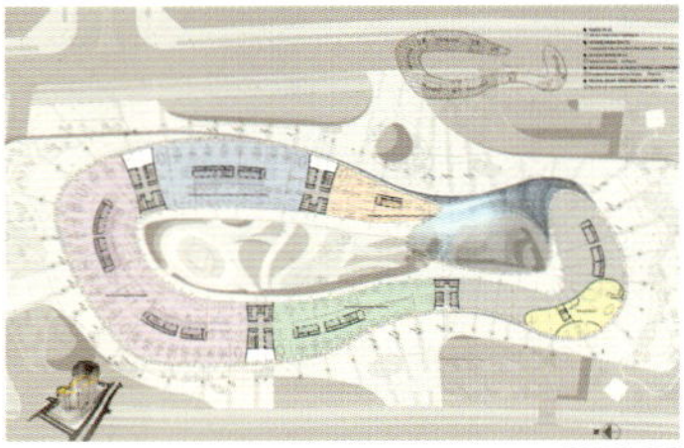　　9F PLAN

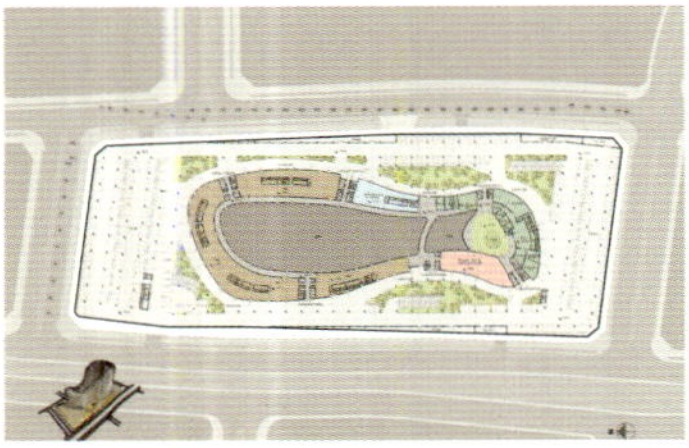

B5 PLAN 　　1F PLAN 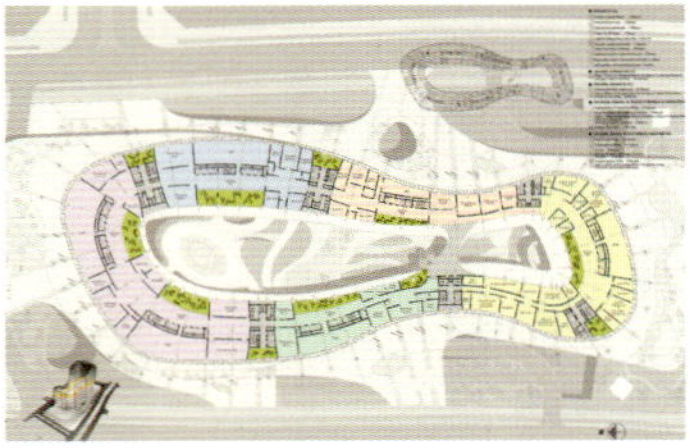　　7F PLAN

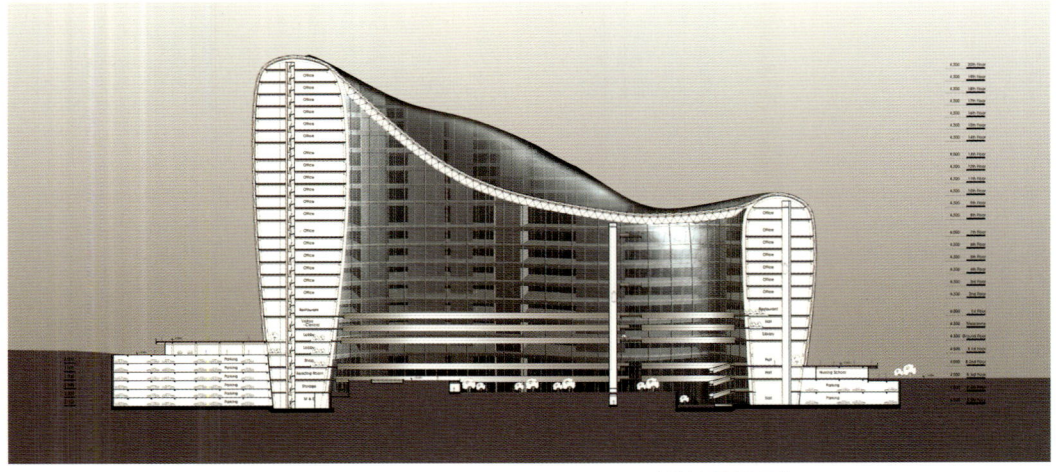

SECTION | 사무소 공간의 조직 SPATIAL ORGANIZATION OF OFFICES

이란석유성 | 희림건축
이영희

기본 구조의 얼개 BASIC STRUCTURE DESIGN

대형 아트리움을 공유하는 사무공간 LARGE ATRIUM PENETRATING OFFICES

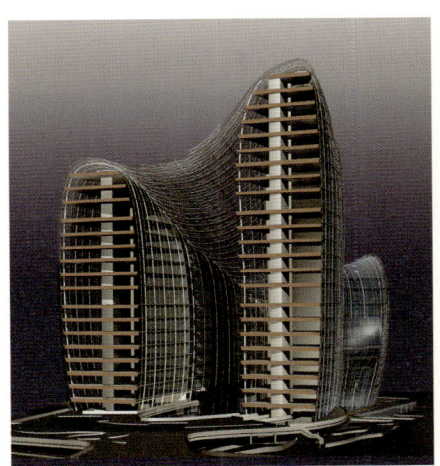

Sukyeon Yoo + Pier Vittorio Aureli, Kersten geers
HNA Ongodang, Korea + Dogma Office, Italy
The First Town
Multi-Functional Administrative City, Korea
Competition Project_Participation

2006

새로 도시를 만드는 것은 도시를 디자인한다는 의미라기보다 강력한 프로토타입 – 재현 가능한 기본 단위의 형태 – 를 정의하는 것에 더 가깝다. 프로토타입은 개념적인 엄격함을 통해 주변 컨텍스트에 영향을 주거나, 보편적인 적용 가능성을 통해 공간의 다양성을 추구할 수 있다는 데 강점이 있다. 이런 점에서, 프로토타입은 언제나 원래 거기 있었던 장소의 속성을 구성하는 역할을 하게 된다.

Making the city doesn't mean to design the city, but rather to define a charismatic prototype in the form of one reproducible unit. The strength of the Proto-type is the ability to influence its context through its conceptual rigidity and to leave space for variation through its universal adaptability. In this way, the prototype always refers to the place where it is located by assuming the role of a frame to what is already there.

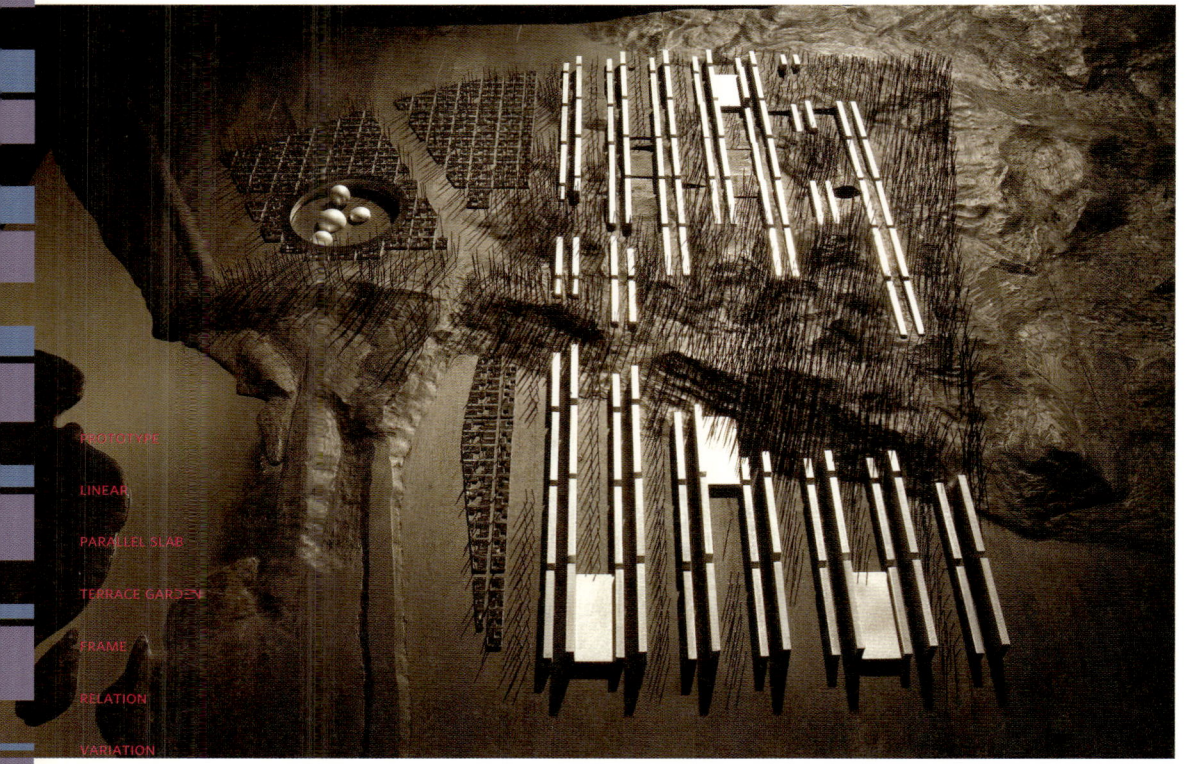

PROTOTYPE
LINEAR
PARALLEL SLAB
TERRACE GARDEN
FRAME
RELATION
VARIATION

첫마을

유석연 + Pier Vittorio Aureli, Kersten Geers
온고당 + Dogma

Concept 1
프로토타운의 형식은 112m 길이와 8m의 폭을 가진 판형의 매우 추상적인 건물 유형인데, 정면적 이미지보다 오히려 평면적인 형태에 집중한 전례없는 선형 배치이다.

The form of Proto-town that we choose is a very abstract building type - a slab 112m long and 8m wide - arranged in an unprecedented linear composition that stresses the planar condition, rather than the frontal image.

Concept 2
각 유닛은 두 개의 평행한 판형 매스로 이루어져 있는데, 그 사이의 길은 공공적인 장소로 기능한다.

Each unit is made out of two parallel slabs that form a passage which is a public ground.

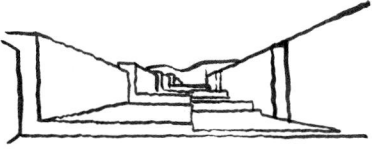

Concept 3
각 유닛 사이의 오픈스페이스 – 테라스 가든 – 는 기념비나 랜드마크로서보다는 장소의 특질을 강조함으로써 마을의 이미지를 형성한다.

The terraced gardens between the units will form the image of the place instead of marking the town with monuments and landmarks.

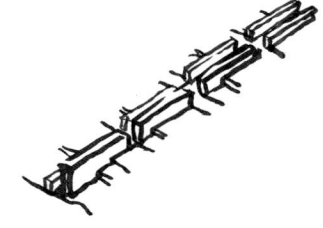

Concept 4
형태는 전략이며, 도시의 발전을 예감하는 장기적 비전이다. 전략을 정하는 것은 본질적으로 결정될 수 없는 것을 결정하는 것이다. 즉 도시가 생기고 번성할 수 있는 뼈대를 만들어가는 과정이다.

Form is strategic, it is a long-term vision that foresees the development of the city without predicting it. To define a strategy is essentially to "decide upon the undecidable", meaning to establish the frames within which cityness can occur and flourish.

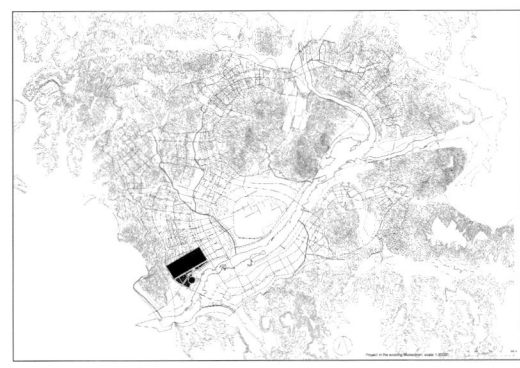

위치와 대지와 개발 분석 ANALYSIS DEVELOPMENT

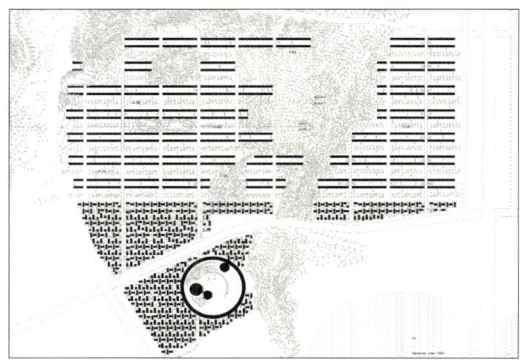

종합계획 MASTER PLAN

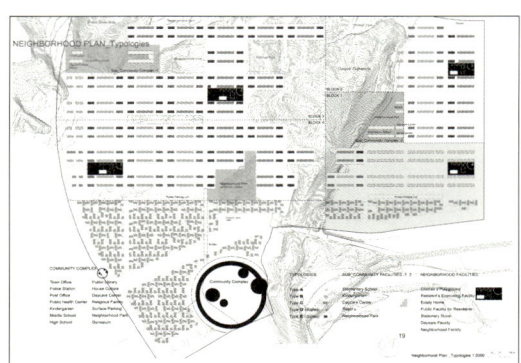

근린계획 NEIGHBORHOOD PLAN

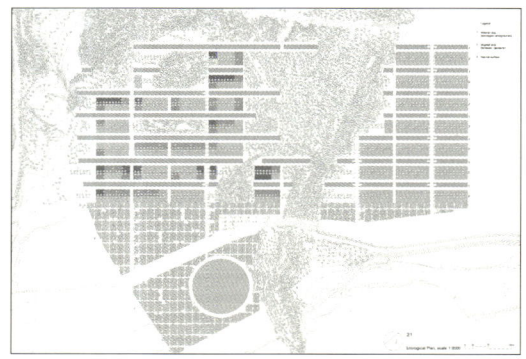

생태 계획 및 주차 ECOLOGICAL PLAN WITH PARKING

주거건축과 자연의 간섭 INTERFERENCE BETWEEN DWELLING AND NATURE

채움과 비움 | 간격의 프로토타입 SOLID AND VOID | PROTOTYPE OF SPACING

사이의 관계 IN-BETWEEN RELATIONSHIP

첫마을

유석연 + Pier Vittorio Aureli, Kersten Geers
온고당 + Dogma

다기능 주거도시, 행정중심복합도시 현상설계 참가작

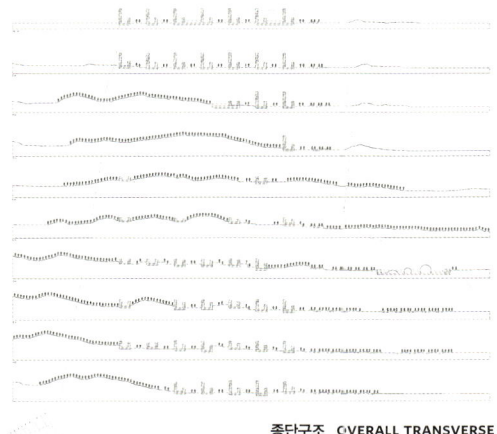

종단구조 OVERALL TRANSVERSE

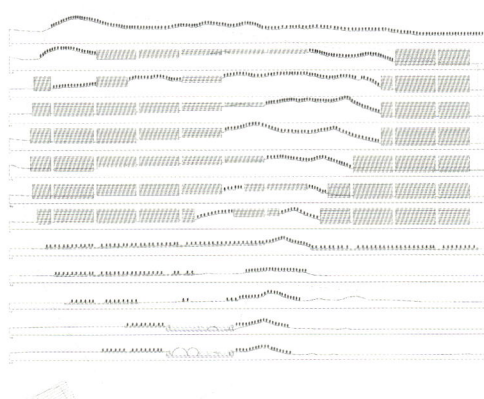

횡단구조 OVERALL LONGITUDINAL

Peter Eisenman, U.S.A. + Haeahn Architecture, Korea
Sewoon District #4 Urban Redevelopment Project
Competition Project_Participation
Seoul, Korea
2004

포화상태에 이른 아시아 주요 도시의 밀집 상태는 현대 도시계획의 새로운 모델을 필요로 한다. 이 계획부지는 기존에는 작은 필지로 잘게 나누어져 저층의 고밀도 건물군이 들어서 있었다. 반면, 부지 주변에는 기존 도시조직과 고층타워가 빠르게 등장하고 있다. 우리는 급속한 재개발과 고층화로 달라지는 이곳에, 중규모의 고밀도 복합공동체라는 실용적인 모델을 제시한다. 유기적 스펀지를 표현한 개념모델로 복합적인 도시 기능을 토대로 한 상징적인 새 도시조직을 형성해주고, 도심 중심부에서 거주에도 좋은 지구로 변화하는 데 대안점이 될 것이다. 즉 중층 고밀도 복합 구성으로 전체 4블록 모두 동일한 기능과 형태를 이뤄 통일성을 갖춘다.

이런 동질성은 저층과 고층건물이 뒤섞인 유형을 지속하기 위한 것으로, 별다른 특징없이 덧붙이기를 계속할 수 있다. 이것은 전 세계적으로 현대 도시의 특징인 개별성 없는 도심을 만드는 것이기도 하지만, 세운상가 부지 주변 다른 지역과의 개발 연계를 염두에 둔 것이다. 이런 중규모는 밀집된 도시가 주는 친밀감으로 도시블록에서 도시가로로, 보행자공간에서 중정공간으로 이어지는 다양한 스케일의 장점을 회복시킨다. 상가 역시 거리 주변을 따라 배치되어 기존 부지가 가진 가로의 역동성 또한 복원될 것이다.

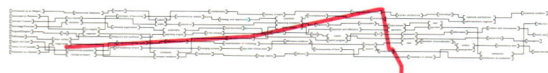
청계천과 새로운 거주공간 CHEONGGYECHEON AND NEW RESIDENTIAL BUILDINGS

세운상가4구역 도시환경정비사업
현상설계 참가작

PETER EISENMAN + 해안건축

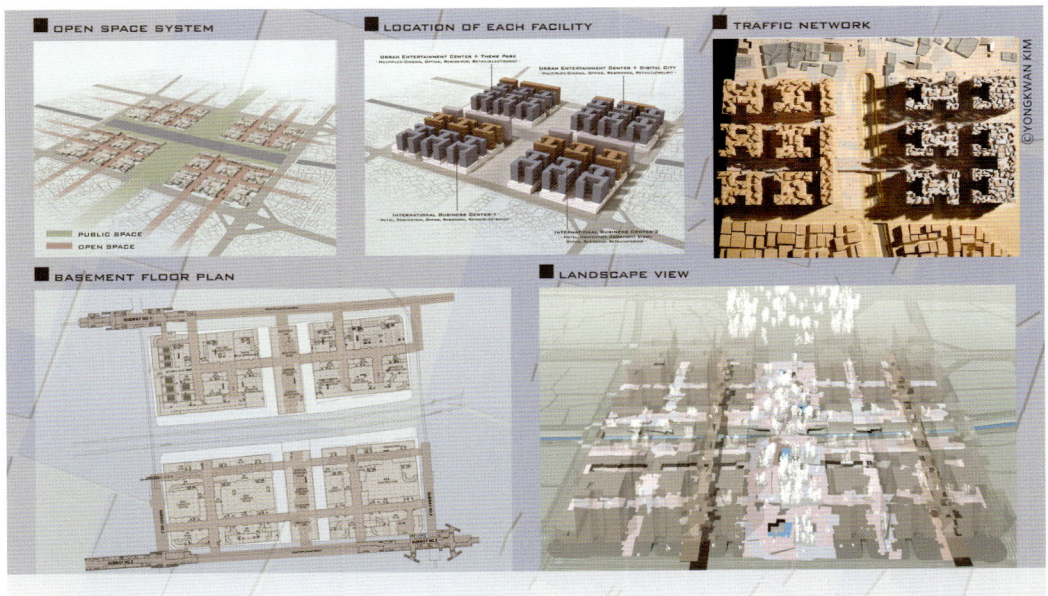

프로토타입의 변형 TRANSFORMING THE PROTOTYPE

  The major Asian cities are dense enough to be saturated, requiring a new model of modern urban plans. The existing site for this project is divided into small lots where low-rise but high-density buildings stand, while its neighbors are being filled rapidly with high-rise towers irrelevant to the existing urban structure. Here, we suggest a practical model called complex community of middle scale but high density. It is a conceptual model like an organic sponge for urban planning, and therefore, it will help to form a new symbolic urban structure playing some complex urban functions. Moreover, it will be a good alternative for a downtown residential zone. The entire four blocks designed to be a middle-rise but high-density complex community will have the same functions and forms to look unified.

  Such a homogeneity will allow low-rise and high-rise buildings to form a well-mixed pattern. Then, buildings will continue to be added easily to the community later. Although such a schema may result in a neutral downtown typical of modern cities throughout the world, it was conceived in consideration of future redevelopments around Sewoon Commercial Quarters. Furthermore, such a middle-scale/ high-density downtown community will be familiar to citizens because of its advantage or diverse scales: urban blocks, street-sides, pedestrians' spaces, patios, etc. The arcades or shops deployed along the street will help to recover the past dynamics of the streets.

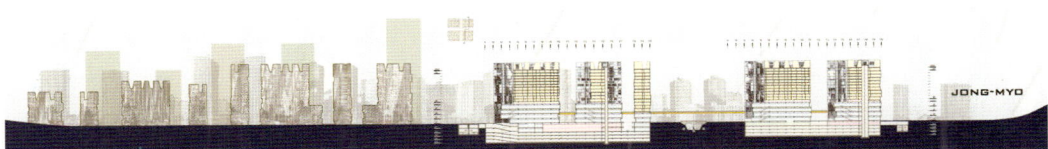

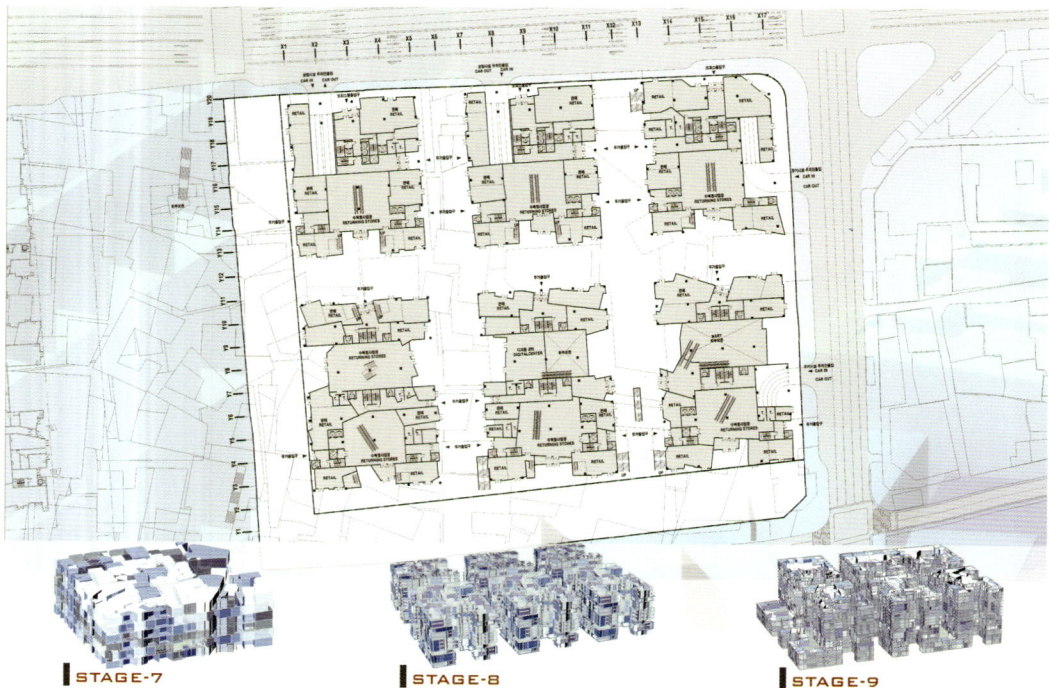

STAGE-7　　**STAGE-8**　　**STAGE-9**

빈번한 외곽의 접촉 MAXIMIZED CONTIGUITY WITH EXTERIOR

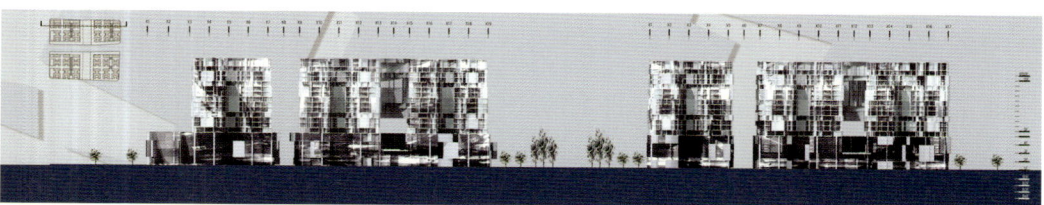

EAST-WEST ELEVATION　　NORTH-SOUTH ELEVATION

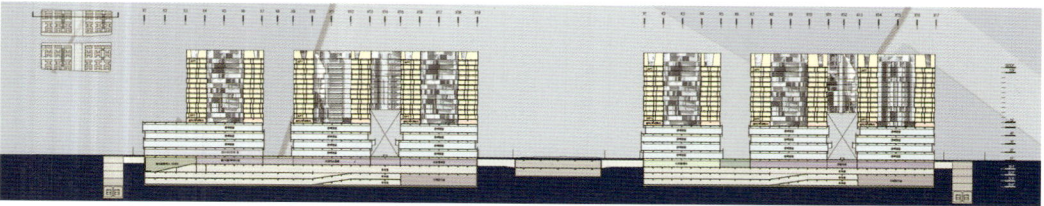

EAST-WEST SECTION　　NORTH-SOUTH SECTION

세운상가4구역 도시환경정비사업
현상설계 참가작

PETER EISENMAN + 해안건축

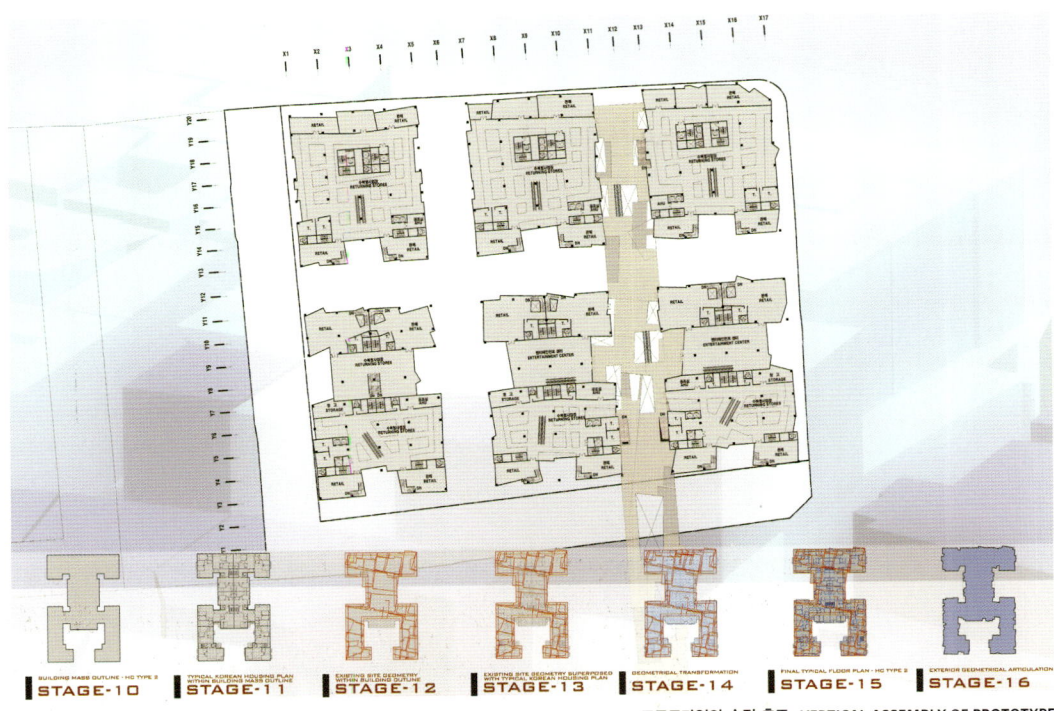

프로토타입의 수직 축조 VERTICAL ASSEMBLY OF PROTOTYPE

Yoongyoo Jang
UnSangDong Architects, Korea
Asian Culture Complex
Competition Project_3rd Prize
Gwangju, Korea
2005

새로운 도시의 스테이지는 길과 광장을 관통하며 만들어내는 비워진 광장과도 같다. 이 작업은 주어진 프로그램과 랜드스케이프를 관통하는 거대한 지도를 만들어주는 것이라 할 수 있다. 광주의 역사를 기록하는 방법으로 기존 도시구조 길의 모습을 입체화하는 구성을 제안한다. 기존 도시의 길을 볼륨화하고 입체적으로 위치시킴으로써 새로운 도시를 구성한다. 계획 대지 내에서 보존가치가 있는 도시구조는 역사적 사건의 기억을 담고 있는 길이라는 공간일 것이다. 도시적 길과 주변 매스의 반전을 통해 길의 형상을 담은 도시 규모의 매트를 만들고 그 매트를 대지적 모뉴멘트로 변형하는 작업을 제안한다. 길이 만드는 매트는 도시 전체를 가로지르는 랜드스케이프로, 다양한 이벤트와 프로그램을 수행하는 광장과 동선 역할을 수행한다.

City Stage

Historical Memories

Cultural Activities

Historical Value

Events

Urban Mat

Earthly Monument

Multi-Dimensional Skin

Landscape

아시아문화전당
현상설계 3등작

장윤규
운생동

City Stage for historical memories and cultural activities Making a city - scale urban plaza of Culture scape Street Mat City - We propose the process of extruding out the existing city structures and fabrics as a way of recording the history of Gwangju existing bands of streets are extruded to become the volume in order to frame the new city condition. This means that the city's existing fabric with historical value is also the spaces and streets with historical memories and events. Through the reversal of old streets and masses, city-scale urban mat is defined, followed by morphing it into the earthly monument. Urban mat framed out by the streets then forms the landscape slicing through the surrounding urban panorama, hosting various events and programs, while providing needed circulations.

Three-dimensional event plaza embedded in the urban mat during the culture stage is inserted in between streets to become the multi-dimensional skin containing cultural programs and activities.

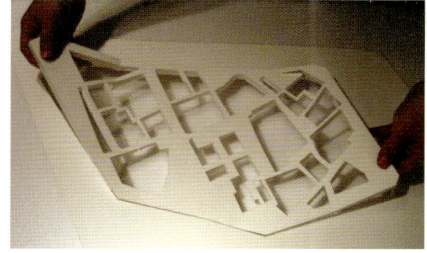

©JAEKYEONG KIM

새로운 지형 NEW TOPOGRAPHY

Deck Program

Deck Slope Level

Stage Ground Level

Direction

Trace of History

skin 2_Street Cell_ Historic Recorder

Combine with Stage

Natural Conta of Step

program zoning

Corridor

skin 1_ City Stage

Multi Stage

Landscape Stage

Glassing-Light Stage

Deck Stage Road

skin 3_ Network Flow_ Hyper-Street Map

Combine with Stage

natural Program Conta

program zoning

Plug in Gardens

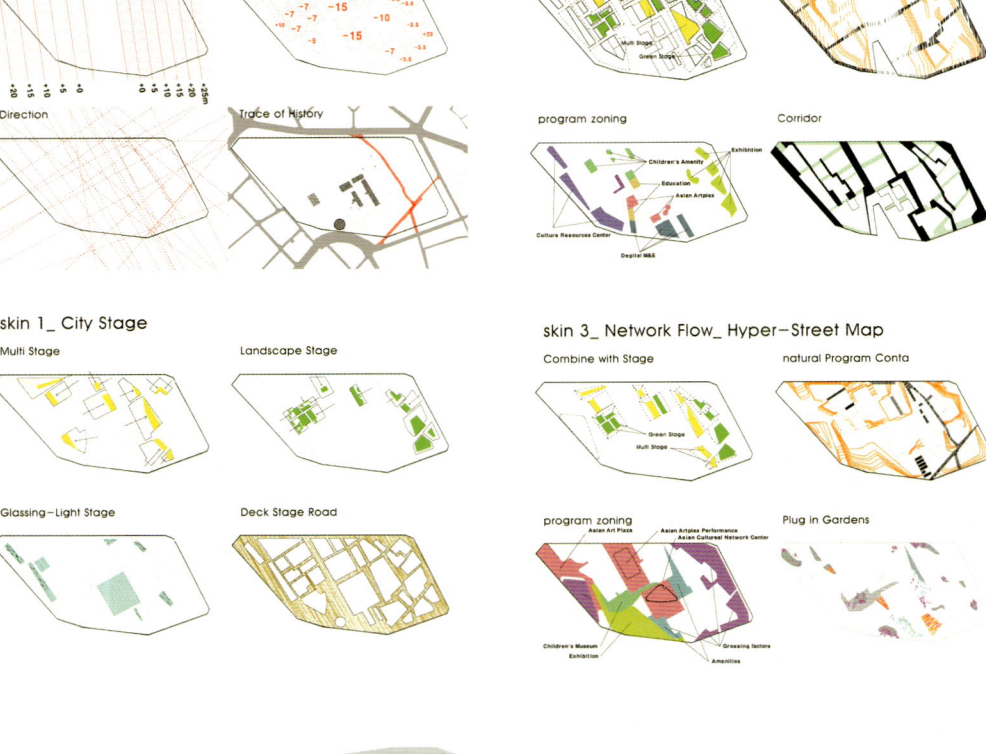

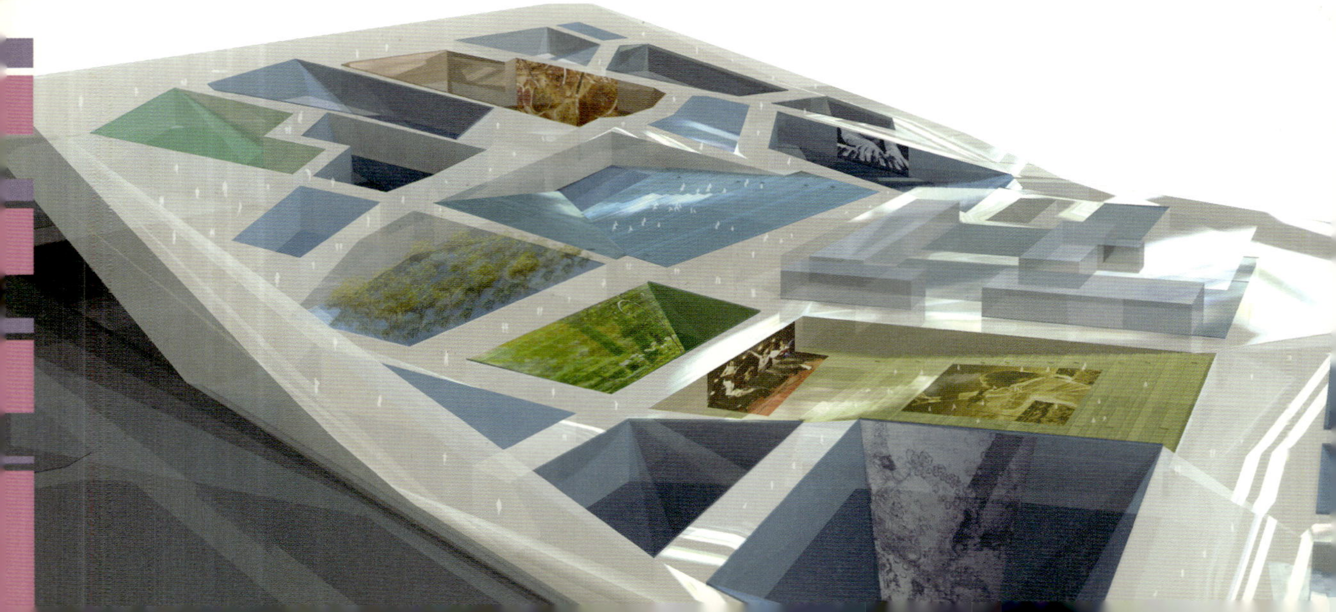

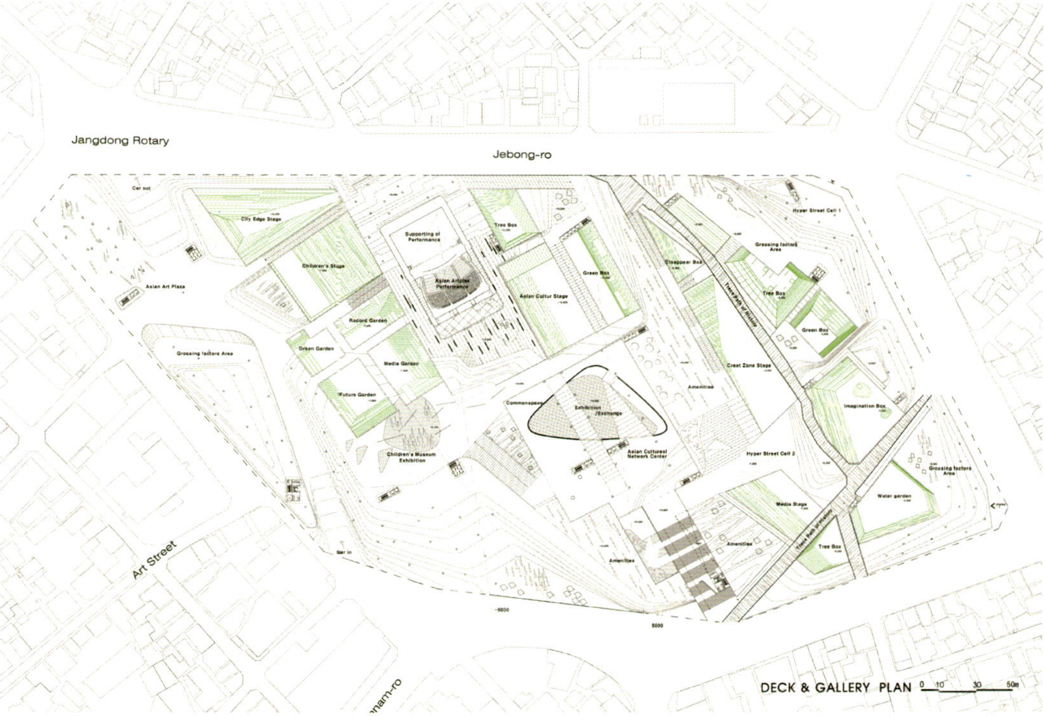

아시아문화전당

현상설계 3등작

장윤규
운생동

DECK & GALLERY PLAN

자유로운 내부의 문화 행위 LIBERATED CULTURAL ACTIVITIES IN INTERIOR SPACE

Yoongyoo Jang
UnSangDong Architects, Korea
Gallery Yeh
Seoul, Korea
2006

이 프로젝트를 통해 도시에 놓인 거대한 '캔버스'를 만들어내려 했다. 기존 캔버스가 평탄한 2차원이라면, 예화랑에 설정된 캔버스는 평면성에 숨어 있는 새로운 코드를 끄집어낸 평면과 공간 사이의 3차원적 스킨이다. 일반적인 건축 요소인 벽이라는 스킨을 공간적인 개념으로 변형하려는 시도라 할 수 있다.

스킨스케이프는 건축 요소인 '스킨'과 총체적인 틀을 의미하는 '스케이프'가 결합했을 때의 가능성을 만들어보는 실험적인 텍스트로, 스킨과 다른 요소들의 결합에 의해서 구성된다. 스킨과 구조의 결합, 스킨과 공간의 결합, 스킨과 프로그램의 결합…. 스킨을 물리적인 코드로부터 변환시키며, '스킨이 공간되기', '스킨이 미디어 되기' 등의 일련의 작업을 대체한다. 들뢰즈가 이야기하는 주름진 표피와 매끈한 표피의 연속성 속에 존재하는 창조적 탄생의 구조를 발견해내는 작업과도 같다.

Enormous urban "canvas" has been attempted through the project Gallery Yeh. If typical canvas can be thought as two-dimensional medium, the canvas we have developed for the gallery is the spatial skin developed out of the new code found between the floor plan and the three-dimensional medium. Two-dimensional aspects of the wall have now become the opportunity to deform into space. Such work is similar to searching for the new generation of space out of structure between the folded and smooth, continuous skin.

Skinscape can be initiated by simply acknowledging urban fabric as rather the envelope structure. Skinscape can be said an experimental text attempted by combining the architectural skin and the loose meaning of the term scape. It is organized through the formula of skin plus other elements and layers as variable? skin plus structure, skin plus space, skin plus program and etc. Then we can imagine its physical deformation. Process of finding the new spatial model is closely linked with Skinscape, with its variables possible to be maximized by removing the excessive spatial elements or adding up more spatial "fat".

URBAN CANVAS

EXPERIMENTAL ARTWORK

SKINSCAPE

URBAN FABRIC

FOLDED SURFACE

CONTINUOUS

SEQUENTIAL

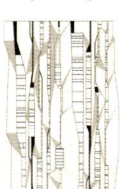

예화랑 장윤규
운생동

새로운 입면 생성과 감춰진 내부 기능 GENERATION OF NEW FACADE AND HIDDEN FUNCTION

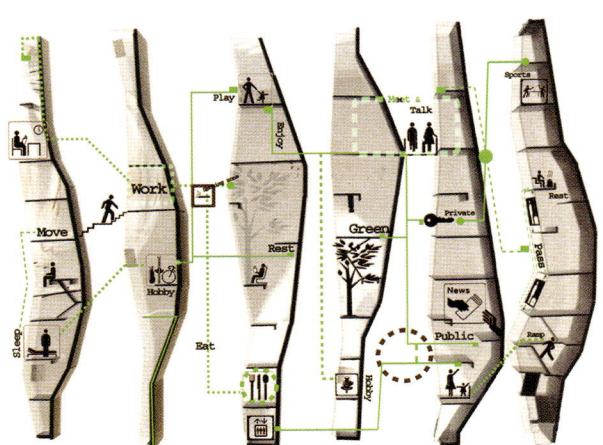

프로그램 다이어그램 – 프로세스 컨셉

장윤규
국민대학교 건축대학

MAP

COMPOUND AND SYNTHESIS

SEARCHING ENGINE

TEXT

BRAND VALUE

PRODUCT AND CONSUMPTION

HYBRID PROGRAM

INFORMATION

DEFORMATION

IN THE PROCESS

> 지도는 온몸을 던져 실재에 관한 실험 활동을 지향하고 있다. 지도는 자기 폐쇄적인 무의식을 복제하지 않는다. 지도는 무의식을 구성해낸다. 지도는 장場들의 연결 접속에 공헌하고, 기관 없는 몸체들의 봉쇄-해제에 공헌하며, 그것들을 고른 판 위로 최대한 열어놓는 데 공헌한다. 지도는 그 자체로 리좀에 속한다. 지도는 열려 있다. 지도는 모든 차원 안에서 연결 접속될 수 있다. 지도는 분해될 수 있고, 뒤집을 수 있으며, 끝없이 변형될 수 있다. 지도는 찢을 수 있고, 뒤집을 수 있다. 온갖 몽타주를 허용하며 개인이나 집단, 사회 구성체에 의해 작성될 수 있다. 지도는 벽에 그릴 수도 있고, 예술 작품처럼 착상해낼 수도 있으며, 정치 행위나 명상처럼 구성해낼 수도 있다. 언제나 많은 입구를 가지고 있다. – 천 개의 고원, 질 들뢰즈 & 카타리

다이어그램을 통해서 말하려는 것은 무수한 고리의 변형을 통해서 도달될 수 있는 경계선을 찾아내는 데 있다. 나는 이따금 그 비유의 하나로 '찾는 기계'를 설정하는 형식을 제안하는데, 원형과 변형 사이에 존재하는 복합과 통합 사이의 무수한 테두리를 찾아내는 설정을 대신한다. 그러면 어떻게 적절한 지도를 설정하여 '복합과 통합' 사이 지도를 만들어낼 것인가라는 질문이 존재할 것이다. 서로 다른 재료를 결합하는 기본 전제는 우선 결합의 소스를 결정하는 데 있다. 우리가 실행해 왔던 도시적 리서치도 기본 재료를 설정하고 움직일 필요가 있다. 가설의 시작은 재료 설정에 있음을 간과할 수 없다. 도시와 건축을 결합하며 설명하던 시대는 오래전이다. 우리 시대는 더 자극적이며 새로운 용기가 될 수 있는 결합을 원한다.

다이어그램은 일종의 지도map를 만드는 작업과 같다. 지도는 방법과 내용을 모두 포함한다. 지도는 프로그램이며 물리적인 구조이고, 상품화된 전략 등을 모두 포함하는 하나의 설정이다. 프로세스와 설계 프로젝트에 따라서 백과사전의 목록처럼 다양하게 끄집어내고 발췌되는 하나의 시스템이라 볼 수 있다. 다이어그램을 공간적인 비유로 대체한다면, 마치 비워진 공간에 숨어 있던 지도를 발견하는 것과도 같다. 비워진 공간에 숨어 있는 하나의 지도로 하이퍼텍스트적 지도를 상상해본 적이 있다. 건축적 텍스트를 공간에 뿌리고, 또 다른 테마의 경로로 연결되는 지도가 숨어 있다고 가정한다. 비가시적으로 숨어 있던 떠다니는 텍스트를 공간에 드러나게 한다. 텍스트들은 상호연관성을 통해 연결점을 찾아낸다. 텍스트는 더 이

Program-Diagram : Process Concept

Yoongyoo Jang
School of Architecture, Kookmin University

What distinguishes the map from the tracing is that it is entirely oriented toward an experimentation in contact with the real. The map does not reproduce an unconscious closed in upon itself; it constructs the unconscious. If fosters connections between fields, the removal of blockages on bodies without organs, the maximum opening of bodies without organs onto a plane of consistency. It is itself a part of rhizome. The map is open and connectable in all of its dimensions; it is detachable, reversible, susceptible to constant modification. It can be torn, reversed, adapted to any kind of mounting, reworked by an individual, group, or social formation. It can be drawn on a wall, conceived of as a work of art, constructed as a political action or as a meditation. A map has multiple entry ways.
- A Thousand Plateaus, Deleuze & Guattari

The intention behind discussing the "Diagram", and countless transforming of its linkages, is to discover a hidden boundary condition. As one example, I often propose formats to establish "search engine" that could replace this setup, discovering its infinite edges and boundaries existing in between the form & deformed, and compound & synthesis. If so, how can we map out the appropriate format in between the "compound & synthesis"? The basic premise ought to involve mixing the two different materials start primarily with determining its source of the mixture, as if the urban research that we have been practicing, must also involve determining and processing of basic ingredients and mixing strategies simultaneously. Such wisdom can provide us a new opportunity and assertion on claiming new mixture that can describe the urban and architectural condition, rather than repeating the past exercise of explaining just simply the city and architecture. Even if this mind set can cause inappropriate accidents and side effects, we must not fear for things to come in our time.

Making diagrams are very similar with making a version of a "Map". A map embraces both method and contents. Map is also a presupposition that includes program, physical structure, and a manufactured strategy. Through processes and design projects, extracting an encyclopedic list of contents must be regarded as one step to establish certain system for further utilization. Comparing a diagram with space, it is similar to discovering hidden maps from the corners of an empty space. One of these cases would be what I call hypertext map presupposition in which architectural text is sprinkled around the space and create a system of movements that can progress toward another theme. This invisible text appears in the space, and of which each subtext associate themselves to establish a web of linkages. Thus the text no longer exists as only one code or singular interpretation. They constantly rearrange themselves in accordance with their changing relationships and linkages a

상 하나의 해석으로만 존재하지 않는다. 무한히 텍스트에 걸쳐진 관계의 지도에 의해서 복합적인 코드로 재편성된다. 완결된 지도를 공간에 대입한다.

건축 작업은 '프로그램 다이어그램'을 만드는 형식과도 같다. 클라이언트에 의해서 주어진 프로그램은 일차적이라 볼 수 있다. 나는 그 일차적인 프로그램을 변화시킬 준비가 되어 있다. 주어진 프로그램을 변화시키는 것이 건축가의 몫임을 우리 모두는 알고 있다. 건축가의 사회적 역할과 직분이 변화되었다. 빠른 사회 변화로 인해 우리의 예측 범위를 벗어난 엄청난 속도를 체득한 지도 오래되었다. 변화 속도는 전통적인 건축가들이 간과해왔던, 혹은 숨겨왔던 상업적 코드의 실체를 드러내게 한다. 철학적인 이야기나 예술 이야기로만 포장되어왔던 건축의 성스러움은 더 이상 옷 걸치기를 원하지 않는다. 프로그램적 다이어그램의 개입은 클라이언트에 대한 전략적인 부분과 맞물린다. 클라이언트의 모습이 변화되었듯이, 사회는 건축가의 유연하고 변화된 적응을 요구한다. 거창하고, 본질적인 건축적 이슈로만 건축가가 움직여서는 경쟁력을 상실한다. 건축 공간도 이제는 '상품'이 되었다. 이제 클라이언트의 욕구와 상상력을 대변하는 노력이 필요하다. 클라이언트의 기본 요구를 변화시키는 작업은 마치 '상품화된 코드'를 근본적으로 포함하기를 원한다고 볼 수 있다.

건축 공간은 사람들의 꿈을 대변하며 혹은 강요하며 구성되어왔다. 사회적 정체성은 생산하는 것보다 소비하는 것과 관련되어 있다. 일반인들이 예술 작품을 체험하는 방식을 '상품화된 공간'을 제공하는 형식으로 채용할 수 있다. 소비자의 욕구를 자극하는 현상으로 드러나는 것이 그 예의 하나라 볼 수 있다. 이제 건축가들도 하나의 브랜드 가치에 기대며 상품을 만들어내는 것처럼 건축을 실행한다. 상품을 만들어내는 방식과 상품을 파는 근본적인 방식에 대한 이해가 요구된다. 디자인을 판다는 것은 단순히 소비자 요구에만 반응하는 것이 아니라 감성을 자극하고 다른 요구를 하도록 상황을 연출하는 데 있다. 이러한 상황을 만들어내고 이끄는 도구로 '프로그램 다이어그램'이 작용한다.

클라이언트의 요구를 변화시키는 작업도 중요하지만, 건축가 자신의 브랜드를 만들어내는 것도 중요하다. 클라이언트는 건축가의 작업을 보고 자신이 구축할 공간과 건축을 예측할 것이다. 정보화 사회에서 건축가는 새로운 브랜드 가치로 무장해야 한다. 자신의 건축을 상품화된 다이어그램으로 만드는 준비를 할 필요가 있다.

complex web of codes and networks. Maps, however, are now ready and complete enough to be altered into a physical space.

In making "Program-Diagram", a list of programs requested by a client could be said as the first set of programs. As I got ready for revising these programs (because we all know that one of architect's role is to transform and make inquiries about any given programs and standards), I was aware that it is also true that architect's social responsibility and role has evolved. Facing the fast-paced social evolution, it has been a while since the realization of the enormous power of speed and unexpected phenomena have exposed discrepancies in architectural practices that have been hidden behind our stereotypes and false impression. The old status of architecture in the sacred world of philosophy and art no longer works today. This matter of harsh reality relates rather directly with the discussion of programmatic diagrams and architect's strategy to deal with a client. The notion of "client" has been changed, so must architect's attitude and reaction respond in flexible manner. There are limits for the star architect to tag out multiple facets of issues in regards to client's realistic demands especially in par with competitiveness of the market economy and reality. Architectural space has now also become a "product", which means that the architect is now granted with a role to represent the desire and imagination of a client. Transforming the very basic demands made by the client is almost similar to fundamentally giving up "manufactured code".

We have been providing architecture on behalf of people's imagination and sometimes by recommending their unconscious dreams to be realized in certain ways. Social identity is more directly related to consumption than production, through which its example can be provided as "manufactured space" that stimulates the consumer's desire and will. Architects today also practice in similar fashion, making production in par with a "brand value". And this is why it is important to understand the fundamental and methodological aspects of making products and selling them. Because selling the design is not only about answering the consumers' demands in one take, but it is as a whole, continuous and fluid process and method of touching their sensibilities and making them to desire for your next product. The project "Everville" gave me a similar case for such challenge making attempts to transform the initial set of programs through the "brand-value" approach that is enormously crucial to this corporation client. We have proposed new set of program-diagrams that can capture new programs for the office environment and spatial possibilities, of which the company can promote as their own cultural characteristics. It was a suggestion of how to establish architectural construction of "branded space" through program-diagram.

Although important to react and be responsible for transforming the client's demands, I feel that it is even more crucial for architects themselves to sculpt their identity as a "brand". After all, it would be the works and identity of an architect that a client will create his image for the future space and architecture he desires. Facing

공간조직이나 도시조직의 본질이 하나의 규범적인 틀로 구성되어 있다고 믿지 않는다. 나무의 구조와 반대 입장을 가진 리좀의 구조는 인터넷이 지금 사회를 반영하는 하나의 조직 모델과도 같다. 통일되고 고정된 것을 벗어나는 저항의 구조를 만드는 데 있다. 사회나 도시를 단순히 하나의 코드로 인식하거나 보이는 것만으로 인식해서는 안 됨을 시사하기도 한다. 어떠한 것도 단일화된 구조나 요소로 구성되지 않으며 지속적인 변형과 변화를 필요로 함이 숨어 있다.

형태나 구조를 변화시키는 요구를 달성하는 것이 아니라 자체의 프로그램을 변화시키는 의미를 가지고 있다. 잡종적 프로그램을 구성할 수도 있고, 형태와 프로그램의 무관한 조직을 하나의 다양체로 결합해내는 시도를 할 수도 있다. 이 다양체는 하나의 프로그램만을 수행하는 시스템이나 장치가 아니며 변종된 요구와 변화를 수용하는 움직이는 프로그램을 수행하는 하나의 틀이다. 프로그램의 변형은 근본적인 건축 모델의 변형을 가능케 하며 근본적인 프로그램 결합방식에 대한 질문을 던지는 데 있다. 서로 결합될 수 없는 코드의 물체들을 결합하게 하는 것이며, 다른 요소로부터 추출된 데이터를 건축적인 공간으로 치환해내는 가능성을 내포한다. 오르페우스의 눈을 율리시즈의 귀로 치환해내는 작업과도 같다. 변형의 범위는 설정을 어디까지 가져가느냐에 달려 있다.

그러나 중요한 것은 도달되는 새로운 공간적 가능성이나 도시적 요구를 새롭게 정의해내는 데 있다. 이러한 설정은 일상적인 코드를 관통할 때 더욱 힘을 발휘함을 잊어서는 안 된다. 우리가 무심코 지나치는 수많은 공간들, 일반적인 주거일 수도 있고, 주거 안에 화장실일 수도 있고, 도시의 여러 공간일 수도 있고, 자주 접하지만 깊게 생각해보지 않던 것들에 대한 새로운 시각과 해석과 구축이 요구되는 것과도 같다. 우리가 수없이 무심하게 만나는 공간이나 장소를 다른 시각의 프로그램으로 대체하는 가능성을 발견하는 작업이며, '의미 지우기'와 '새로운 의미 부여하기'의 틀을 통해서 새로운 프로그램을 삽입하는 작업을 설정하는 것이다. 텍스트의 구축을 의미하는 것처럼 본래 텍스트에 매달린 군더더기 텍스트를 제거하고 본질의 텍스트만을 남겨, 본질의 텍스트를 다른 차원으로 변형시키는 설정이다.

even more intricate world of information, architects must know and act now on how to arm themselves with the new yet their own brand value, and diagrammatize their architectural production and prepare for the race.

I do not believe that spatial organizations or urban fabrics are constructed as one specific framework. The structure of rhizome, an opposite structure of a tree, coincides more with the structure of the internet that reflects the way our society works these days. The issue here is making structure that resists such unified or standardized form or mode of thinking. It addresses for us to be against the notion that a society or a city is comprised as one singular code. Nothing in this world is not constructed in such way, and instead, continuous needs and desires for the transformation and change is hidden somewhere.

There can be a discussion of altering programs rather than satisfying the demands to directly change the form or structure organizing a set of hybrid programs, or making attempts to create compound structure that is not associated with its formal language. This compound structure then becomes not just one of the frameworks that exercise only one programmatic or systematic element. Rather, it becomes a mechanism to accommodate any programmatic elements that can come into the system the variable. Such possibilities of programmatic deformation allow architecture to be transformed physically, therefore brings up a series of questions of its fundamental methodology of assembling programs. It also conveys the possibilities how the outside data can be substituted into architectural spaces. This by the way, is quite similar with substituting the "Orpheus's Gaze" with "Ulysses's ear".

The range of deformation depends on where to stretch a given knowledge. More important issue is to take the given spatial possibilities or urban demands and proceed with new definitions. Such establishments of knowledge can be especially powerful if they are in conjunction with our everyday lives and modes of thinking. Examples can be - countless spaces being ignored in our consciousness, typical lifestyles, a bathroom in a house, many places in our cities, or nearby spaces that we have not thought deeply at all. The work is about setting up a new interpretation and perspective, and then altering them with new breed of programs and opportunities, such as the framework we have worked through the "delete meaning" and the "insert new meaning". Just like constructing the meanings of the text, it is about removing all of the unnecessary extras and deform the text into other dimensions and definitions.

I do not constitute that there is a large gap between the diagram and architectural spaces. I do want the diagram itself to deform into architectural space. If we make a kind of diagrams and models in the process, I would like for them to directly become a physical space or a piece of drawing.

Moongyu Choi
Yonsei University, Korea / GA.Architects
M³_Questions on the Space
2006

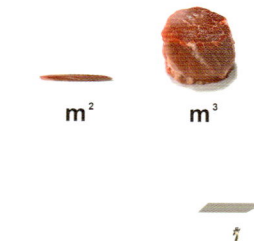

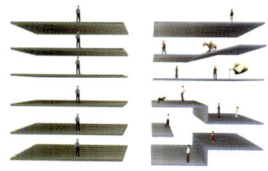

3-DIMENSIONAL VOLUME

CUBIC METERS

SIZES OF EACH HUMAN BODY

PSYCHOLOGICAL DEMEND

INNER SPACE TO ITS SKIN

DIFFERING CEILING HEIGHT

DYNAMIC PERSPECTIVE SPACE

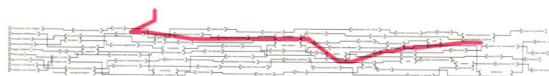

"만약 소고기를 부피로 판다면, 물을 먹이지 않고 공기를 불어넣을 것이다."

건물을 사고팔 때 부피 단위로 판다면 층고를 줄이기 위해 이렇게 다양한 방법이 사용되지 않을 것이다. 우선 공간을 측정하는 단위가 세제곱미터임에도 불구하고 그 단위 중 하나는 사라지고 제곱미터(평)가 의심없이 통용되는 세상은 이상하지 않은가? 부피를 측정하는 단위로 면적이 쓰이는 세상. 한 층이라도 더 짓고 싶은 경제적 이유에 일조권이나 도로 사선제한 그리고 인동간격 등의 법규 제한이 더해져서 층고는 낮아지고 같아진다.

결국 건물은 비슷한 층으로 구분되고 이러한 층들은 같은 높이로 이루어져 결국 기준층이라는 결과를 만들어낸다. 경제성과 설계와 시공의 편리함은 만족되더라도 의문은 남는다. 각각의 인간이 물리적 크기가 다르듯 심리적 요구 공간이 다르다면 다양한 공간을 만드는 방법에 대해 관심을 갖고 탐구하는 것은 건축가의 임무가 아니겠는가.

공간에 대한 질문들은 부피를 건축화하는 작업들을 통해 만들어질 수 있다. 프로그램을 만들 때 평면이 아닌 부피로 시작해서, 이를 한정된 공간 안에 배치하는 방법이 그 하나로, 서해문집과 G29의 초기 계획안과 태학사가 이런 경우다. 볼륨으로서 공간은 결국 외피로 확장되어 층으로 구분될 수 없는 건축이 가능해진다.

다른 경우는 프로그램에 따라 각 층의 높이를 변화시켜서 같은 높이의 층들의 적층으로부터 탈출하는 것으로 노원구 도서관 현상설계안이나 서해문집 그리고 아름드리미디어가 이에 속

"If we buy & sell a chunk of meat by the volume of it, they will inject air instead of feeding water."

If they trade buildings by cubic meters not by square meters, there will be no more trying to make stacks of slabs. Questions on space were started from this. The beginning was why we buy and sell space by the unit of 2-dimensional area, not by 3-dimensional volume without questioning it. As we consider space as volume, shouldn't we use cubic meters instead of square meters? I fully understand economic reason that they should build as many as possible profit-wise as well as restrictive building codes.

These factors have made us to count on our buildings commonly with numbers of floors that are identically same height. As biological forms and sizes of each human body are different, psychological demands on space of each one might be also different. This is why architects should explore space and try to come up with new spatial ideas. At first, the questions on space lead to the possibility of space of certain program and its arrangement in space. When the program is given, each program can be made based on the volume of the space but the area. Considering the matter and the possible solution I have mentioned above, the first proposal for *G29* and *Booksea Publishing*

한다. 이와는 별도로 투시도적 공간에 대한 질문 – 오렌지 튜브 – 과 깊은 공간의 가능성 그리고 하나의 볼륨으로 만들어지는 단일 공간의 크기와 관계는 아직 많은 연구의 여지를 두고 있다. 유형화가 될 수 있어 보이는 이러한 공간에 대한 관심과 연구는 건축공간에 새로운 가능성을 열 것이다.

were designed. This idea of volumetric architecture extends its inner space to its skin. It frees us from false measurement of space.

Another method could be differing ceiling heights of floors according to its program and this helps to change the notion of stacking floors consisting the same height. This idea was reflected on *Nowongu Information Library* competition project, *Booksea Publishing* and *Arumdri Media*. *Distinctive* to those matters above, questions on dynamic perspective space - Orange Tube, *Namjun Paik Museum* competition - deep space, omni-volumetric space such as *Taehaksa* and changeable space - *Cheese* - are still on progress at the same time. I believe that our interests and evolving questions on space that could be organized and categorized will open up a new chapter in architectural field.

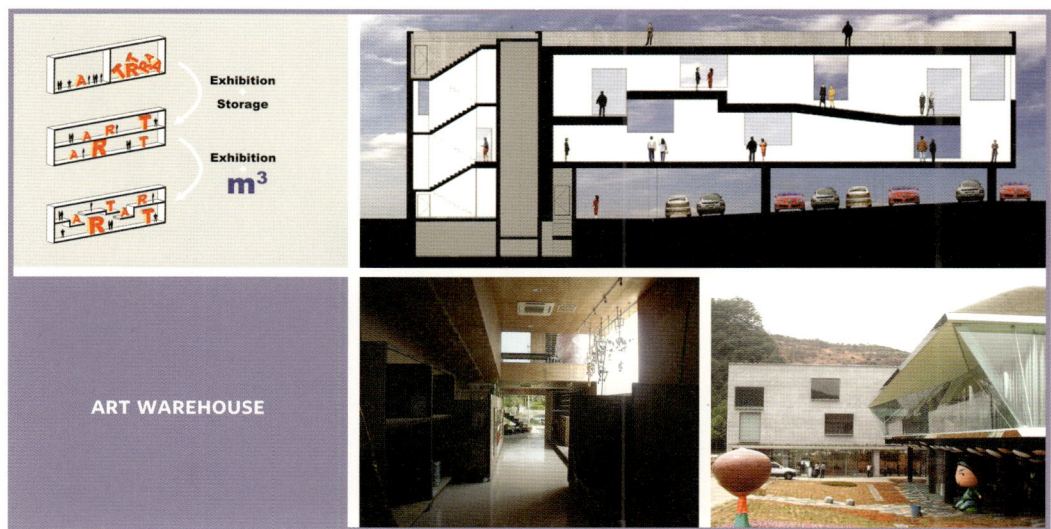

ART WAREHOUSE

ARUMDRI MEDIA

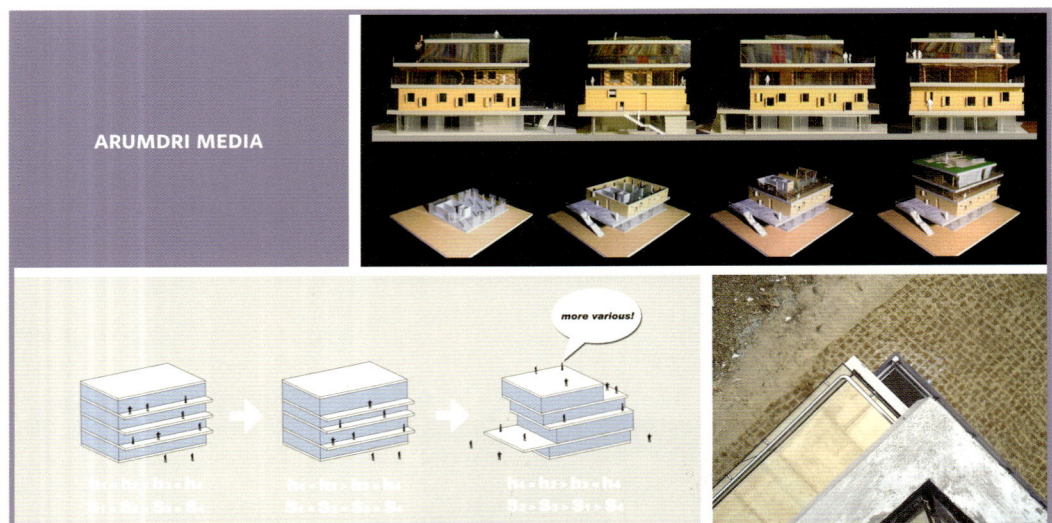

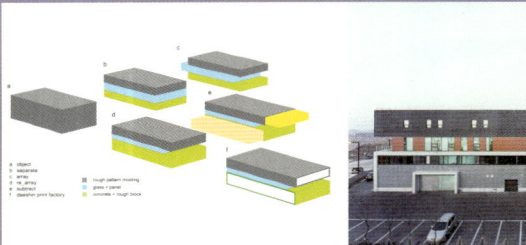

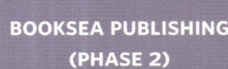

DAESHIN PRINT FACTORY

BOOKSEA PUBLISHING (PHASE 2)

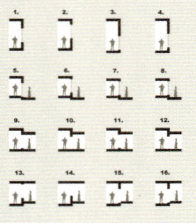

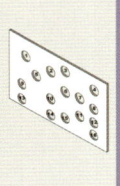

M³_공간에의 질문 | 최문규
연세대학교 | 가아건축

NOWONGU INFORMATION LIBRARY

BOOKSEA PUBLISHING (PHASE 1)

G29

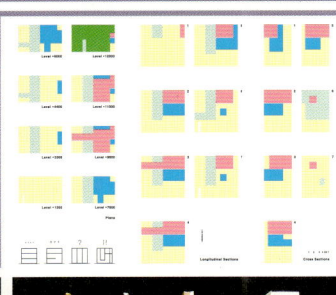

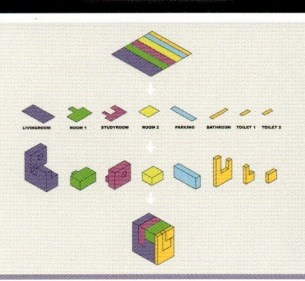

M³_공간에의 질문   최문규
연세대학교 | 가아건축

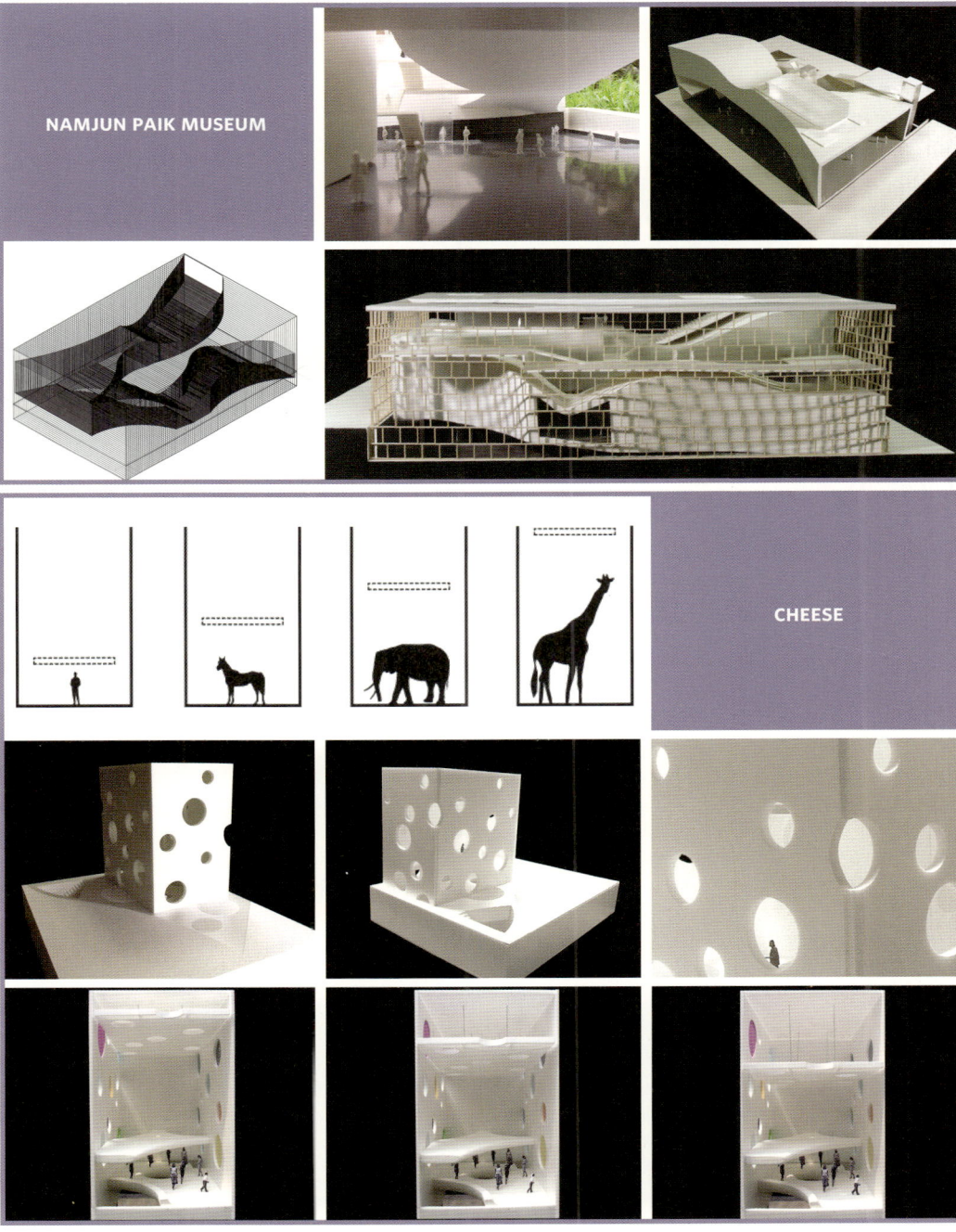

NAMJUN PAIK MUSEUM

CHEESE

Minsuk Cho
Mass Studies, Korea
Handsome Hotel
Namyangju, Gyeonggi-do, Korea
2005-

확산/수렴

부지는 백두대간의 한북정맥이 무수히 확산되어 천마산을 거친 마지막 잔가지와 북한강의 한 지류가 동시에 북한강에 도달하는 지점에 위치한다. 그런 의미에서 이곳은 지형적으로 시작과 끝이 공존하는 경계적·전이적 성격을 갖는다.

우리의 건축적 제안은 기존의 주변 인공환경을 고려하기에 앞서 이 개발의 대전제 조건인 주변의 '자연환경', 즉 광활하며 동시에 섬세한 주변 산세, 그리고 힘차게 흐르는 한강을 닮고자 한다. 이는 주변 자연환경과 연계해 새로운 '유기적 질서'를 재정립하려는 시도다. '경계 지역'이라는 지형적 특성을 최대한 장점으로 활용하면서, '전이적' 성격을 가진 이 장소의 특성에 부합하는 새로운 공간 질서를 건축물로 부여하는 일이 우리가 이 부지에서 해결해야 할 과제이다.

제안하는 건물은 네 개 요소가 북한강을 향하여 남북 방향으로 펼쳐져 있고 모두가 다양한 레벨의 내부 또는 외부에서 연결되어 하나의 전체를 이룬다. 남단 5층의 호텔동과 북단 2층 2개 동의 볼룸/맨션은 각기 진입로와 단차로 분리된 영역을 가지며, 독립적으로 이용될 수 있다. 호텔동과 볼룸/맨션동 사이에는 이 두 건물을 기능적으로 보조해주면서 전체가 하나의 건물로 역할할 수 있도록 연결하는 어메티니/지원시설동이 위치한다.

Divergence/Convergence

After dispersing and passing through Cheonma Mountain and numerous other points on the map, the last remnant of the majestic Baekdu Mountain ridge finally meets the Bukhan River at one of its tributaries-this is the site's propitious location. Beginning and end coexist topographically, making the location both a demarcation and a transition.

Our architectural proposal is to communicate with the natural surroundings rather than with the existing man-made environment. Our project intends to approximate the development's main inspiration and raisons d'etre: the area's expansive yet delicate mountain range and powerful Han River. It will be an attempt to connect with the natural environment to establish a "new" organic order. The primary issue that we must resolve is how best to utilize the topographic characteristics of a "boundary zone" while architecturally devising a new spatial order that is consistent with the distinctive "transitional" nature of this location. The proposed building's four components will face the Bukhan River, spreading out north to south, interconnected by internal or external features at different levels to form a cohesive whole.

The five-story hotel on the southern end and the pair of two-story buildings to the north, housing the ballroom and mansion, will each establish its own domain through a dedicated entrance and differing levels and may be used independently. An amenity/support facility will stand between the hotel and the ballroom/mansion buildings, providing functional support to both sides while connecting them as one entity.

DIVERGENCE

CONVERGENCE

COMMUNICATE

NATURAL ENVIRONMENT

NEW ORGANIC ORDER

DEMARCATION

TRANSITIONAL

BOUNDARY ZONE

RIVER VIEWS

WATERFRONT PARK

한섬호텔 조민석
매스스터디스

물, 대지와 형태 동조 SYMPATHY AMONG WATER, SITE AND THE BUILDING FORM

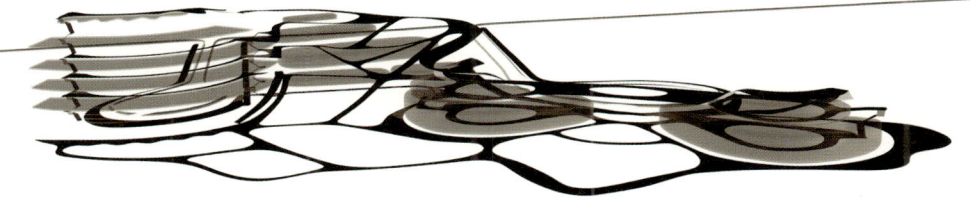

동선체계 CIRCULATION SYSTEM

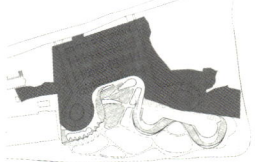 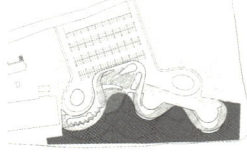 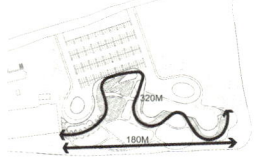 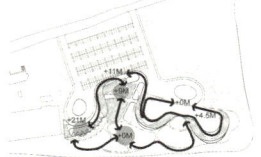

DIVISION 1: SITE NORTH **DIVISION 2: SITE SOUTH** **EXTENSION 1: WATER VIEWING SPACE** **EXTENSION 2: WATER ACTIVITY SPACE**

분절 1: 대지 서측
분절된 서측의 대지는 진입로가 될 국도와 면해 있으면서 남북 양단에 각각 호텔과 연회장으로 접근하는 두 개의 진입로를 갖는다.

분절 2: 대지 남측
자유롭게 남북 방향으로 흐르는 듯한 선형 건물은 북한강에 접한 부지 남측의 외부 공간을 품으면서 부지 내 주된 실내외 활동공간을 대지 주변의 난삽하게 개발된 환경으로부터 시각적으로 분리시키며 동시에 자동차를 위한 공간과도 분리한다.

DIVISION 1: SOUTHERN LOT
The property's southern section faces the freeway that serves as its approach and has two separate driveways that each leads to the hotel and reception hall at the southern and northern ends, respectively.

DIVISION 2: WESTERN LOT
The building's striking outlines appears to flow freely along a north-south axis, cradling the site's southern outdoor space abutting the Bukhan River. The structure thereby visually delineates the complex's primary internal/external spaces from the garish clutter of neighboring development while simultaneously functioning as a divider from the parking space. This separation allows the site's waterfront space to maintain a relaxing park?like atmosphere reserved for pedestrians.

연장 1: 수변 조망공간
개발의 가장 큰 동기인 '북한강의 조망'이 가능한 부지는 강변 방향을 따라 유연하고 자유롭게 움직이듯이 배치된 건물로 인해 조망이 가능한 건물의 길이가 두 배 이상으로 늘어난다. 결과적으로 모든 주된 활동을 위한 단지 내 실내 공간들은 북한강으로 열려 있는 조망을 갖는다.

연장 2: 수변 활동공간
북한강에 인접한 단지 동단의 외부 공간에서 시작되는 수변공간은 지면 레벨에서 건물의 실내와 접하는 다양한 데크로 연장되며 계단, 램프, 경사면 등을 통해 건물의 옥상면과 연결된다. 건물 자체가 하나의 수변공원의 산책길이 되어 수변 활동공간이 연장되며 수평·수직적 확장을 하게 된다. 입체적인 그물구조와도 같은 이 산책길은 남단에서 북한강과 맞닿으면서 시작하여 지면 레벨의 몇 개의 인공연못을 거쳐 호텔 옥상레벨의 야외수영장에 이르기까지 모든 층의 다양한 레벨에서 산책, 야외 식음 및 식사, 각종 옥외 연회 및 행사, 피트니스, 야외스파, 야외수영장 등 다양한 수변 활동공간을 연속시키게 된다.

EXTENSION 1: WATERFRONT VIEWS
The Bukhan River view is the strongest natural asset to this development and is accessible along 200 meters on the eastern end of the site. With the supple, dynamic placement of buildings, the area with river views increases to over twice the length.

EXTENSION 2: WATERFRONT PARK
The outdoor space adjacent to the river in the eastern section is extended by a variety of ground-level decks that leads inside, while stairs, ramps and slopes connect the park to the rooftop level. The building itself becomes a part of the park trails, expanding space both horizontally and vertically. The net-like composition of paths begins at the river on the southern end, continuing on past several artificial ponds at ground level to finally reach the hotel's rooftop pool on the sixth floor. The network of paths thus creates a continual sequence of waterfront activity, connecting indoor and outdoor activities.

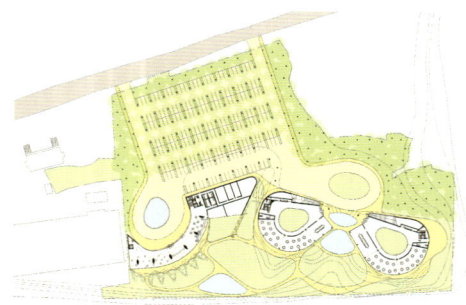

1F PLAN

수변을 향한 또는 수변에서의 조망

189　　　　　　　　　　　　　　　　　　　　　　　　　한섬호텔　　조민석
　　　　　　　　　　　　　　　　　　　　　　　　　　　　　　매스스터디스

수변 활동을 위한 내부와 외부의 유기적 결합　ORGANIC CONNECTION BETWEEN IN AND OUT OF SPACE

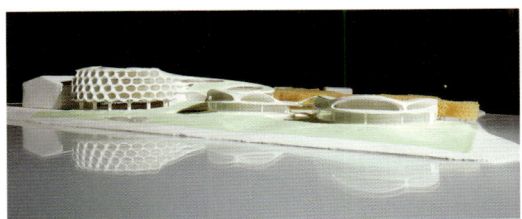

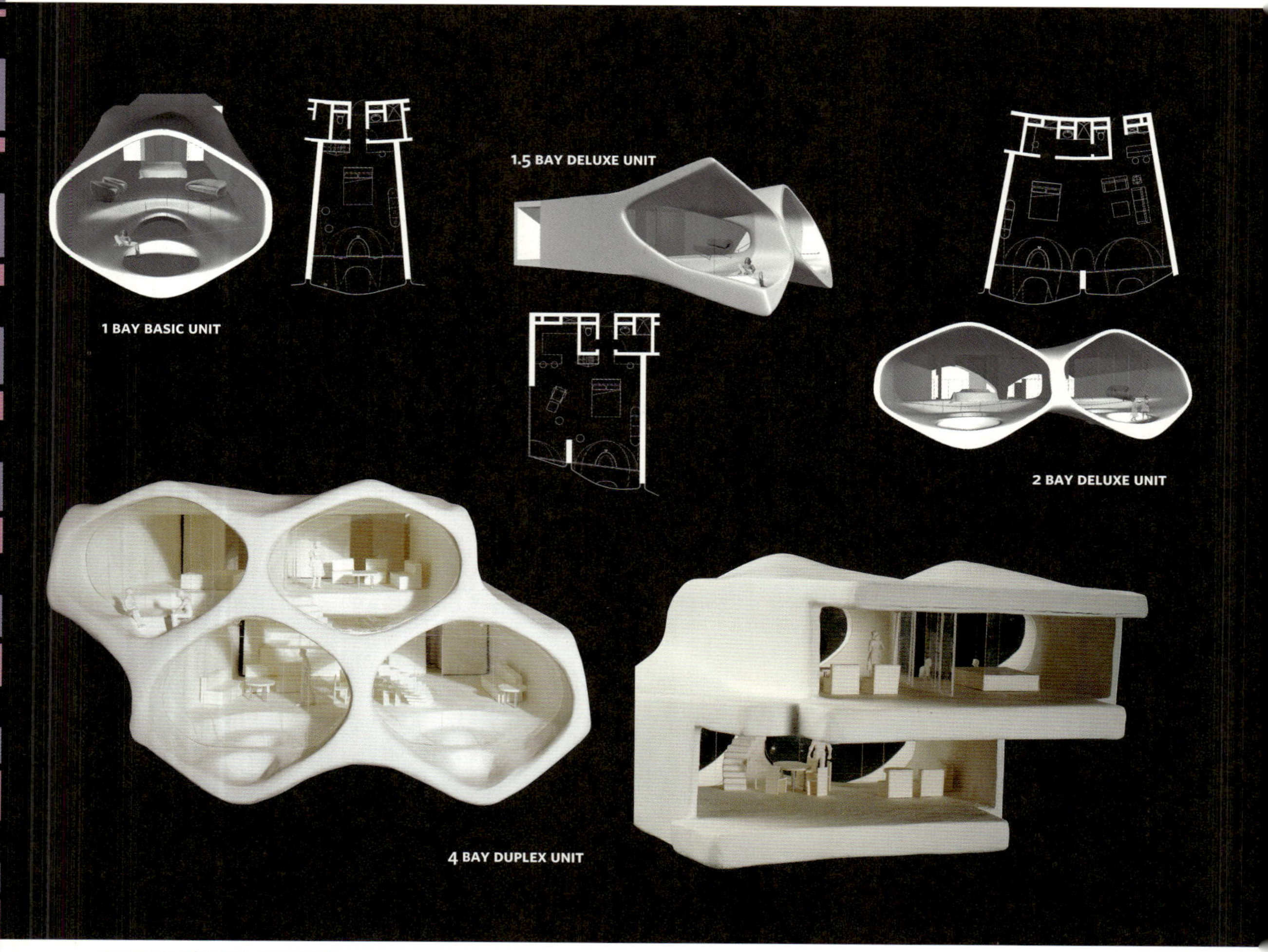

1 BAY BASIC UNIT

1.5 BAY DELUXE UNIT

2 BAY DELUXE UNIT

4 BAY DUPLEX UNIT

객실 유닛의 조직과 전개 UNIT TYPES

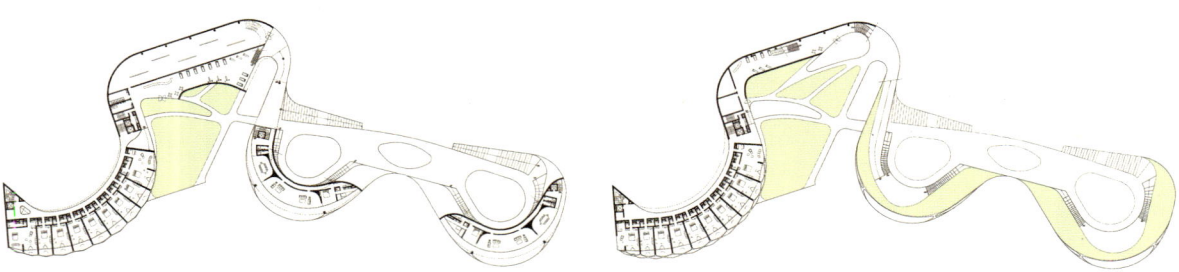

2F PLAN

3F PLAN

한섬호텔 조민석
매스스터디스

복층형 유닛에서의 수변 풍경 VIEWING WATER SPACE, 4 BAY DUPLEX UNIT

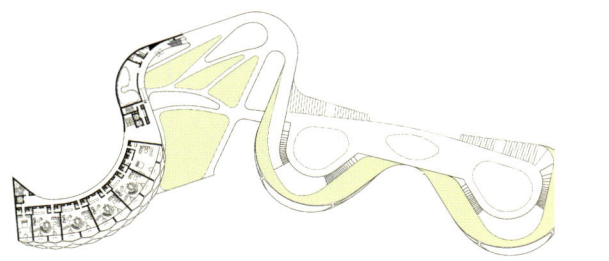

4F PLAN

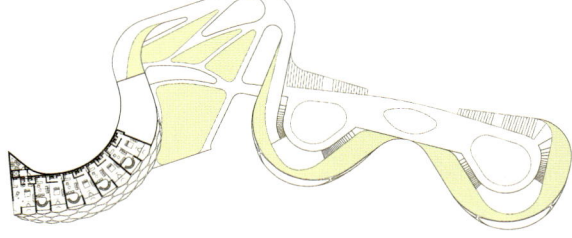

5F PLAN

Alejandro Zaera Polo + Fashid Moussavi
FOA, U.K.
Ewha Campus Center
Competition Project_Participation
Seoul, Korea
2004

지금까지 이화여대 캠퍼스는 일정한 기준없이 건물이 세워져 전반적으로 혼란스러운 배열을 보여왔다. 이에 대응해 위치 상으로도 캠퍼스의 중심점이 될 캠퍼스센터는 지형적 특성을 방사선 구조로 풀어 다양한 경사로로 계획, 주변 시설들과 긴밀한 연계성을 가질 수 있도록 했다. 전체적으로 링 형태를 띠어 경사로와 바닥, 뜰에 이르는 모든 동선이 연속적인 경로를 가진다. 기존 지형과 위치적 특성을 살린 초점화 전략은 캠퍼스가 가진 에너지를 한곳에 모아 전체적으로 통일감있는 대학 캠퍼스를 보여주게 될 것이다.

교육용 건물들은 대개 매우 밝은 분위기를 갖기 때문에 외부와의 연결 동선이 중요하다. 반면, 캠퍼스센터는 지하공간에 조성되므로 무엇보다 일광에 관한 전략이 관건이었다. 강당과 주차장을 제외하고 단지 내 모든 공간은 입구부가 수평으로 놓여져 외부 공간과 방이 시각적으로 시원스럽게 연결되어 자연광을 유입할 수 있다. 특히 연속적으로 돌다가 뻗어나가는 링 형태에 내부 공간을 담아내고, 그 선형구조 주변에 다양한 기능의 프로그램으로 크고 작은 뜰을 조성하여 건물에 충분한 일광과 환기를 제공할 수 있게 하였다.

캠퍼스 입구의 해체 DE-COMPOSITION OF CAMPUS ENTRANCE

이화여대 캠퍼스센터 FOA
지명현상설계 참가작

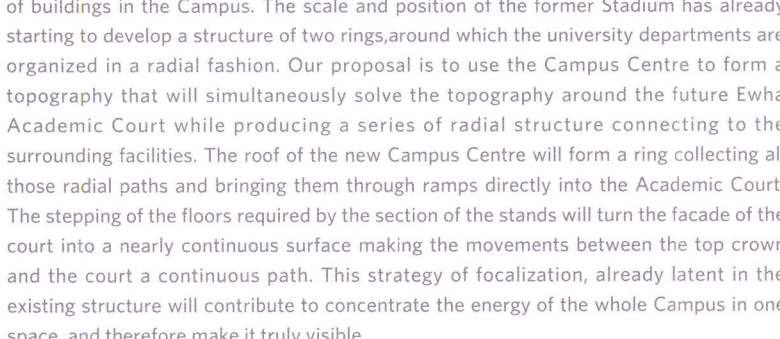
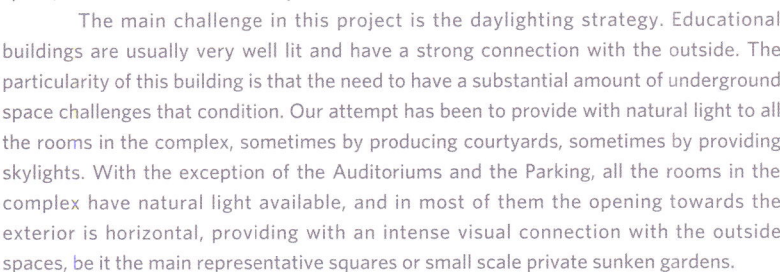
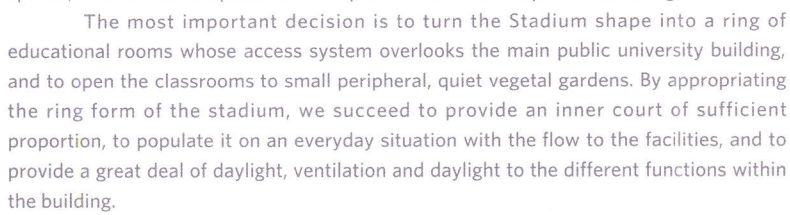

The biggest potential of Ewha Stadium is its baricentric position within this arrangement of buildings in the Campus. The scale and position of the former Stadium has already starting to develop a structure of two rings, around which the university departments are organized in a radial fashion. Our proposal is to use the Campus Centre to form a topography that will simultaneously solve the topography around the future Ewha Academic Court while producing a series of radial structure connecting to the surrounding facilities. The roof of the new Campus Centre will form a ring collecting all those radial paths and bringing them through ramps directly into the Academic Court. The stepping of the floors required by the section of the stands will turn the facade of the court into a nearly continuous surface making the movements between the top crown and the court a continuous path. This strategy of focalization, already latent in the existing structure will contribute to concentrate the energy of the whole Campus in one space, and therefore make it truly visible.

The main challenge in this project is the daylighting strategy. Educational buildings are usually very well lit and have a strong connection with the outside. The particularity of this building is that the need to have a substantial amount of underground space challenges that condition. Our attempt has been to provide with natural light to all the rooms in the complex, sometimes by producing courtyards, sometimes by providing skylights. With the exception of the Auditoriums and the Parking, all the rooms in the complex have natural light available, and in most of them the opening towards the exterior is horizontal, providing with an intense visual connection with the outside spaces, be it the main representative squares or small scale private sunken gardens.

The most important decision is to turn the Stadium shape into a ring of educational rooms whose access system overlooks the main public university building, and to open the classrooms to small peripheral, quiet vegetal gardens. By appropriating the ring form of the stadium, we succeed to provide an inner court of sufficient proportion, to populate it on an everyday situation with the flow to the facilities, and to provide a great deal of daylight, ventilation and daylight to the different functions within the building.

PLAN

이화여대 캠퍼스센터
지명현상설계 참가작
FOA

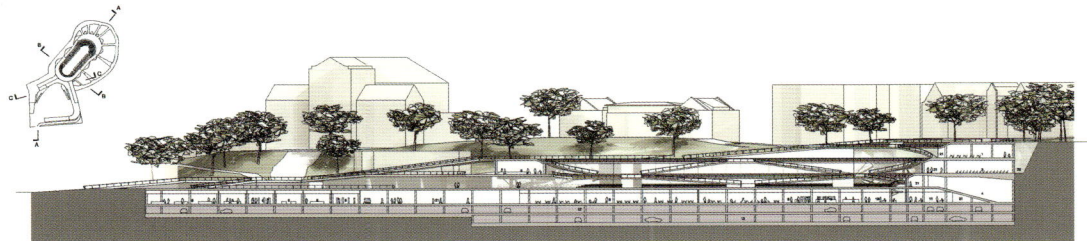

A-A SECTION

B-B SECTION

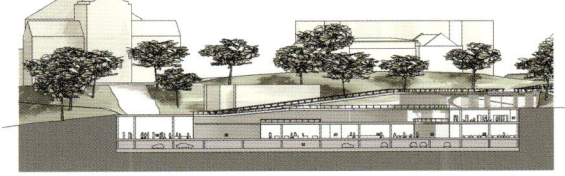

C-C SECTION

캠퍼스 외곽에서의 접근 경로들 **PATHS FROM OUTSIDE OF CAMPUS**

LEEMJONG JANG
YONSEI UNIVERSITY, KOREA
田 + FISH & FISH
SEOUL, KOREA
2004

田 九宮格　붓이 빈 공간을 물 흐르듯 움직이며 의미있는 선의 세계를 펼치듯 구궁격, 아홉의 맞닿아 있는 정방형이 서로가 하나를 이루고 그 면들은 자기 증식을 통하여 시간이 지나면서 스스로의 모습을 변형시키게 된다. 이는 서서히 생명력 있는 모습으로 공간을 형성하며 3차원의 존재가 되며, 다시 건축적 의미공간이 되게 하는 실험 작업이다.

FISH BEING A HOUSE　물고기는 움직인다. 물고기는 물과 자유롭게 교유하며 자신의 몸을 변형시킨다. 물고기는 섬세하게 그 몸을 감싸는 비늘을 갖고 있다. 그리고 매우 유연한 몸의 구조를 갖고 있기도 하다. 물고기가 집이 된다면? 집이 물고기가 된다면?

田 THE CALLIGRAPHIC SPACE MAKING　A brush runs as water flows, creating a meaningful lines into a shape. A white plane becomes a transforming entity.

FISH BEING A HOUSE　A fish swims. A fish gently but sometimes abruptly runs through the water adjusting her body by herself. The fish has also delicate skin and a perfect structure. How about a fish being a house? How about a house being a fish?

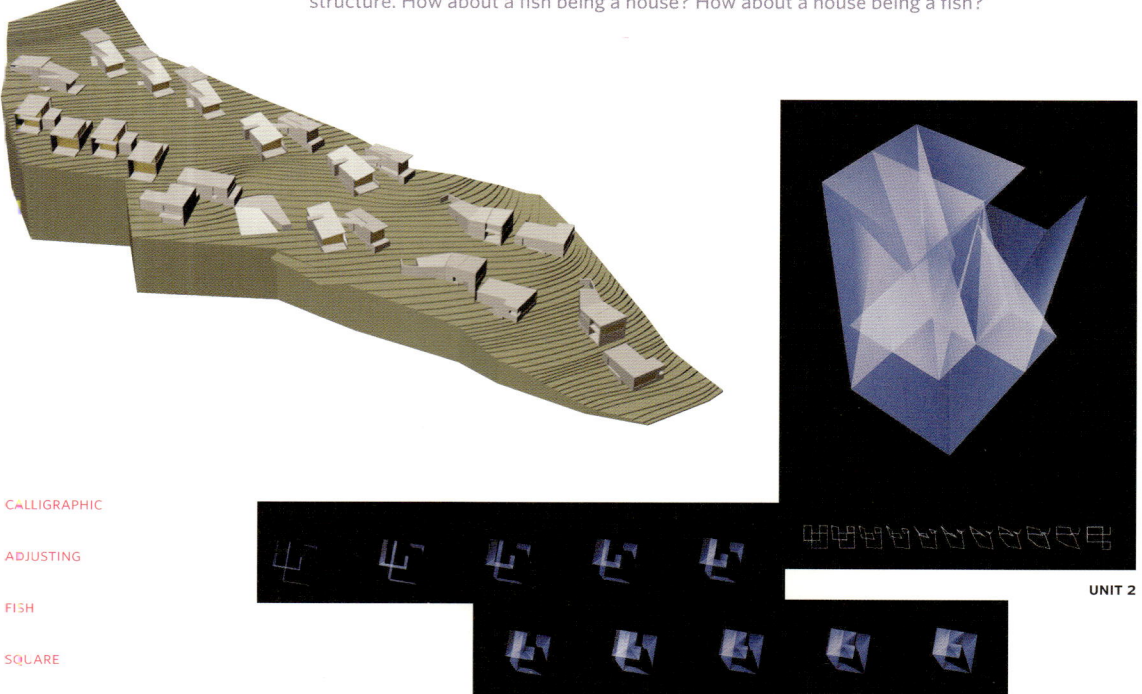

CALLIGRAPHIC

ADJUSTING

FISH

SQUARE

TRANSFORMING ENTITY

UNIT 2

UNIT 1

田 + FISH & FISH

장림종
연세대학교

FISH 1

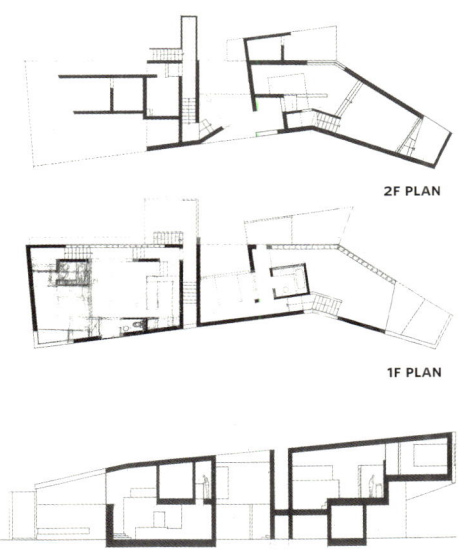

2F PLAN

1F PLAN

SECTION

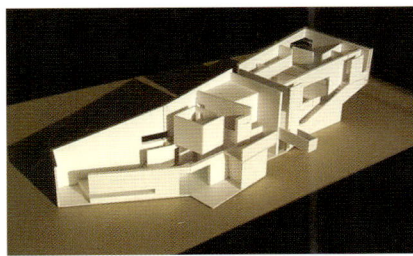

FISH 2

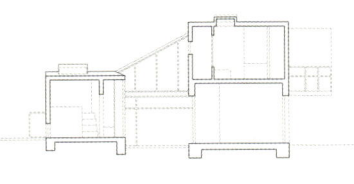

NORTH-SOUTH SECTION

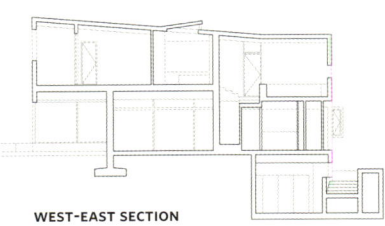

WEST-EAST SECTION

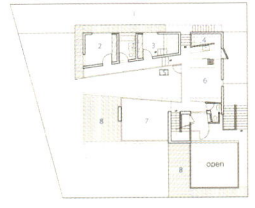

1F PLAN

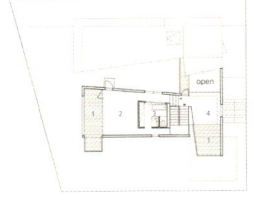

2F PLAN

Leemjong Jang
Yonsei University, Korea
Jeongok Prehistory Museum
Competition Project_Participation
Gyeonggi-do, Korea
2006

멀리서 보면 마치 수평선처럼 계곡 양편의 산등성이 위에 얹히면서, 지형의 흐름을 존중하고 땅의 모습을 훼손하지 않도록 새로운 박물관을 계획하였다. 이곳을 찾는 방문객들은 흘러 올라가는 흐름과 연속적이고 순환적인 동선을 따라 골짜기를 걸어오르며 발굴 터(전시공간)를 발견하게 된다. 또한 비물질화된 지붕은 하늘이 반사되어 하늘 속으로 사라지고, 그 위를 방문객이 주변 경관을 둘러보며 거닐도록 하였다.

The new museum starts with the natural site. The main spaces of the building are concealed underground, leaving the natural contours of the land undisturbed. But the natural is enhanced by a strategic and multi-faceted treatment of the landscape and building surfaces, creating an environment that is more natural than nature itself. The resulting museum offers visitors a simultaneous experience of the prehistoric and the present, the natural and the otherworldly.

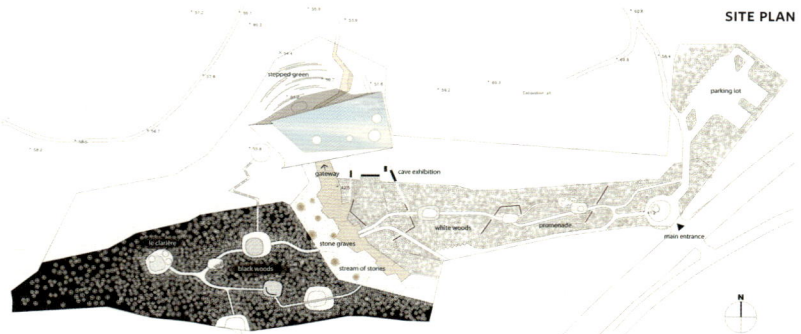

SITE PLAN

전곡리 전경에서의 박물관 VIEW TO THE MUSEUM IN THE LANDSCAPE

MORE NATURAL THAN NATURE ITSELF
CONTOURS
CONCEALED UNDERGROUND
MULTI-FACETED
NATURAL AND THE OTHERWORLDLY
LANDSCAPE
SIMULTANEOUS EXPERIENCE

전곡선사박물관 장림종
현상설계 참가작 연세대학교

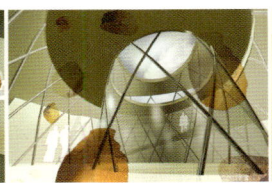

수평적 구조를 만들기 위한 '길게 끔' "ELONGATION" TO MAKE HORIZONTAL STRUCTURE

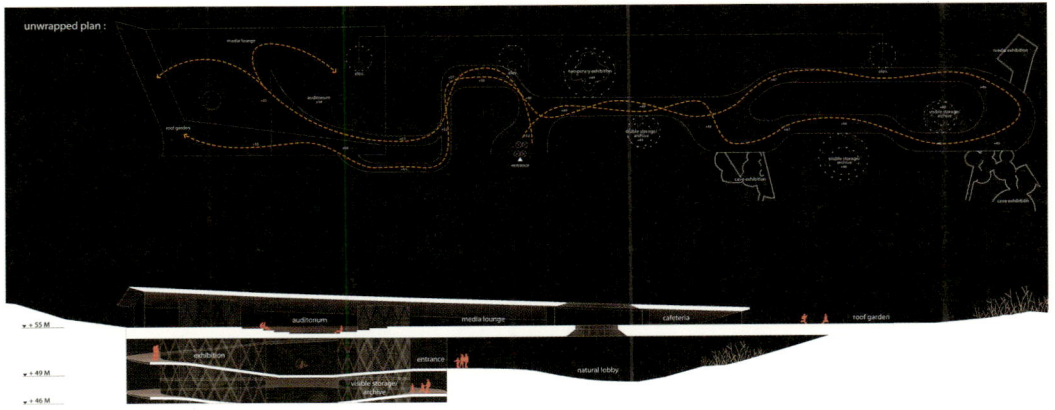

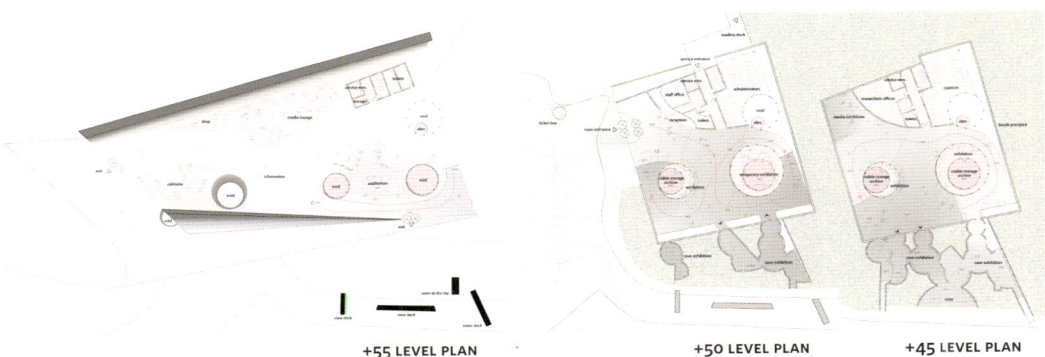

+55 LEVEL PLAN   +50 LEVEL PLAN   +45 LEVEL PLAN

YOUNGJOON KIM
YC2 ARCHITECTS, KOREA
HERYOOJAE WOMEN'S HOSPITAL
ILSAN, GYEONGGI-DO, KOREA
2001-2004

복합 프로젝트 시작의 개념은 복합이었다. 오피스와 오피스텔, 상가와 병원이 있었다. 공공업무지구의 지침이 개발의 순리와 만난 어쩔 수 없는 선택이었지만, 그것은 일면 우리 현실이었다. 다양한 프로그램이 고밀도로 얽혀 있는 적층의 논리를 생각하였다.

불확실 많은 변화가 있었다. 대지가 상업지구로 바뀌었고 더불어 프로그램이 바뀌었다. 경기가 부침을 거듭했고 더불어 규모가 바뀌었다. 불확실하다는 것, 확실히 우리 시대 중요한 특성의 하나다. 그래도 변하지 않는 것에 주목하였다.

병원 몇 번의 경험으로 병원건축을 복합의 관점으로 해석한 바 있다. 업무공간도 상업공간도 주거공간도 등가의 영역이 병원에 있었다. 그것을 엮는 방안이 중요했다. 병원건축은 독자 영역을 집적하는 방식에 해결안이 있었다.

보이드 영역간을 조정하는 보이드를 중시했다. 우리네 고밀의 환경에서 보이드의 중요성은 두말할 나위가 없다. 밀집과 적층의 현실에서 프로그램의 일부로, 프로그램의 조정자로, 다양한 성격의 보이드를 활용하였다. 보이드의 많은 부분을 외부공간으로 규정하였다.

저층부 지표면은 건축의 바탕이자 도시와 만나는 접점이다. 고밀의 도시에서는 저층부의 느슨함이 삶의 숨통을 트는 출구가 될 수 있다. 저층부는 고층부와 다른 처리가 필요하다고 생각하였다. 그래서 필로티, 오픈스페이스, 내부 가로, 스탠드형 계단 등 다양한 시설을 도입하였다.

4층 지표면의 기능을 4층에서 한 번 더 반복하였다. 4층은 저층부의 처리가 반복되는 장소이고, 지표면과 동일하게 공공에게 주어지는 영역이면서, 집회와 관람의 유사 기능이 집중되는 개방의 장소로 생각하였다. 지표면과 4층이 직접 연결되는 수직 동선이 보완되었다.

외부 공간 외부 공간과 내부 공간이 함께 어우러지는 적층구조가 결론이었다. 도시 곳곳에 외부 공간이 존재하듯, 병원 곳곳에 외부 공간이 지면처럼 존재하는 단면이다. 외부 공간은 도시의 오픈스페이스를 유추해보는 정원보다는 쓰는 공원을 지향하였다.

포커스 외부 공간을 인테리어 디자인의 포커스로 생각하였다. 단순한 마감의 중간중간에 외부 공간 디자인의 이미지가 실내 정경으로 유입되는 구상이다. 절제된 내부 공간과 풍요로운 외부 공간의 구도를 결정하였다.

동선 다양한 성격의 공간이 집적된 구조에서는 동선 처리가 상대적으로 중요한 변수이다. 하나의 시퀀스로 몰아가는 단순함보다 연결의 다양한 선택을 생각하였다. 지면과 최대한 밀착하기 위해서, 도시와 건축의 원활한 교류를 위해서, 분산된 연결 동선을 제시했다.

집적 결국 집적의 대안으로 프로젝트 성격을 생각하였다. 고밀의 현실에서 복합의 현실에서 느슨하고 여유로운 대안이 목표였다. 지면이 겹쳐지고 독립된 연결고리로 내·외부 공간이 혼재되는 제안이었다. 도시구조를 마음에 두고 건축구조로 다가서려는 시도다.

형태 단순히 영역을 표시하는 매스로 형태를 결정하였다. 내부 기능을 대변하기보다는 내부 혼돈을 정리하는 질서를 외관으로 결정하였다. 질서와 기능이 대비되며 만들어낸 결과는 불규칙적인 다양성이다.

DIFFERENT COMBINATIONS
MULTIPLE CHARACTERISTICS
UNCERTAINTY
LAMINATION
VOID
OUTDOOR SPACE
IRREGULAR MULTIPLICITY
FRAMEWORK

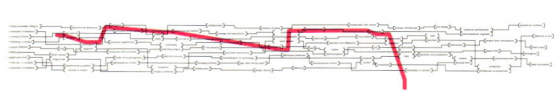

허유재병원   김영준
김영준도시건축

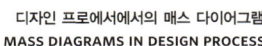
디자인 프로세스에서의 매스 다이어그램
MASS DIAGRAMS IN DESIGN PROCESS

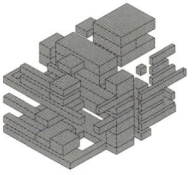
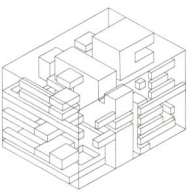
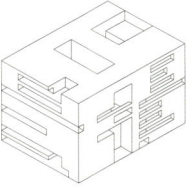

This project began with programmatic exploration of multiple characteristics embedded in functions of office and hospital. Complying with the urban design guidelines for the City of Ilsan, we have come across to the solution of mixing an array of various programs into different combinations. (The program of office was later removed, leaving only the hospital part to be developed as a partially independent mass). These combinations should neither collide with each other nor generate a conflict, suggesting that they should stay within its independent territory and characteristics.

    We have imagined a condition where different programs operating simultaneously on the same site. Each function for the hospital has been separated into specific categories and re-grouped into several programmatic fragments conveying multiplicities in function. Considering each of them as a unit, we have conducted infinite number of possibility studies in connecting them among each other. Architectural suggestion for this project, therefore, was to rethink the organization methods and urban fabrics by which the outdoor spaces become a combining agent to connect each pieces into the whole mass.

    The main task was to divide and distribute the outdoor space in appropriate manner. As similar to the past urban landscape, where each building scatters throughout the field of ground, a complex of architectural set-up that is distributed around the artificial surfaces require some extent of control carried by the outdoor space. The outdoor space became the tool of creating a framework to embody complexity within the building.

    As an image, we have captured an organization that transcended the interpretations of individual program. It was a system where the outdoor space, as an object, and the program, as its counterpart, were assembled in rather the simple framework. By acknowledging the role of architecture as defining a loose footprint of territory, more so than devoting to resolve each program's requirements, the new architectural suggestions can be made.

일산, 도시에서의 정면 FACADE TO THE CITY, ILSAN

삽입된 여러 개의 아트리움 INSERTED ATRIUMS

삽입된 정원 INSERTED GARDEN

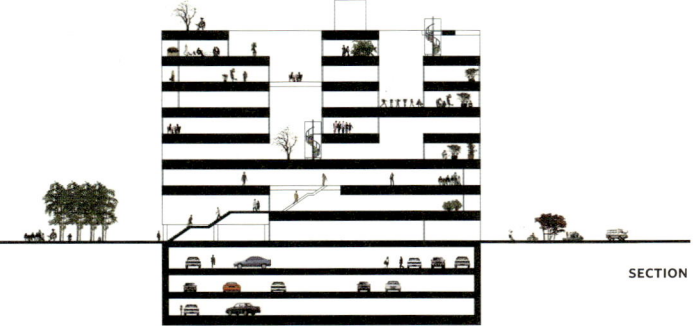

SECTION

허유재병원　김영준
김영준도시건축

2F PLAN

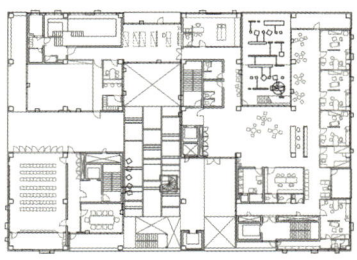
4F PLAN

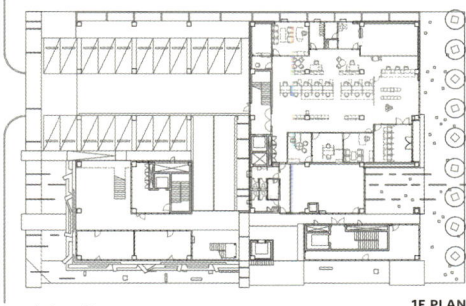
1F PLAN

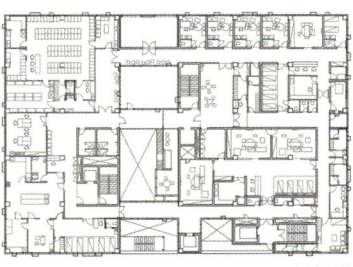
3F PLAN

각층 기능으로의 내부 액세스 INTERIOR ACCESS TO THE PROGRAM ON DIFFERENT FLOOR

Youngjoon Kim
YO2 Architects, Korea
Hamburg Architectural Olympiad
Wandsbekerzoll Strasse, Germany
2006

이 프로젝트는 주어진 현실에서 유추되는 가로 활성화 전략을 목표로 하였다. 전략은 다음과 같은 단계의 결과에서 유추되었다.

첫째, 기본적으로 도시 풍경의 완성을 목표로 한다. 둘째, 가장 중요한 수단으로 주변 현황에서 도출된 점을 제안의 근간으로 삼았다. 점은 부가되는 조경 요소의 위치 정보인 동시에 다양한 대응을 위한 형태의 기본 요소이기도 하다. 셋째, 새로운 개발이 준비된 지역의 건축적 과제는 건축 유형의 제안으로 정리하였다. 건축 유형은 가로와 대응하는 방법에서 유도되었다. 가로 영역을 확장하는 유형으로 제안하고, 건축 지침으로 정리하였다. 넷째, 조경의 점을 보완하는 바닥 패턴을 부가하였다. 새로운 개발 지역보다는 기존 컨텍스트를 유지하는 지역에서 상호연결하는 수단으로 활용하였다. 다섯째, 중요한 위치의 점을 발전시켜 파빌리온으로 특화시켰다. 점의 집합은 가로에 수평 수직으로 대응하는 방식에 따라 세 거점의 질서로 통합되어, 가로공간 순환의 매듭으로 자리매김하였다.

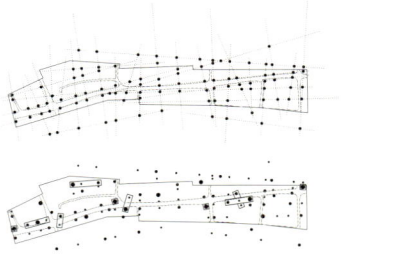 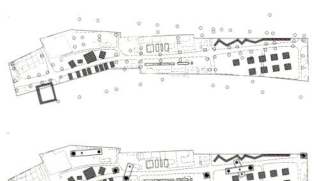

점들로 이루어진 다이어그램으로부터 평면 다이어그램까지의 프로세스
PROCESS FROM POINTS DIAGRAM TO PLAN DIAGRAM

CITYSCAPE

COMPLEXITY OF MULTIPLE LAYERS AND ORGANIZATIONS

UNPREDICTABILITY

CONNECTION

NETWORK OF POINTS

EVOLVING FORM

SCENARIO

MAPPING

LESS HIERARCHICAL

MORE HETEROGENEOUS

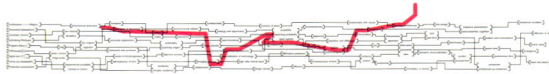

©YONGKWAN KIM

Main objective for this project is to generate strategy out of given standards and existing apparatus as the following:
1. Basic framework of urban landscape shall be established.
2. As most important source for the proposal, several "points" located out of the existing condition has been considered as the physical information of landscaping elements, and the diverse array of form-generation and alternation.
3. Building types have been suggested for the areas ready for the development. These types have been derived out of the manner to deal with the street in contact as they are regarded more as the elements that can potentially expand the territory of the street. All of its related criteria have been suggested as a guideline.
4. Rather than designating the new development location, pattern of street pavement have been utilized as one of ways to maintain the existing context and to set up reciprocal relationship within the site.
5. Number of critical points have been developed as pavilions, as the network of points are organized as three different groups of conditions - hot point, strip, and wedge. They have been located along the circulation of street called "node".

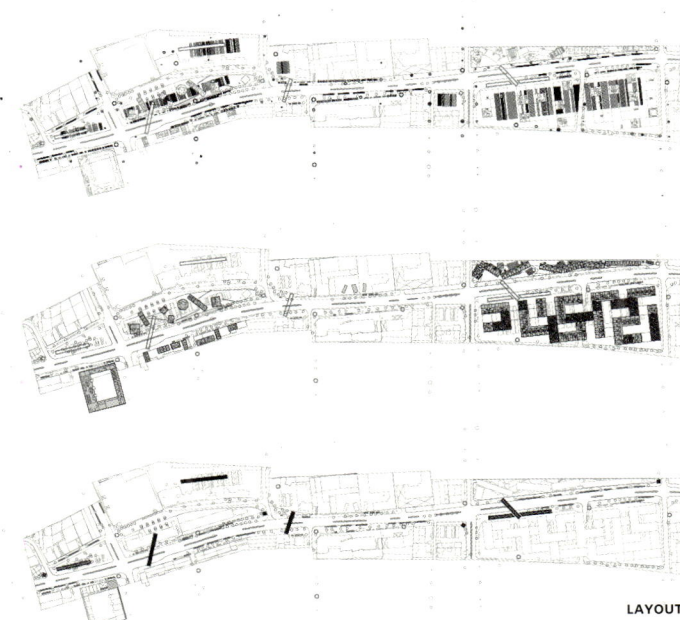

LAYOUT

Tesoc Hah
iARC Architects, Korea
Systematized Prototype
2003/2006

시스템적 프로토타입은 일반 시스템으로서 가상이다. 이것은 현실화됨으로써 특정성을 갖는다. 비형태적 물질과 비형적 기능이 프로토타입의 추상기계(이것은 가상에서 다이어그램적으로만 작동한다)를 만들듯이 분화된 형태(양상)들이 이것이 현실화되면서 생성된다. 대량생산 소비재의 물질-형태의 관계는 추상기계-양상의 관계로 대체된다. 이 양상은 정해진 형태 없이 형태가 변수로 존재하는 것이다(이 미분 프로토타입은 이것의 변화 가능성으로 인해 열린 시스템이지만 동시에 스스로의 자기조직적인 면 때문에 닫힌 시스템이기도 하다).

시스템적 프로토타입은 끊임없는 변이의 평면이다. 이러한 변이들은 연속적인 분화된 내용과 표현이 있는 하나의 고원으로 여겨질 수 있다. 시스템적 프로토타입은 추상기계의 사실상의 형태(양상)이다. 매스 커스터마이제이션의 건축에서는 이 시스템적 프로토타입 개념이 적절한데, 체계적으로 스스로를 분화시킬 수 있기 때문이다. 즉 프로토타입은 매번 거주자의 다양한 요구를 충족시키기 위해 반복될 때마다 또 다른 모드로 존재하게 된다. 이것은 또한 특정의 단일한 디자인보다는 다수성의 시스템을 요구한다.

PROTOTYPE

COMMUNITY

GEOMETRY

MULTIFORMITY

STRATUM

SHELVE

DIFFERENCE

WAVING WALL

ACCOMMODATION

A systematized prototype is virtual as generic system, only when it actualized it is become specific. As non formal matter and unformed function make abstract machine of prototype (it only works as diagrammatic in virtual), differentiated forms (modality) are generated when it actualized. Thus "matter-form" relation of mass produced consumer products is changed to "abstract machine-modality". It is modality where no fixed form plays significantly only form exists as variables (open system as its possible worlds but close system as its self-referentiality).

A systematized prototype is a plane of continuous variation; each actualization can be considered a "plateau" of variation that places differentiated contents and expression in continuity. It is an abstract machine in a de facto form (modality). A desiring machine trigger the actualization of a prototype. In mass-customized architecture, the concept of systematized prototype is viable since it has the capacity to differentiate itself systematically, within defined range parametrically. In other words, a prototype could exist different mode every time it repeats in order to meet diverse demands of its contextual conditions. Rather than being specific for a single design, it demands system of multiplicity.

시스템적 프로토타입 하태석
아이아크

S LIBRARY PROTOTYPE
2006

이 프로토타입은 커뮤니티 기능과 중학교 학생이나 성인을 위한 도서관 등 다양한 요구를 충족하는 어린이도서관을 위해 디자인한 것이다. 인간의 성장 과정에서 나타나는 다양한 키 높이에 따라 도서관은 여러 지층으로 조직된다. 서로 다른 요구에 따라 각 선반 지층이 쌓이고 서로 다른 용도에 따라 배열이 달라진다. 이 프로토타입은 각각의 상이한 학교의 상황에 따라 다른 양상으로 적용된다.

The prototype is designed for children's library with various demands including community function and junior high school students and adult library. In order to accommodate these demands, it is organized as strata to match various heights of human growing. Each shelve-strata is stacked for different demands and differentiated its geometry for different use. The prototype is intended to occupy different school in different modal conditions.

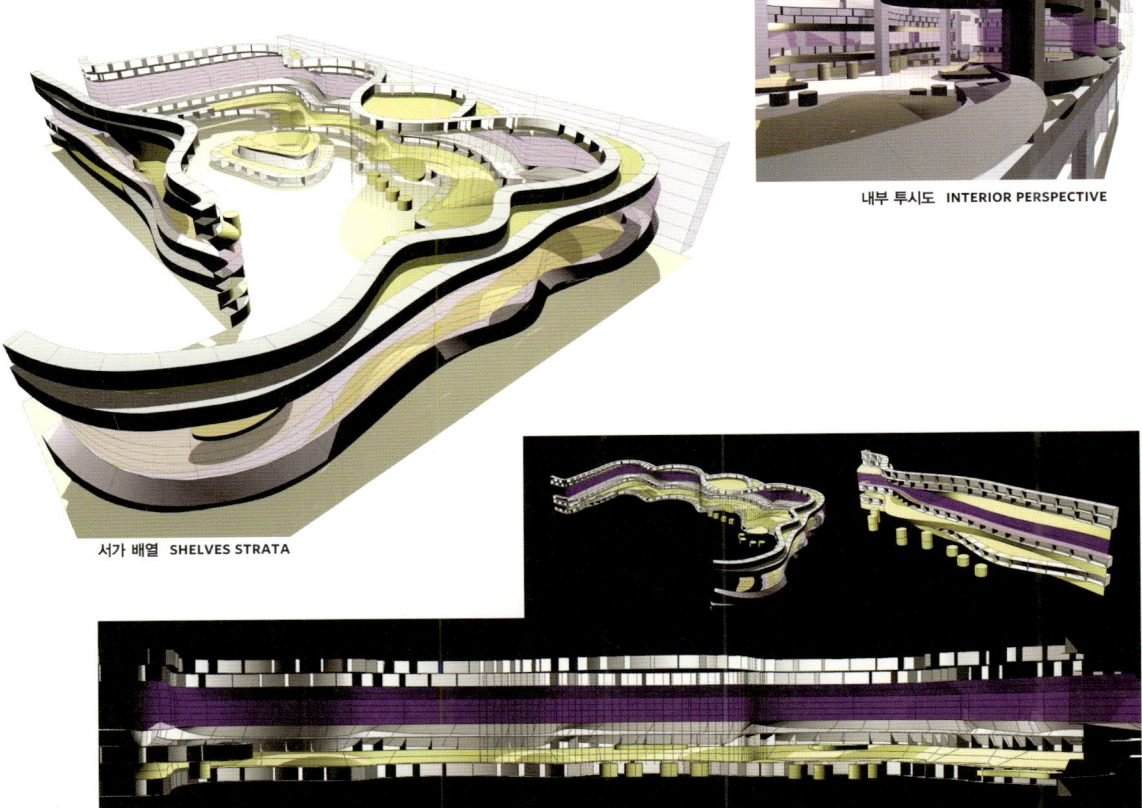

내부 투시도 INTERIOR PERSPECTIVE

서가 배열 SHELVES STRATA

서가 배열 프로토타입 PROTOTYPE COMPONENT FOR SHELVES STRATA

M Loft Prototype
2006

M 로프트는 다양한 노드의 건축을 구성하는 네트워크 아키텍처이다. 노드 아키텍처가 서로 연결되어 전체 네트워크를 구성한다. 대상 부지들은 교통량이 많은 도로와 연결되어 교통이 편리하고 가시성이 높다. 네트워크 연결이 차량으로 이루어지는 만큼 자동차가 전체 시스템을 운영하는 중요한 요소가 된다. 프로토타입은 쉴드, 주거유닛, 상업유닛, 자동차 리프트의 4개 요소로 구성된다. 부지 고유의 정보가 여러 다양한 맥락 요건을 충족하기 위한 프로토타입의 실현을 촉진한다.

M loft is a network architecture which constituted various nodal architecture. Each nodal architecture connected each other to create a networked whole. The generic sites have a common aspect which all the sites are located in next to busy road with good traffic connectivity and visibility. The connection of network is provided by car. The car became significant element to operate the whole system. The prototype constituted with four components; shield, housing units, commercial unit and car lift. The site specific information is triggering the actualization of the prototype to suit different contextual demands.

전경 CONTEXTUAL APPLICATION

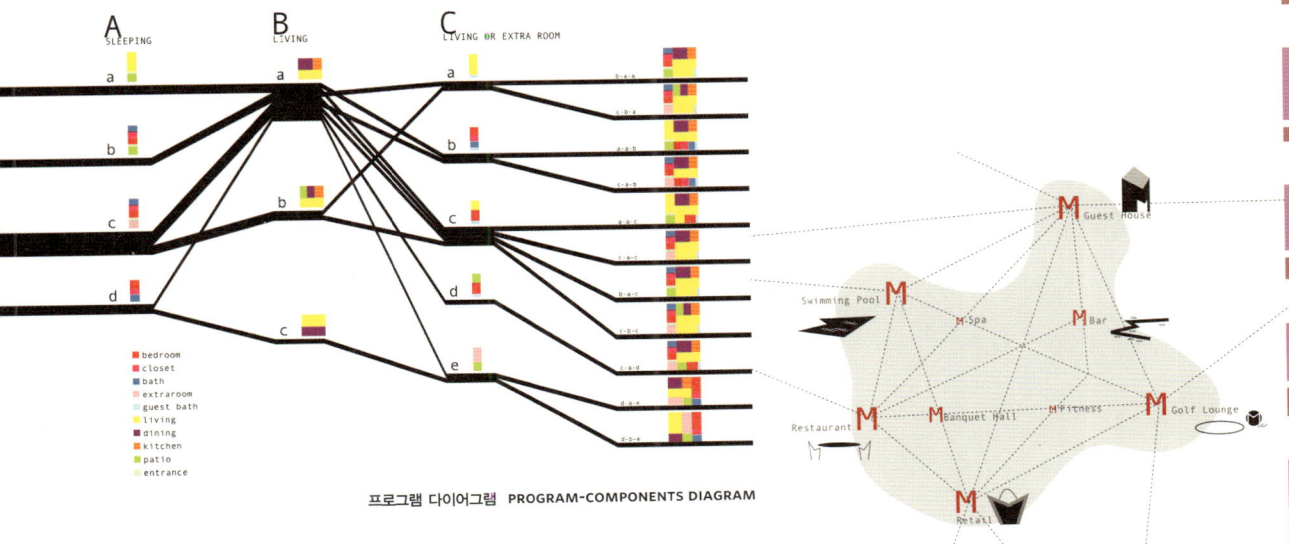

시스템적 프로토타입 하태석 아이아크

프로그램 다이어그램 PROGRAM-COMPONENTS DIAGRAM

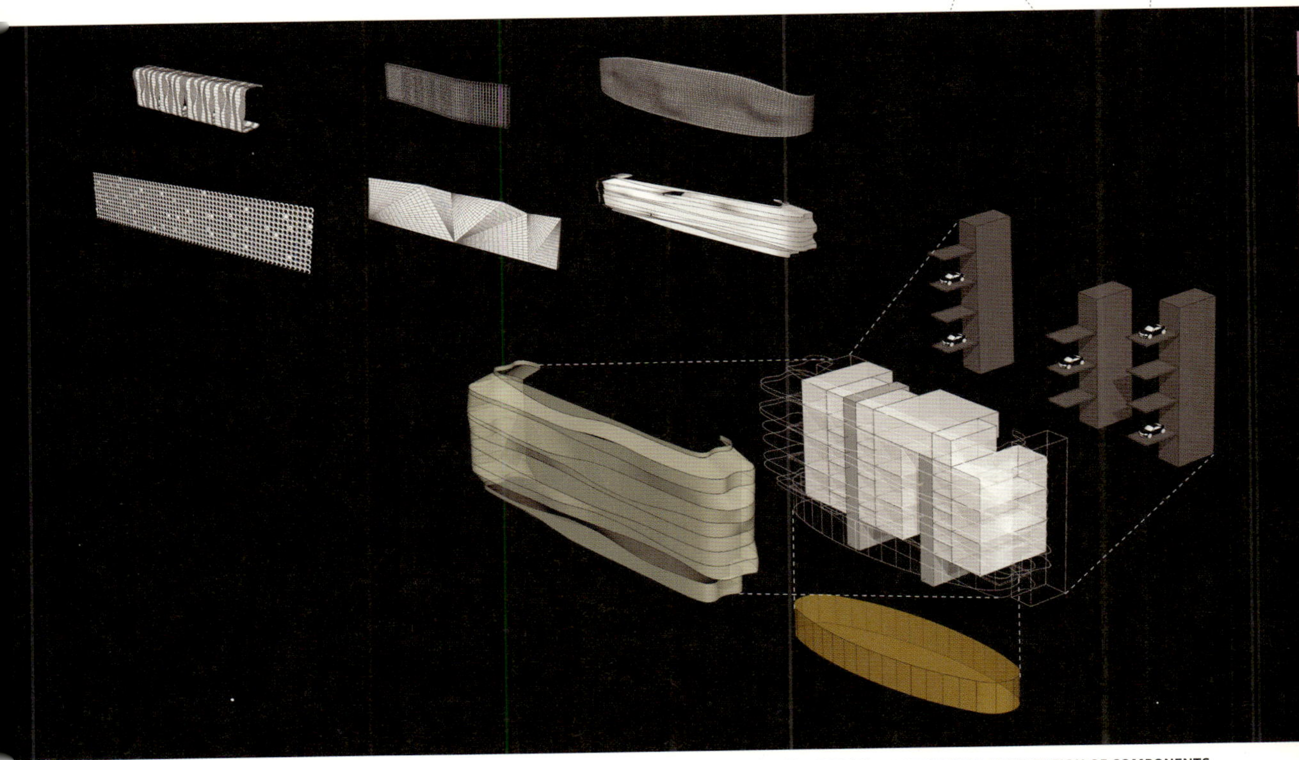

유닛 구성 프로토타입 PROTOTYPICAL FABRICATION OF COMPONENTS

FLOWCITY
Izola, Slovenia
2003

계통화된 프로토타입과 처방된 배치체계에 따라 조직된 저층주거 콤플렉스. 각 주거유닛은 서로 다른 세대, 라이프스타일, 클러스터 유형에 따라 다양한 유닛으로 분화된다. 이들 유닛이 모여서 다층주거 블록을 이루고 있는데, 연속 지붕 표면으로 물 부족을 해소하기 위해 빗물을 모은다. 지붕 스트립이 조경 스트립에 연결되어 온전한 빗물 수집 유닛을 구성한다. 즉 각 지붕+조경 스트립 유닛이 연결되어 전체 빗물 채취 시스템 네트워크를 이룬다.

The low rise housing complex is organized itself according to its systematized prototype and prescribed deployment system. Each housing unit is differentiated according to different household, lifestyle and clustering types. These units are clustered together to form a multi-stories housing blocks which have a continuous roof surface in order to harvest rain water for its water lacking environment. The roof strips are connected their landscape strips to form a complete rain water harvesting unit. Each roof+landscape strip unit is connected together to form the whole as a networked rain water harvesting system.

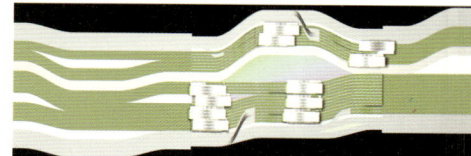

PROTOTYPE

COMPLEX

SYSTEM

UNIT

CLUSTER

CONTINUOUS ROOF SURFACE

LANDSCAPE

STRIP

NETWORK

DIFFERENTIATION

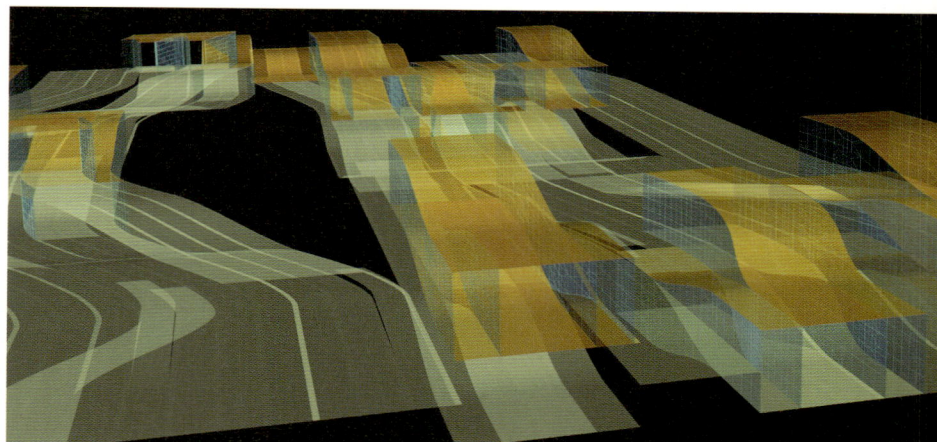

지붕 스트립 시스템 ROOF STRIP SYSTEM

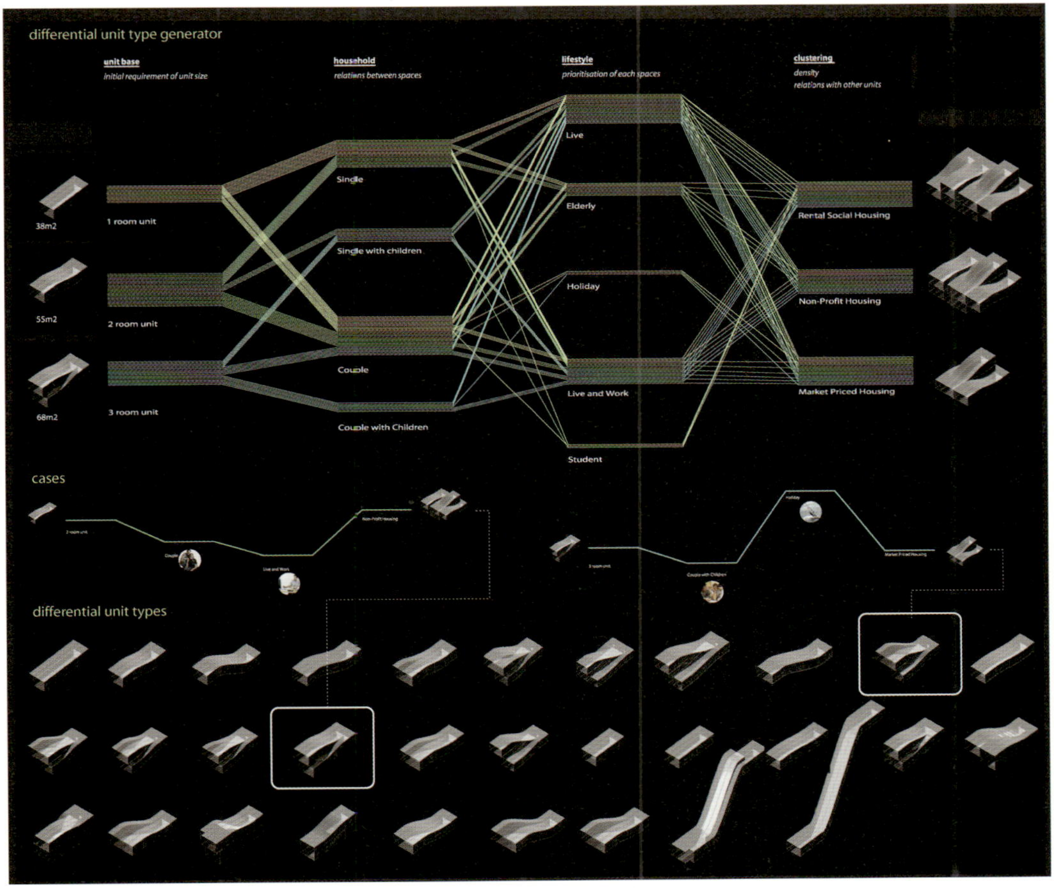

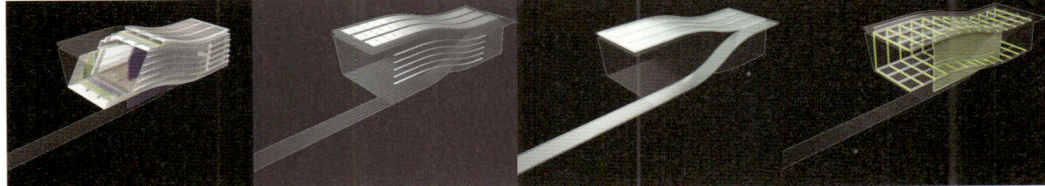
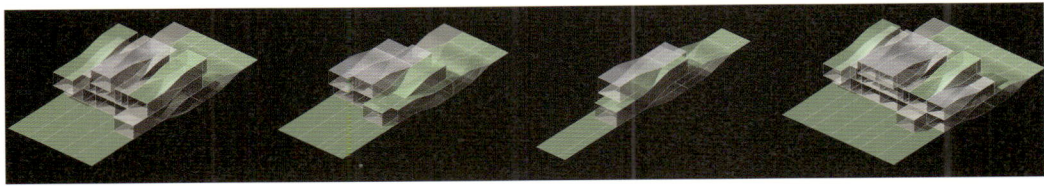

유닛 블록 구성방식 DIFFERENTIAL UNIT TYPE GENERATOR

Ciro Najle
GDB, Argentina
Protostructures
Generative Normativities, Life Prototypes and Complex Systems

MATERIAL

DESIGN

PROTOTYPE

INFRASTRUCTURE

LANDSCAPE

TECHNOLOGICAL

CREATIVE

HETEROGENEOUS

VIRTUAL SYSTEMS

COMPLEXITY

프로토스트럭처는 소재 탐색, 빌딩 디자인, 프로토타입 개발, 인프라와 조경에 걸쳐 연속성을 개발하려는 시도로 일관된 건축 및 구조 매체 내에서 일련의 건축 관습, 빌딩 전통, 기술과 예산에 대한 제약을 통합한다. 일련의 물질적 제약을 결합과 조정의 기술로 사용함으로써 다양한 생산조건과 시나리오에 체계성과 학문적 엄격함을 대비시킨다.

프로토스트럭처는 복잡하고 과도한, 풍부하고 왕성한 물질 조직을 다양한 시스템의 창조적 관리 관점에서 유형적인 것을 시스템적인 것으로 바꾼다. 이것은 영역들의 상호교류, 다양한 척도에 걸친 정보 교차, 서로 다른 전문지식 영역과 전통을 구분하는 재정의를 가능하게 한다.

프로토스트럭처는 추상적인 프리아키텍처 구조물로, 단독 가상시스템에 따라 물질적 행동과 조직의 역동성을 통합하며, 건축적 점유에 열려 있다. 자체 수정을 하고 규정을 포함하는 이종적이고 일관된 자급 시스템으로, 단독 규범으로 합성되고 추상적이자 구조적인 논리에서 처리된다. 프로그램적으로 확산되어 있지만 건축적으로는 엄밀하다.

프로토스트럭처는 결정들이 축적되어 생성되고 이러한 의미에서 건축에서 일반성의 결정적 수단을 구성하지만, 동시에 질서의 중립적이고 동질적이며 무표정하고 메마른 형식에서 그 은신처를 우회한다. 가장 다양하고 마이너한 합리적 행동에 체계적 일관성을 부여하는 보강재로 작용하는 구조적이고 법적인 제약을 갖고 있다. 흡수, 조율, 복잡성의 구조물이자 장엄함의 신비스럽고 난해하고 관능적인 형식을 갖고 있다.

In the attempt at developing continuity across material exploration, building design, prototype development, infrastructure and landscape, Protostructures integrate in a consistent tectonic and structural medium a series of architectural conventions, building traditions, technological and budgetary constraints. They confront systematicity and disciplinary rigor against different scenarios and conditions of production by engineering a series of material constraints using them as techniques of incorporation and mediation.

Protostructures involve a shift from the typological into the systemic in view to the creative management of a variety of systems through complex excessive, exuberant and robust material organizations. They enable cross-fertilization of domains, transposition of information across scales, and largely a redefinition of the division between fields and traditions of expertise.

Protostructures are abstract pre-architectural structures that integrate material behaviors and organizational dynamics according to single virtual systems, open to architectural occupation. They are heterogeneous, consistent and selfcontained systems of regulation and selfcorrection, sinthetized in a single normativity and engineered in an abstract constructive logic. They are programmatically diffuse yet tectonically rigorous.

They are generated through accumulation of determinations, thus constituting the ultimate means of generality in architecture, yet bypassing its retreat in neutral, homogeneous, inexpressive and sterile forms of order. Protostructures are embedded with constructive and legal limitations operating as armatures that give systematic coherence to the most diverse and minor rationalities. They are structures of absorption, coordination and unfolding of complexity. They host cryptic, intricate and voluptuous forms of the sublime.

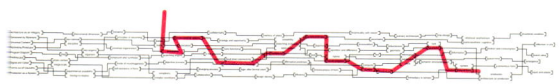

SPbranching / 가구 시스템

SP브랜칭은 하나의 가지가 두 갈래로 나뉘는 분열증식 원리로 이루어진 조직 시스템으로, 이전에 프라이 오토 가 진행했던 실험 양식을 발전적으로 승계하여 상상력 의 범위를 확장시키고자 한 것이다. SP브런칭은 구조 의 유동성을 강화하고, 하부구조의 잠재력을 다각도로 향상시키며, 구조적인 패턴의 배열을 확대함으로써 이 러한 모형의 분화된 형태를 제시했다. SPb000884는 SP브랜칭을 최초로 적용한 사례로 2001년 런던 AA스 쿨 전시를 위해 설치되었다. 철근을 구부려서 만든 기 본 구조들을 플라스틱 끈으로 연결하여 이차적인 구조 를 형성하고 구조 변형을 통해 기하학적 다양성을 연출 했다. 구조 윗부분의 표면은 간단한 조작을 통해 여러 다발의 기본 구조들을 지탱하도록 설계되었다.

SPbranching / Furniture System

SPbranching is a system of structural organization based on the proliferation of a branching principle from one to two components. SPbranching is the continuation and further evolution of experimentation previously developed by Frei Otto with the purpose of expanding their scope beyond their ideality. SPbranching proposes a differentiation of this model, providing it of higher tectonic fluidity, enhancing its infrastructural capacities at many scales and generating a wide array of organizational specimens. SPb000884, the first sample of the system was built at the Architectural Association in London for an exhibition in year 2001. It consisted of a primary system of steel bars bent in space and bundled locally by a secondary system of plastic ties, which adapt to the variations in geometry resulting from structural deformation. Its upper surface was designed to support a series of "floating" books with easy and soft manipulation.

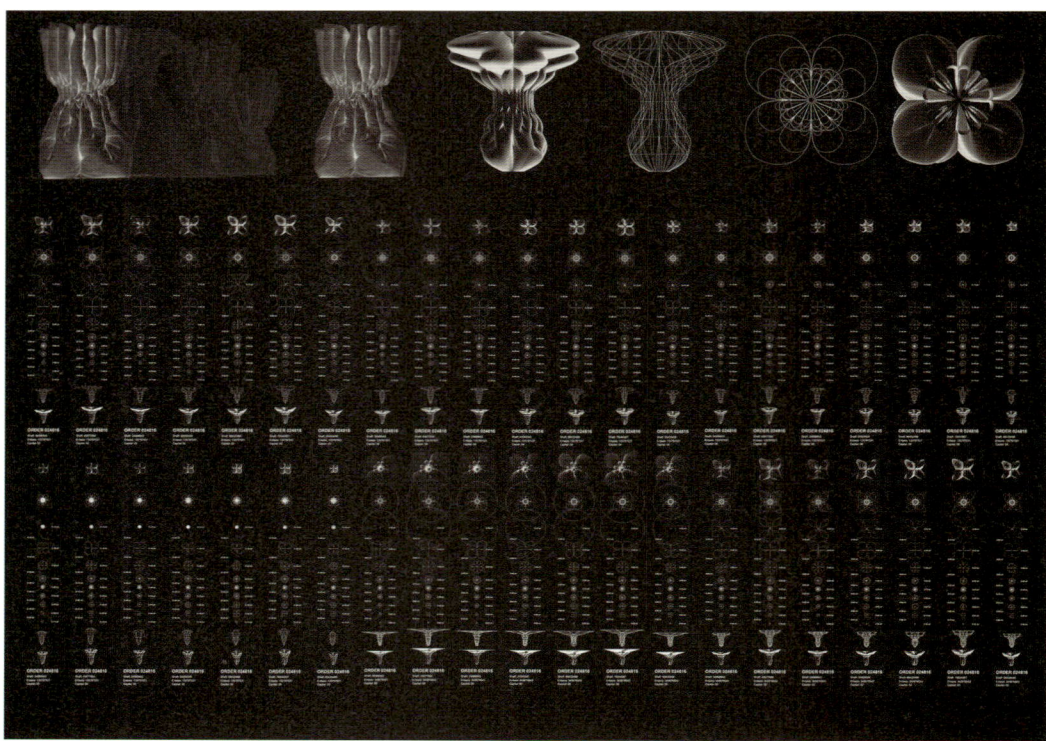

Thrust Order / 기둥 시스템

트러스트 배열은 이상적이고 정적이며 구상적인 성격의 고전적인 방식을 대체하는 복합적이고 역동적이며 실용적인 기둥 시스템이다. 트러스트 배열은 수직으로 쌓아나가는 움직임의 방향에서 발전시킨 간단한 구조 원리들을 변형한 것이다. 트러스트 배열은 압력에 약한 물질 매체가 분리된 물질 덩어리로부터 영향을 받는 역학체계에서 생성된 변형의 결과물로, 건축적인 표현에 대한 구조적인 토대를 형성한다. 이 체계는 직경 9m, 높이 3m, 두께 4m인 원형의 검정색 구조물에 최초로 적용되었으며 2003년 6월 프라하아트비엔날레에서 설치되었다. 설치가 간단할 것, 조립과 설치에 있어 전문지식의 필요성을 최소화할 것 그리고 이동이 용이하고 가볍게 만들 것 등이 설계의 주요 고려사항이었다.

Thrust Order / Column System

Thrust order is a system of prototypical columns assumed as a complex, dynamic and operative order alternative to the ideal, static and representational orders of the classical tradition. Thrust order is a system based on the variation of simple structural principles that evolve from behavioral trends in a vertical accumulation. Thrust order is a lineage generated from the variations in a regime of forces where a material medium unstable to tension is affected by a mass of disaggregated material, configuring a tectonic substrate for architectural expression. The first simple of the system was constituted by a circular black piece 9m diameter, 3m height and 4m volume and was built as an installation in the Prague Art Biennale in June 2003. Simplifying the construction on site, minimizing the amount of assembly and construction expertise and requiring light and easy transportation were main determinants of the design.

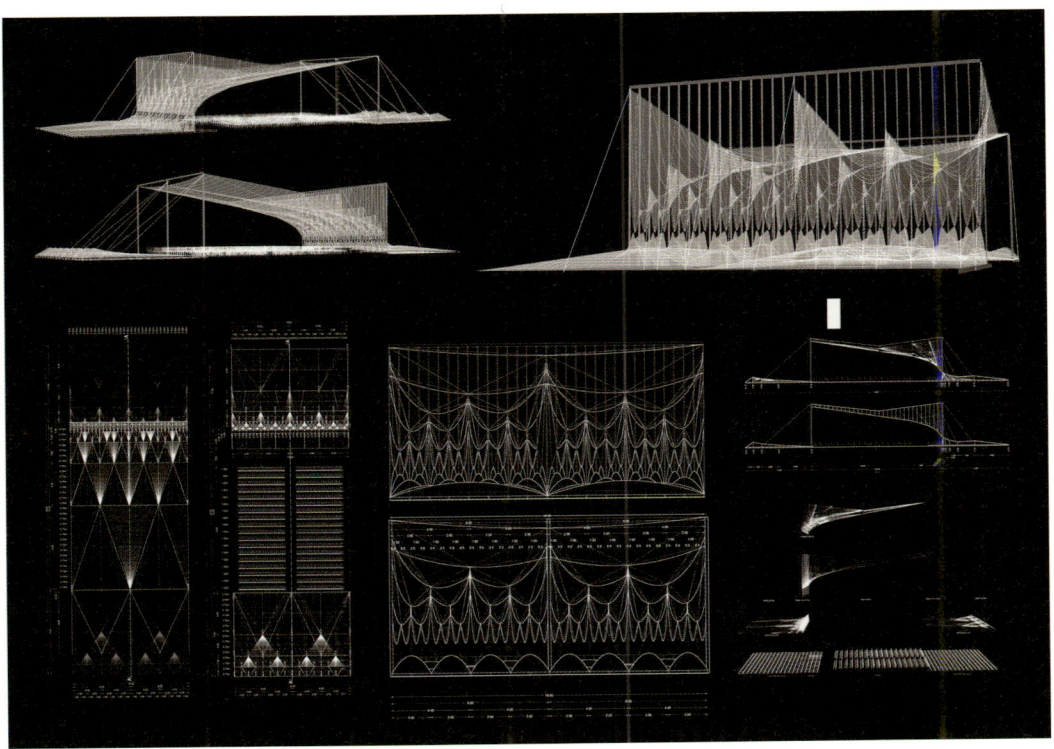

Wmembrane / 지붕 시스템

W멤브레인은 기념식이나 개막식, 일시적인 행사 등을 위해 고안된 지붕 시스템이다. W멤브레인의 구조공학과 용도 그리고 재료 사이에는 긴밀한 관계가 형성되어 있다. 구성요소간의 상호작용은 부수적인 시설을 병합하는 역할을 한다. 기본 설비와 배수시설, 무대와 라운지의 관계에서 이러한 상호작용을 확인할 수 있다. W멤브레인은 복잡한 쇠사슬구조를 따른다. 지붕면은 케이블을 이용한 삼각형 지대에 의해 지탱된다. 이 지지대는 일단 힘을 측면으로 분산시킨 다음, 샌드백 완충 효과를 지닌 지면으로 보낸다. 따라서 지붕면의 크기와 각도에 따라 부지의 경계가 제한될 수 있다. W멤브레인은 런던 AA스쿨의 의뢰를 받아 2004년 런던 베드포드 광장에서 열린 연말축제에 처음으로 설치되었다.

Wmembrane / Roof System

Wmembrane is a roof system conceived as a structure for ceremonies, openings and temporary events. Wmembrane introduces continuity between structural engineering, complex use and material behavior, where the interactivity between components is used as a medium for the absorption of ancillary systems like furniture and drainage systems, stages and lounges. Wmembrane is constituted by a roof surface of overabundant material that folds down in a complex catenary surface whose pattern is followed by water circulation and spatial organization. The roof surface is held by a triangulated cable system that transmits forces laterally to a system of sand-bag cushions on the ground, whose variation in size and angle qualifies the use of the perimeter. The first sample was commissioned by the Architectural Association in year 2004 for its end-of-the-year ceremony at Bedford Square in London.

Ivault / 전시 천장 시스템

I볼트는 전체적인 전시를 최소 단위의 구성요소로 분해하고 그 구성요소들을 근접성과 인접성의 증감에 따라 공간에 분산시키기 위해 고안된 천장구조다. 이러한 천장구조는 가시성 조절을 통해 노출된 사물을 모호하게 함으로써 내용의 가독성을 문제화한다. I볼트는 철제 케이블 체인이 가로 방향으로 반복되면서 오목한 A4 패널을 잡아주는 구조다. 패널의 위치와 접히는 정도를 달리하면, 패널의 배열 방향에 변화를 줄 수 있다. 패널들 사이의 거리는 완전히 겹쳐지는 경우도 있고 완전히 분리되는 경우도 있을 정도로 서로 다르다. 회오리나 미풍과 같은 공기의 움직임이 전달되면 부분적으로 경직되면서 전체적으로 생동감 있는 분위기가 조성된다. I볼트 구조는 벽과 바닥이 없는 소규모 계단부의 천장에 처음으로 적용되었다.

Ivault / Exhibition Ceiling System

Ivault is a ceiling system for exhibition purposes that breaks down the integrity of an exhibition to its smallest components and distributes them in space according to gradients of proximity and adjacency, problematizing the readability of content by controlling the difficulty in its visibility producing ambiguity in the object being exposed. Ivault is a collective of chained catenaries that transfer local variations transversally and hold a series of concave A4 panels. The relative changes in location and extension of folds enables variation in rotation in the two directions of the vault. A gradient of adjacencies between panels ranges from superimposition to detachment, creating local stiffness and global mobility to transfer movements created by swirls, eddies and breezes. The first sample of the system was developed as a vaulted ceiling at a stairwell with no walls, no ground and small volume.

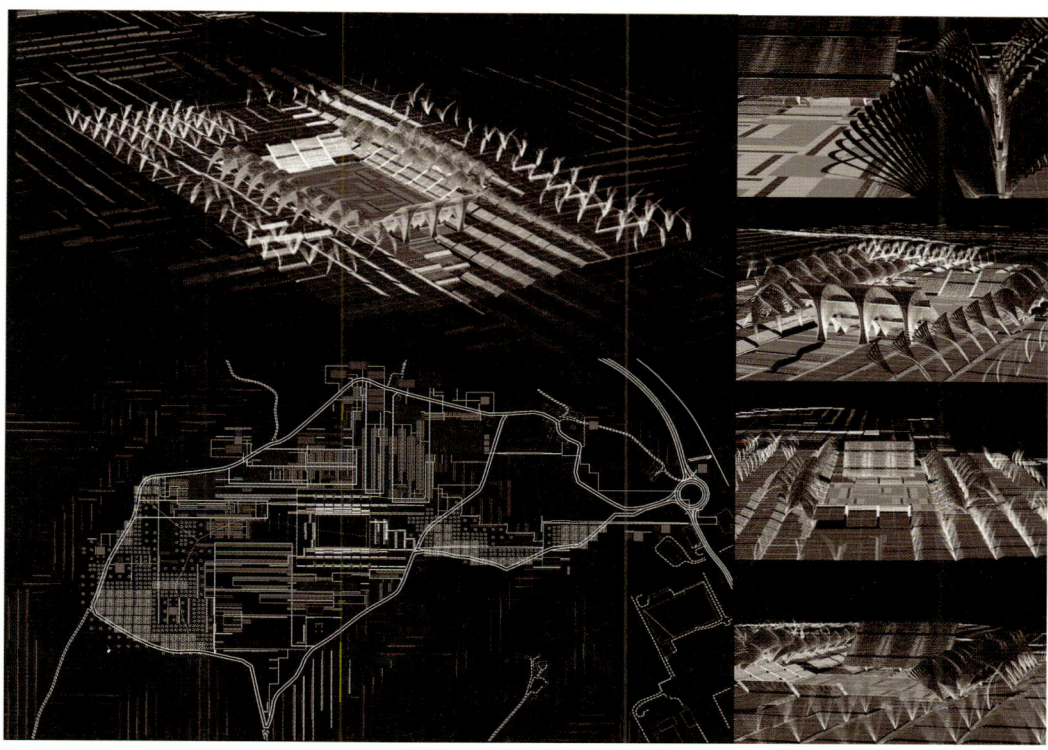

Mille Plateaux / 종합 스타디움

시에나의 새 스타디움은 스타디움이 갖춰야 할 엄격한 요구사항들을 충족시키는 설계 작품을 완성하는 것 이상의 기회를 제공한다. 이 스타디움은 복합적이고 체계적인 구조를 바탕으로 지역사회의 거대한 레저공간으로 기능하면서 지역주민들의 일상을 변화시킬 것이다. 스타디움의 규모를 보면, 정기적인 축구경기나 단발성 행사를 유치하는 공간으로만 보기는 힘들다. 스타디움이 창출할 수 있는 가치는 그보다 훨씬 크기 때문에, 잠재적인 가치에 대한 인식과 함께 개발 노력이 필요하다. 제시된 구조물은 복합적이고 견고하게 구성된 스타디움으로 다양한 스포츠 활동과 이벤트를 수용할 것이다. 이 스타디움은 환경적인 조건, 건축 기술, 기반시설의 네트워크, 기본 설비 및 시설을 하나의 시스템으로 통합하여 시공·관리하며 이를 통해 매개변수에 따라 통제되는 정사각형의 종합 스타디움으로 완성된다.

Mille Plateaux / Stadio Complex

The new stadium for Siena offers opportunities far beyond the satisfaction of tight programmatic necessities and their competent design accomplishment. It involves a complex systemic challenge, available as a device to transform the every day activities of the locals by breeding large leisure potentials in the region. Because of its size it cannot be treated merely as a building receptacle for regular football events or for sporadic spectacles. Its effects are wider and problematic and need to be recognized and exploited. The proposed matrix of organization is understood as a complex and consistent field to host a changing multiplicity of sport activities and events, integrating environmental conditions, construction technologies, infrastructural networks, furniture and services in a single system of organization, construction and management, coordinated in a parametrically-controlled square-tiled stadium-complex.

패러-스케이프 : 도시, 조직 그리고 건축

김영
국민대학교 건축대학

COMPLEXITY IN CITIES

도시와 건축에서 컨텍스트의 복잡성에 어떻게 대응해야 하는가에 대한 문제의식과 그 결과적 실패는 근대건축을 인식론적·실천적 대응에 대한 가장 심각한 비판에 직면하게 만든다. 세기 후반에 제기된 이 문제에 대해 대체로 건축가는 양 극단에 해당하는 대안을 통해 문제점을 피해가고자 하였다. 그 하나가 일종의 도시기호학적 처방으로 컨텍스트를 읽어내는 데 익숙하고 오래된, 그리하여 약호화된 의미를 도시에 새로 인입된 물적 구조물을 덧씌우는 것(Contextualism, New Urbanism)이었다면 나머지 하나는 도시의 컨텍스트를 독해할 수 있다는 믿음에 타격을 가하는 일종의 미학적 선언을 양식화하는 것(Desonstruction)이었다. 양자의 대안이 도시건축을 구축하는 데 사회·미학적으로 일정한 자기 영역을 확보한 것과 무관하게, 건축적으로 보면 전자가 자기기만적 이미지의 보수주의적 봉합이었다면, 후자는 파편화된 이미지를 통한 아방가르드적 자기부정이었다.

근대 이후의 건축적 지형에서 벌어진 두 가지 해프닝을 실천적 대응의 오류로부터 비판의 출발점을 삼는 것은 또다른 오류의 양산을 예시할 뿐이다. 오히려 도시와 건축의 복잡성을 인지하고 이해하는 데 필요한 인식론적 기반의 문제점을 제기하는 것으로부터 출발해야 할 것이다. 도시와 건축은 단지 독해를 위한 주체와 분리된 객체로서 존재하는 것이 아니다. 각각의 조직체계가 총체적으로 분절되었기에 궁극적으로 읽어낼 수 없는 대상만은 아니기 때문이다.

우선 도시는 그 자체로서 자기완결적 조직이 아니라는 점이다. D. 하비, M 드 세르토 등의 지적과 같이 도시는 정치경학적 존재로서 그 유지와 발전을 위해서는 자원의 입출력이 가능한 열린 상태를 유지해야 하는 개방적 시스템이다. 또한 앞서와 같은 이유로 도시는 전체를 관장하는 통합조직의 규칙이 있을 수 없으며 오히려 도시 환경을 영위하는 각 개체가 맺는 관계와 조직의 구성방법이 전체 도시의 성격을 좌우하게 된다. 따라서 도시와 건축의 공간은 개체이기 이전에 관계의 집합일 뿐이다. 도시의 유동성과 비결정성은 이것으로부터 연유하는 것이다.

- URBAN CONTEXT
- COMPLEXITY
- OPEN SYSTEM
- LOCAL INTER-CONNECTIVITY
- HYBRID
- FIELD
- FACE+TIME
- PARAMETRIC
- ORGANISM
- EMERGENCE
- FORMAL/SPATIAL MATRIX

PARAscape:
City, Organization and Architecture

Young Kweon
School of Architecture, Kookmin University

Complexity in Cities

Regarding the questions of how to response the complexity of the context in architecture and urbanism, modern architecture faces the most crucial criticism in its epistemological commitment. The question, raised at the end of the 20th century, is avoided by two irreconcilable attitudes: One is an urban-semiotic remedy that to overlap the old and familiar meanings (so that they are even coded) to the newly inserted physical structure into city, in reading the context of city. Contextualism, New Urbanism. The other is that to stylize aesthetic manifestations attacking the belief in the possibility to read urban context. Regardless the fact, both standpoints assure their own territories in constructing urban architecture. Architecturally speaking, however the former is nothing but a conservative compromising of self-deceptive images, and the latter is no more than avant-guardian self-negation with fragmented images.

Any viewpoints based on the fallible responses of the two different commitments at post-modern architectural topography, can not initiate a valuable criticism but only to lead to another fallacies. It should rather begin with questioning the epistemological foundation which is necessary to recognize and understand the complexity of architecture and city. Because, city and architecture do not exist as comprehensible object isolated from its subject, nor are ultimately unreadable mystified object of which respective system is totally fragmented. First of all, city is not a self-sufficient system. As D. Harvey and M. de Certeau have highlighted, city is an economic-political entity. City is a system required to be open for the continuous flow of resources. For the same reason, it is also impossible that any rule exist to control overall synthetic organization in a city, rather the relationships and the organizations of each individual decide the characteristics of the city. The space of city and architecture is only the collection of relationships prior to an entity. The notions of flexibility and the indeterminacy of city space derive from these reasons.

Emergent Spatial Sensibility

Architecture, especially that in after modern era, is indebted to technical apparatus of

EMERGENT SPATIAL SENSIBILITY

다른 어떠한 문화적 실천보다도 건축, 특히 근대 이후의 건축은 기하학이라는 기술적 장치에 빚을 지고 있다. 계몽기 이후 데카르트적 유산으로부터 기원하는 기하학에 대한 전통은 투시도법의 발명과 바로크 공간의 탄생, 액소노메트릭을 통한 근대적 공간의 탄생 등 건축에서 공간적 발전 양상이 가능하게 하였다. 그러나 특정 시기 이후 그동안 유효했던 기하학이 특정한 공간적 감수성을 나타내기에는 부족하고 부적절하다는 것이 인지되기 시작한다. 그것은 대체로 근대적 인식론 비판의 등장과 더불어 근대건축 이후 양식의 고갈을 드러내는 시기와 때를 같이한다고 보인다. J. 키프니스는 이토 도요의 '유동적인 공간Liquid Space', 바흐람 시르델의 '심연Deep' 과 같이 명확하지는 않지만 새로운 공간적 감수성에 대해 구체적 언급이 다양한 방식으로 이루어진 시점에 주목한다. 이와 비슷한 시기 클레어 로빈슨, OCEAN, 아미 랜데스버그, 브루스 마우, 그레그 린, 세지마 가즈요, 샌포드 퀸터, 레이저+움에모토, FOA 등의 저작과 초기 작품들이 이와 같은 새로운 감수성의 등장을 예고하고 있다고 한 바 있다.

어떤 경우든 건축에서 진정 새로운 감수성이 발현하고 있는 것으로 여겨지며, 이는 1990년대를 전후하여 건축계에서 유형적 분류로 신미니멀리즘 혹은 신표현주의 정도로 이름 붙여졌다. 위에 언급한 건축가들은 기하학이 놓치고 있는 부분에 대해 다른 속성들을 적극적으로 도입함으로써 그들이 추구하는 새로운 감수성에 다가간다. 예를 들어 J. 키프니스와 B. 시르델의 작업에서 그들이 포착하고자 했던 공간은 물고기 떼처럼 거대하고, 복합적이며, 순간적으로 끊임없이 움직이면서도 서로 다른 형태를 만들어내는 것이었다. 물고기들의 그칠 줄 모르는 움직임이 만들어내는 특정 순간의 조직, 구성은 바로 그 순간에 서로 충돌하는 힘과 영향, 흐름들이 우발적으로 해소되는 과정의 순간 포착이었다. 이때 만들어진 형태는 선험적 주체에 의해 주어진 기하학적 '재현' 이 아니라, 조형의 주체가 부재한 채 형태를 구성하는 요소들의 상호관계와 그것이 작동하게 하는 규칙의 수준에서 '발현' 하는 것이다.

EMERGENCE : FIELD CONDITION

이머전스 사이언스Emergence Science는 I. 프리고진 등에 의해 주창된 열역학 이론의 하나인 복잡성 이론을 보다 다양한 분야에 적용하기 위해 파생된 새로운 학제적 연구 영역의 하나이다. 1990년대 중반을 기점으로 스탠 알렌, C. 젠크스,

geometry. The tradition of geometry has been originated from the Cartesianlegacy after Enlightenment, and has enabled various spatial developments in architecture; such as the invention of perspective, Baroque spaces, and modern spaces by axonometric. It started to be perceived, however, that geometry is inappropriate and insufficient in terms of representing spatial sensibility, despite of its obsolete validity in specific eras. It seems to coincide with the exhaustion of styles after modernism in architecture, following the criticism on modern epistemology. J. Kipnis has noticed the vague instant when new sensitivity of spaces has been expressed by Toyo Ito with "liquid space" and by Bahram Shirdel with "the deep". Claire Robinson, OCEAN, Amy Landesberg, Bruce Mau, Gregg Lynn, Kazuyo Sejima, Sanford Kwinter, Reiser+Umemoto and FOA are the architects illustrating the emergence of the new sensibility of space. It is conceived that totally new sensitivity has been manifested and merely categorized as "New Minimalism" or "Neo-Expressionism" reflecting the typology in contemporary architecture. Introducing what have been deficient by geometry, above mentioned architects have approached new sensible architecture. For example, in the work by J. Kipnisand B. Shirdel, what they were trying to apprehend is a school of fish creating various forms by massive, complex and ceaseless moving. The structure and organization in a specific moment, created by continuous movement of fish is the representation of forces colliding in that moment and contingent resolution of movements. The forms are not geometrical representation by a priori subject, but the emergence of interrelationship of the elements in the level of rules, absenting the subject of form-making.

EMERGENCE : FIELD CONDITION

Emergence Science, suggested by I. Prigogine, is a realm of inter-disciplinary research to adopt the Complexity theory in Thermodynamics into other fields. Architectural critics, such as S. Allen, C. Jencks, M. Weinstock and J. Kipnis have shaped meaningful level of architectural discourse around mid-90's. Emergence -Field condition in Stan Allen's term- is a formal, spatial matrix with the ability to integrate various elements rather than individual element itself or its property. It is a loosely organized set by local interconnectivity, therefore, the rules connecting parts are more important and overall appearances and forms are flexible. Emergence is a phenomenon of upward, that is, it is not defined by any dominant geometrical rules, it is rather determined by complicated local relationships. The notion of "form" in this context is meant to be the form between objects rather than that of objects. In the same time, Emergence is also an independent and objective process not controlled

M. 와인스톡, J. 키프니스 등에 의해 건축적으로 유의미한 수준의 담론이 형성되기 시작하였다. 이머전스(스탠 알렌이 말하는 Field Condition)는 일반적으로 개별 요소 그 자체의 속성이라기보다는 그 다양한 요소들을 통합하는 능력을 지닌 형태적·공간적 매트릭스이다. 이것은 상호연관성에 의해 규정되는 느슨하게 조직된 집합이며, 따라서 각 부분을 연결하는 규칙이 가장 중요하고 전체적 형태와 모습은 매우 유동적이다. 이머전스는 보텀업Bottom-up 현상이다. 즉 지배적·기하학적 규칙에 의해 정의되지 않고 오히려 복잡하게 얽힌 지엽적 관계에 의해 결정된다. 이때 형태는 '사물의 형태'라기보다는 '사물들간의 형태'라 할 수 있다. 동시에 이머전스는 인간의 주관적 사유와 행위에 통제되지 않는 독립적이면서 객관적인 과정이다. 따라서 '통제 불가능성'은 그 자체로 이머전스의 속성을 특징 짓는 단어가 된다.

PARAscape Architecture on In-between Threshold

복잡한 조직체로서의 도시, 기하학에의 새로운 감수성, 사건과 사물의 과정적 관계에 대한 새로운 시선으로서 이머전스는 이전과는 다른 건축의 종 출현을 예시한다. 기존 건축이 결정론적 시공간에 고착된 존재였다면, 새로운 건축에서는 열린 시간의 차원이 결합된 통시적 존재로 등장한다. 새로운 건축은 형태 생성의 결과가 아니라 과정이며 열린 시공간적 조건 하에서 끊임없이 변모하며 적응한다. 그 내부 요소들은 개체의 독립적 성질보다는 서로간의 관계에 의해 규정된다. 계량화된 매개변수의 작용과 반작용이 지배하는 기하학의 세계는 시간과 공간을 동시에 포괄하며, 인간의 주체적 작용을 배제한다. 산재된 혼종성이 만들어내는 정보공간은 따라서 생물학적 과정을 비의도적인 방식으로 모방한다.

건축은 현대 자본주의가 만들어낸 시공간적·사회경제적 복잡성의 경계로부터 출현한다. 그 경계로부터 탈출이 아니라 그 복잡성에의 긍정과 진입을 통해 건축 형식의 내파를 꾀한다. 더불어 새로운 건축의 종이 전개될 경계없는 지구 환경 하에서 조직체로서 도시와 환경의 지리적 구분은 무의미해져가고 있다. 문화의 경계는 지리적 거리로부터 발생한다기보다는 타자적 시선의 관계 속에서 발현하기 때문이다. 일곱 작가의 작품을 통해 느슨해진 문화적 경계 아래 페러스케이프 건축의 실천 전략을 감상하자.

by human subjective thought and activity. "Uncontrollability" is, therefore, the term characterizing the property of Emergence.

PARAscape Architecture on In-between Threshold

Emergence is a new perspective towards the city of complex organization, new sensitivity in geometry and the progressive relationships in events and objects, proclaiming new specie in architecture. While traditional architecture has been fixed by determinist's space-time, new architecture exists as a diachronic being combined with open time dimension. New architecture is not a result of form-making, but a process. It adopts itself under the condition of open space-time by constant transformation. Internal elements are defined by the interconnectivity of individual elements, not by their own property. The sphere of geometry is governed by the action and the response of quantified parameters, and excludes human subjectivity while including space and time simultaneously. The information space creates the distributed hibridity to imitate biological organism without intention.

Modern Architecture	PARA-scape Architecture
Space	Space + Time
Materiality	Softeriality
Geometric	Parametric
Mechanism	Organism
Hierarchy	Distribution
Identity	Interconnectivity
Humanity	Cybernetics
Pure	Hybrid
Minimal	Optimal
Object	Field
Mass production	Mass customization
Transparency	TransPRESENCE
Form	inForm
Resistance	Response
Zeitgeist	Datageist

PARAscape Architecture emerges from the threshold of the space-time, socio-economic complexity produced by contemporary capitalism. It intends the implosion of architecture not by escape from the threshold, but by affirmation and penetration into the threshold. It is, furthermore, getting meaningless to divide city environment geographically where new architectural species are being developed. Cultural threshold transpires from the relationship of the other's views rather than geographical distances. Examining the works of seven architects, enables us to enjoy the strategic commitments of PARAscape architecture under loosened cultural threshold.

Jooryung Kim
Urbano Architectural Research Group, SAIA+, Korea

Namjune Paik Museum
Competition Project_Participation

Suwon, Gyeonggi-do, Korea

2003

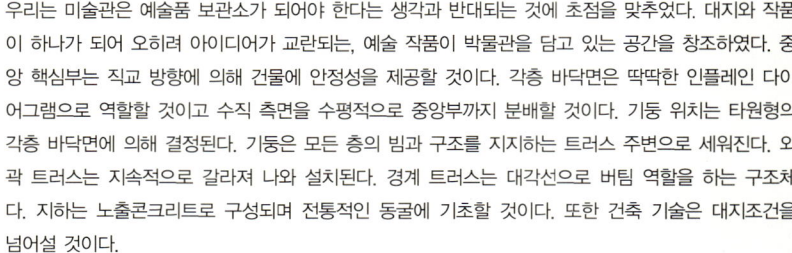

우리는 미술관은 예술품 보관소가 되어야 한다는 생각과 반대되는 것에 초점을 맞추었다. 대지와 작품이 하나가 되어 오히려 아이디어가 교란되는, 예술 작품이 박물관을 담고 있는 공간을 창조하였다. 중앙 핵심부는 직교 방향에 의해 건물에 안정성을 제공할 것이다. 각층 바닥면은 딱딱한 인플레인 다이어그램으로 역할할 것이고 수직 측면을 수평적으로 중앙부까지 분배할 것이다. 기둥 위치는 타원형의 각층 바닥면에 의해 결정된다. 기둥은 모든 층의 빔과 구조를 지지하는 트러스 주변으로 세워진다. 외곽 트러스는 지속적으로 갈라져 나와 설치된다. 경계 트러스는 대각선으로 버팀 역할을 하는 구조체다. 지하는 노출콘크리트로 구성되며 전통적인 동굴에 기초할 것이다. 또한 건축 기술은 대지조건을 넘어설 것이다.

The Nam June Paik Museum is to serve as a repository for the Artist's creative oeuvre, and a home for his future artistic activities.

Building vs. Artwork: Our strategy focused on reversing the notion of a museum being a container for art, but rather the art contains the museum, disturbing the isomorphic idea between the ground and the figure.

Stability System: The central core will provide all stability to the building in both orthogonal directions. The floor plate will act as stiff in-plane diaphragms and distribute lateral loads horizontally to the core

Gravity System: Column locations have been determined with respect to each ellipsoid floor plate. Columns are located around the perimeter truss which supports all floor plate beams and structure. Perimeter truss is located a constant offset from perimeter. The perimeter truss is a diagonally braced structure.

Relation

Contain

Landscape

Gravity

Void

Boundary

Geometry

Isomorph

Stability

3D Visual Morphology

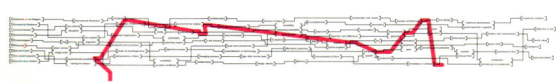

백남준미술관 김주령
현상설계 참가작 도시건축연구소 SaiA+

미술관 내부의 동적 속성 DYNAMIC FLOW OF THE MUSEUM SPACE

JOORYUNG KIM
URBANO ARCHITECTURAL RESEARCH GROUP, SAIA+, KOREA
EYES OF PORDENONE
EUROPAN 6 / PORDENONE, ITALY
2002

도시 외곽을 마주보고 있는 끊어진 연결고리 세 곳을 웨이브 패턴을 활용하여 다시 연결시키고자 한다. 빽빽한 주거 지역인 서부와 도시 중심부, 팽창하고 있는 소수민족 주거 지역인 동부와 공공장소 그리고 상업 지구인 북부와 더 멀리 있는 농촌 지역이 그 세 곳이다.

한 층에서 또 다른 층으로 이어지는 프로그램에 따른 밀집 주택의 연속적인 단면을 제안한다. 풍경 안에서 단면적인 연속성은 방해하는 것이 아니라 기대되는 확장과 분리의 흐름을 소생시킨다. 프로젝트 주변은 미래 확장을 위한 도시의 주된 공원광장이 되고, 이는 저소득층 주거 지역의 무책임한 지형적 점령이 아닌 높은 수준의 공동체를 위한 투자 기회를 제공한다.

오목한 지역에서 우리는 거주지와 건강관리시설 프로그램을 충족시켰다. 그 프로그램들은 막대한 개인 안뜰을 가질 수 있다. 남쪽으로부터 언덕과 길 아래 알맞게 포개진 밀집된 프로그램들의 완전한 영역을 볼 수 있다.

CONCEPTION Converting urban void into landscape complex. By exploiting the wave pattern, we try to re-connect currently missing network forces in between three faces of the edge city: namely, west side of the dense housing block and city centre, east side of growing multi ethnic housing force, and public services, and the north side of commercial bands and further on rural landscape.

APPROACH Amalgamation of the urban edge through sectional exploitation. We suggest a sectional continuity in programmatic density of housing in one level and on another level, sectional continuity in landscape of scenery in order to revitalize and not to disturb the flow of the anticipated expansion and segregation. So that the periphery of the project becomes the prime urban park square for the expanding city in future and thereby provide a good opportunity to invest on high quality neighborhood rather than occupational. topography of derelict low income housing.

OCCUPATION In concave area we implemented the required program of dwellings and healthcare facilities so that those programs can get the vast private courtyard gardens in front of them. From the south side you can see the full scope of the dense urban programs nested underneath of the mounds and pathways.

WAVE PATTERN

NETWORK

AMALGAMATION

SECTIONAL EXPLOITATION

SECTIONAL CONTINUITY

PROGRAMMATIC DENSITY

FLOW

OCCUPATION

PROGRAM

URBAN PROGRAM

4D VISUAL MORPHOLOGY

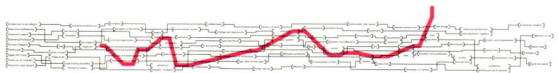

포르데노네의 눈 김주령
도시건축연구소 SaiA+

밀집된 도시주거 지역을 연결하는 연속 패턴 WAVE PATTERN NETWORK RE-CONNECTING DENSE URBAN HOUSING BLOCK

Shaun Murray
U.K.
Disturbing Territories_Camargue Ecology
Rhône Valley, Southeast France

론 계곡과 인접하는 강은 작가에게 끊임없는 정신적 방황의 대상으로 당신이 읽고 있는 이 글과 모델, 그림의 원인을 제공했다. 그중 하나는 조르주 뒤 피에에서 부유하는 오브젝트와 관련있으며, 매년 9월 한 주 동안 발생하는 홍수기에 그 활동이 두드러진다. 론 강 유역에 위치한 네 개 장소에서 물, 대지, 불, 바람의 네 요소에 따른 건축을 통해 사용자는 환경을 읽을 수 있다.

이 프로젝트는 피드백 루프와 리트로센싱 장치의 복잡한 웹 안에서 상호연결된 일련의 '분리된' 사이트와 관련있다. 남프랑스 론 강의 독특한 환경조건 안에서 자연현상, 복잡한 생태네트워크를 이용하는 데 초점을 맞춘다. 케이지드 라이트와 magnetorheorological compound 등 현대기술을 사용해서 이 자연적 유동 시스템을 확대하고 왜곡하여, 자체 조절하는 쌍방향 구성을 형성하고, 응용과 변형의 무한한 가능성을 제시한다. 연구 목적인 생태 정보 수집 및 대조뿐 아니라, 이 네트워크는 사용자가 독자로서 '자연에 빠져들' 수 있는 방법을 제공한다. 다양한 중요한 장소에서 사용자는 네트워크의 또 다른 구성요소가 되고, 환경에 영향을 주는 동시에 다양한 요소에 영향받는다.

이러한 건축은 홍수를 유발하는 골짜기들, 염전 또는 프랑스에서 가장 얇은 지각판 등 지형과 생태적으로 민감한 특별한 장소에 걸쳐 위치한다. 모두 카마르그(론 강 하구 삼각주 본류와 오른쪽 지류인 프티론 강 사이 위치)에서 끝나고, 미스트랄 북서풍의 피해를 자주 입는 론 강 유역에 위치한다. 이러한 구조 안에서 복잡한 메카니즘이 초미학적 결합, 증대되고 왜곡된 지각, 관광객의 개인적 대화를 위한 immersive intensity orator를 형성한다. 이들 장소를 통합하는 것은

FEEDBACK LOOPS
RETROSENSING DEVICES
SELF-REGULATING INTERACTIVE CONFIGURATIONS
PLUGGED INTO NATURE
COMPONENT OF THE NETWORK
HYPERAESTHETIC
BETWEEN THE NATURAL AND THE ARTIFICIAL
LABYRINTH
INTERVENTION
CONDENSATION
BLURRING THE THRESHOLD
FLUID WHOLE
MACHINIC PHYLUM
DIALOGUE WITH THE ENVIRONMENT

The Rhone Valley and adjacent river is for me the subject of certain chronic mental wanderings, clinical enough in nature to produce these drawings, models and explanatory text you are now reading. One of them concerns a vacillating object floating in the Gorge du Fier and becomes animated during annual flood conditions one week in September. There are four locations stretched down the Rhône, each of the locations enable the user to read the environment through the architectures which are in accordance to the four elemental forces of Water, Earth, Fire and Wind.

This project reverberates around a series of "split" sites linked to each other within a complex web of feedback loops and retrosensing devices. It centres on the harnessing of natural phenomena, complex ecological networks within the unique environmental conditions of the Rhone valley in southern France. By making use of modern technologies, including caged light and magnetorheorological compounds, these naturally fluid systems will be amplified and distorted to form self-regulating, interactive configurations, with endless possibilities for adaptation and transformation. As well as collecting and collating ecological information for research uses, this network provides the means for the user as reader, to be "plugged into" nature. At various critical locations the user becomes yet another

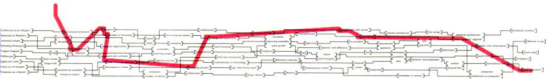

Shaun Murray

영역 교란_카마르그 생태학

당장의 위치를 넘어서서 민감한 풍경에 광범위하게 영향을 끼치며, 자연과 인공 사이 경계를 모호하게 한다.

이 프로젝트에서 논의할 주요 초점은 우선 물질성을 통해 각 노드를 도입하고, 각 다섯 장소에서 그 자연적 계기와의 관계를 도입하는 것이다. 다음으로 드로잉 환경에서 사용자, 즉 독자와 공간의 관계를 논의하고, 자연과 인공 사이의 상태를 변경해서 어떠한 영역이 교란되는지 설명한다. 각 노드별로 설명하고, 다시 한 노드가 어떻게 다른 노드에 건축적으로 또한 공간적으로 영향을 끼치는지 종합적으로 설명하고자 한다.

물 / 조르주 뒤 피에 / 프랑스

길이 300m, 수심 60m의 레 조르주 뒤 피에는 알프스산맥에서 가장 흥미로운 요소 중 하나이다. 그 협류 중에 거대한 물웅덩이가 있는데, 입구에 르 피에 강이 침식하는 미로, 라 메르 데 로세르가 인상적이다. 과거 홍수 발생시 기록된 강 수위를 보여주는 눈금표가 있는데, 1960년 9월에는 평상시 수위보다 27m나 높게 올라간 바 있다.

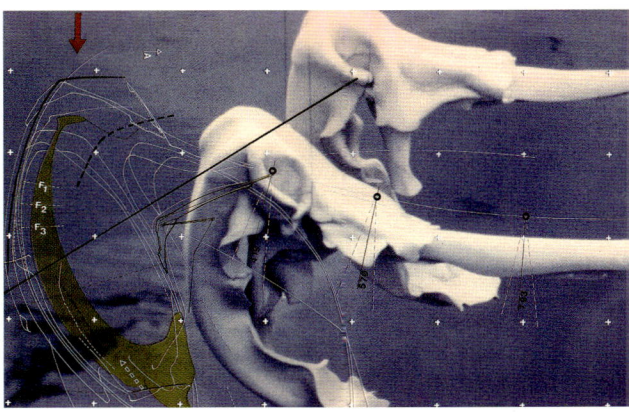

component of the network, influencing the environment; and simultaneously being acted on by diverse factors. These architectures are distributed across sensitive geomorphologically and ecologically special locations, such as gorges that flash flood, salt pans or the point of the France's thinnest Earth's crust, all of which are along the Rhone corridor that terminates at the Camargue and is prone to the ravages of the mistral wind. Within these frameworks intricate mechanisms will form hyperaesthetic connections, augmented, distorted sensations and an immersive intensity orator for a personal dialogue with the modern tourist. The synthesis between the locations will have an extensive impact on this sensitive landscape beyond their immediate location, disturbing the threshold between the natural and the artificial.

The key points to be discussed in the following project are firstly to introduce each node through materiality and the relationship with its natural trigger in each of the five locations. Following that I will discuss the relationship between user - reader - space in the drawing environment and describe what territories are disturbed by altering the states between the natural and the artificial. This will be explained through each node individually and then collectively how one node will affect the other tectonically and spatially.

Water / Gorges du Fier / France

Les Gorges du Fier, at 300 meters long and 60 meters deep, is classified among the largest curiosities of the Alps. Inside these throats one can admire a giant pot of water. To the exit is the impressive aspect of la mer des rochers, a labyrinth eroded

대지 / 크루아 / 프랑스

대성당 주변에 모인 오래된 석조 건물들은 강 유역 커다란 시멘트공장에서 나오는 하얀 먼지로 덮여 있다. 원자력발전소의 대형 탑에서 멀지 않은 곳으로 커다란 연기 기둥이 솟아오른다. 크루아 근처의 대지는 프랑스에서 지각이 가장 얇은 곳으로 앞으로 심각해질 수 있는 지각 움직임이 종종 관찰되는 곳이다.

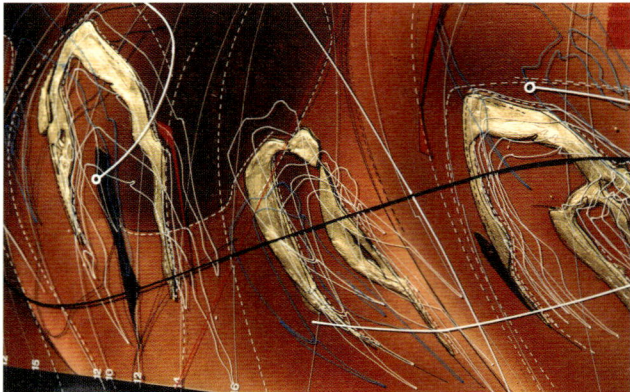

불 / 쇼베 동굴 / 프랑스

1994년 12월 24일, 세 명의 동굴학자가 20,000년 동안 인간이 들어가본 적 없는 것으로 추정되는 동굴을 발견했다. 이 쇼베 동굴에는 사자, 곰, 무소 등의 동물을 그린 구석기 시대의 벽화가 있다. 탄소 측정에 따르면 동굴 벽화 중 일부는 30,000년도 전에 그려진 것으로 추정된다. 이는 기존에 알려진 가장 오래된 동굴 벽화보다도 3,000년이나 앞선 것이다. 아르드슈 지역의 쇼베 동굴은 방대한 갤러리(500m)와 실의 네트워크로 이루어져 있다.

by the river le Fier. Along the vertical faces is a scale showing the heights reached by the river in flood, including September 1960, when it rose 27m above its normal level.

Earth / Cruas / France

Old stone houses huddled around an abbey, all covered with a fine white dust from the large cement works by the river. Not far away the great towers of the nuclear power station pour forth huge plumes of smoke. The land around Cruas is the thinnest part of the Earth's crust in France, and is subject to frequent movements, which might or might not become more serious.

Fire / Chauvet Cave / France

On 24th December 1994, three speleologists discovered a cavern they believed hadn't been entered by humans in around 20,000 years. The Chauvet cave, as it is called, contains Paleolithic paintings of lions, bears, rhinos and many other animals. Carbon dating has shown that some of the paintings in the Chauvet cave are over 30,000 years old. That makes them 3,000 years older than the oldest cave paintings previously known. The physical intervention is sited in the Chauvet cave in the Ardeche region, where there is a vast network of galleries and rooms (about 500m galleries).

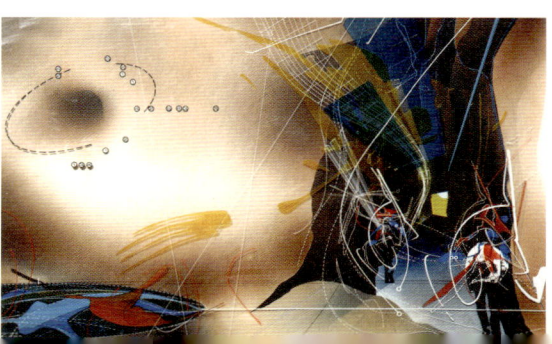

공기 / 레 보 드 프로방스 / 프랑스

고립되고 돌출된 암석 위로 알필(산맥)이 튀어나온다. 양쪽이 가파른 계곡인 레 보 드 프로방스는 극적이고 아름답고 불길하고 우울한 그림자를 이룬다. 여전히 복잡하게 얽힌 폐허로, 암석에서 자라난 것 같은 성 유적이 그 정점을 이룬다. 복수를 부르짖는 반시(아일랜드와 스코틀랜드 전설의 여자 요정)처럼 무너진 성벽 사이로 미스트랄 바람이 윙윙거린다.

AIR / LES BAUX DE PROVENCE / FRANCE

On top of an isolated spur of rock, thrust out from Les Alpilles, and with steep ravines on either side, Les Baux de Provence is dramatic, beautiful, a shade sinister and melancholy. It is still tangled ruins, crowned by the wreck of a castle that seems to grow from the rock itself. The Mistral wind howls through the broken ramparts like banshees calling for vengeance.

압축 / 카마르그 / 프랑스

이 노드는 다른 노드들의 데이터를 대조하고 압축해서, 생태 연구 프로그램의 허브로 작용한다. 론 계곡 주변에서 발생하는 지각, 움직임, 감정, 다양한 노드를 통해 수집된 데이터가 이 주요 매개를 최종적으로 통과하게 된다. 각 노드 사이에는 시각적이고 동적인 일련의 상호작용이 존재하며, 자연과 인공 환경 사이의 경계를 모호하게 한다. 이 간결한 리스트에서는 물(N1), 대지(N2), 불(N3), 공기(N4) 노드 내 네트워크의 가능성을 점쳐본다. N1 갤러리의 난간은 N2에서 발생하는 지진 활동에 대한 반응으로 진동한다.
N1에서 발생한 텍스처 필드는 N2 전시공간의 바닥구조 일부에 반영된다. N4 리본에서 발생한 풍압은 N2 구조에 맞춰진 일련의 미니어처 풍력 터빈으로 재생산되어, 전시공간에서 풍동風洞을 만든다.

CONDENSATION / CAMARGUE / FRANCE

This node collates and condenses the data from the other nodes to become the hub of the ecological research programme. The sensations, movements and emotions generated around the Rhone valley, and the data collected in the various nodes: These are ultimately channeled through this primary vessel.

There are a series of interactions, visual and kinetic between each of the nodes, blurring the threshold between the natural and artificial. This brief list views the potential of a distributed network within Water (N1), Earth (N2), Fire (N3), Air (N4) nodes.
The handrail on the gallery on N1 vibrates as a response to any seismic activity occurring on N2.
The generated texture field on N1 is animated within parts of the floor structure in the exhibition space on N2.
The wind pressure generated on the ribbons on N4 is regenerated with a series of miniature wind turbines fitted into the structure of N2, creating wind tunnels in the exhibition space.
Some of the sensors within the cave on N3 pick up acoustics, which are played real-time in the gallery on N1.

N3 동굴 내 일부 센서에서 음향을 탐지하고, 이를 N1 갤러리에서 실시간으로 들려준다.

N4의 센서 일부가 울부짖는 바람 소리를 탐지해서 N3 동굴에서 실시간으로 들려준다.

N2의 센서 일부가 대화를 탐지, N1과 N3에서 실시간으로 들려준다.

N3의 라이트 세일light sail에서 흡수된 빛의 양이 N1 선가cradle의 실리콘 광채의 강도에 영향을 끼친다.

N2의 센서가 바람의 강도를 탐지하고 이 정보는 N1에 디지털 방식으로 전달되어 바람의 흐름을 컴퓨터 조작으로 시각적 이미지로 전환하고 N1에서 포노스코프로 표현한다.

N4와 N2의 전경을 녹화한 비디오가 N1에 보내져서, 진동 오브젝트 내 디지털 인터페이스를 통해 실시간으로 보여진다.

N4에 전개된 센서가 풍압을 탐지하고 이를 N1에 디지털 방식으로 전달해서, 이 정보가 진동 오브젝트 등 받침에서 스마트 젤 내 일련의 움직임으로 표현된다. 이러한 상호작용은 기존 수압과 간헐적으로 발생하게 된다.

N4와 N2의 전경을 녹화한 비디오가 N3에 보내져서, 동굴 벽에서 실시간으로 보여진다.

Some sensors on N4 pick up the howling of the wind, which is played real-time in the cave on N3.

Some of the sensors on N2 pick up conversations, which are played real-time on N1 and N3.

The amount of light absorbed in the light sails on N3 effects the intensity of the silicon glow in cradle on N1.

Sensors on N4 pick up the changing intensity of wind flows, this information is digitally sent to N1 and the wind flows are translated into a visual image through a computer manipulation programme and presented down the phonoscope on N1.

Video recordings of the panoramic views on N4 and N2 are sent to N1, to be played real-time through the digital interface within the vacillating object.

Sensors spread out on N4 detect wind pressure, which is sent digitally to N1, where the information is translated into a series of movements within a smart gel in the back rest of the vacillating object. The interaction will occur intermittently with the existing pressure of the water.

영역 교란_카마르그 생태학   SHAUN MURRAY

자연은 자체 변동 시스템을 발전시키고, 이 안에서 다양한 생태 현상이 결합해서 완전히 새로운 유체를 발생시킨다. 작가가 달성하고자 했던 바는 마누엘 드 란다가 표현했던 대로 기계적 퓔룸 machinic phylum으로서의 건축이다. 이에 따라 자연은 현대기술의 문화적 가능성의 일부가 된다. 건축은 환경에 반응하고 대화하는 법을 배우게 된다. 이러한 이해를 통해 사용자는 노드, 그 핵심 요점, 개입의 중심에서 시스템에 몰입할 수 있게 된다. 하지만 동시에 시스템이 갑작스럽게 중대한 경계에 도달할 때 충격적인 대조를 경험할 수 있다. 이러한 환경적 힘의 에너지는 건축 축조 및 공간 변동의 과격하고 흥분시키는 교란에서 그 모습을 드러낸다.

이 프로젝트에서 논의된 노드들은 피드백 루프와 릐트로센싱 장치의 복잡한 웹 안에서 상호 긴밀하게 연결되어 있고 매우 구체적이다. 그 환경 및 인공적 교란은 각 노드간 영역을 형성한다. 최종적으로 이들은 혼합된 배아기의 촉매로, 사용자는 이를 통해 공간적인 장소와 목적지에서 강렬한 대안공간적 경험을 얻을 수 있다. 현실에 존재하는 환경이 아니며, 기억된 오브젝트도 아니며 세 번째로 상상된 것일 뿐이다. 건축은 더 이상 인식의 대상이 아닌 커뮤니케이션 시스템이 되어 제 목소리를 내게 된다.

Video recordings of the panoramic views on N4 and N2 are sent to N3, to be played real-time on to parts of the cave walls.

Nature evolves its own fluctuating systems, in which diverse ecological phenomena combine to create a fluid whole. What I have aimed to achieve is to develop an architecture as a machinic phylum, as expressed by Manuel De Landa. Thus, the natural becomes apart of the cultural potential of modern technologies. The architecture learns to respond and conduct a dialogue with the environment. This understanding enables the user to be plugged into the system at the nodes, the focal point, and the hub of the intervention. Elemental forces caress and tease the user with gently fluctuating movements. Yet, a shocking contrast is experienced when the system suddenly reaches critical threshold. The potential energy of these environmental forces manifests itself in violent and exhilarating disturbances in the tectonic and spatial fluctuations of the architecture.

The nodes discussed in this project are highly specified and intimately linked to each other in a complex web of feedback loops and retrosensing devices. The environmental and artificial disturbances create windows between each node. Ultimately, they are hybridized embryonic catalysts, whereby the user gains an intense alternative spatial experience within special locations and destination. It is not the real environment; not even with the object remembered but third imagined. It gives voice to architecture because architecture is no longer an object of perception but rather a system of communications.

Shaun Murray
J.K.
Disturbing Territories_Archulus Ecology
Aldeburgh Coastline, Suffolk, England

아쿨루스는 올드버러 지역 강이 바다와 만나는 곳에 새로운 풍경을 가정한다. 강은 작은 곶으로 바다와 분리되는데 곶에는 부식이 빠르게 발생하고 있다. 이 곳은 곧 사라질 것이고 해안은 범람하게 될 것이다. 이 제안은 파열의 순간과 위치에서 금방이라도 활동을 시작할, 일련의 거의 구부러지고 가득 찬 단편들을 만들자는 것이다. 그 결과 동식물을 위한 새로운 땅을 손에 넣을 수 있을 것이다. 범람 방파제는 공간을 장식하고 엮는 마법의 베틀로 서로 연결된 풍경에서 공명할 것이다. 사용자, 독자 공간의 과도기적 영역을 통해 아쿨루스는 자신의 관심사를 명시적으로 주입하고 고취한다. 파도, 조수의 불균형, 해류의 움직임은 끊임없이 변하는 아쿨루스의 종단면도를 결정짓는다. 곶의 좁은 목 부분에 생기는 강과 바다의 수심 차이가 힘의 원천으로 작용한다.

올드버러의 해안은 자연적 프로세스와 인공 프로세스의 역동적 행동이 만나는, 미묘하게 균형 잡힌 인터페이스이다. 이러한 프로세스는 단기 변동에서 수천 년에 걸친 장기 변화까지 여러 시간대에 걸쳐 작용한다. 아쿨루스는 서포크 동쪽 해안에서 반복되어온 범람의 결과로, 지난 400년 동안 매년 1m 이상씩 침식되었다.

진짜 세계의 모습을 가상 세계에 담아내는 수많은 기술을 사용하고, 진짜 세계의 풍경(온도, 습도, 염도 등)과 가상 세계의 데이터와 프로세스에서 유래된 값의 집합을 사용해서 환경과 드로잉 간의 모델을 구성했다. 이 모델에 시간 기반 데이터를 입력하고, 이 데이터를 통해 형식이 활기를 띠게 되고 진동하는 건축이 가능하다.

NATURAL AND MAN-MADE PROCESSES

BALANCED INTERFACE

RANGE OF TIME SCALES

ENVIRONMENT AND THE DRAWING

LANDSCAPE

DATA AND PROCESSES

REAL WORLD

VIRTUAL WORLD

FED TIME-BASED DATA

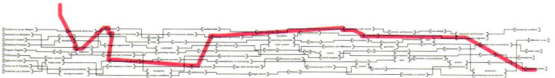
자연풍경과 인공세계의 중첩을 보여주는 이미지
SUPERIMPOSITION OF NATURAL LANDSCAPE AND MAN-MADE WORLD

영역 교란_아쿨루스 생태학   SHAUN MURRAY

현실세계의 데이터와 연동하는 가상공간의 모델링
MODELING OF CYBERSPACE BASED ON THE DATA FROM REAL WORLD

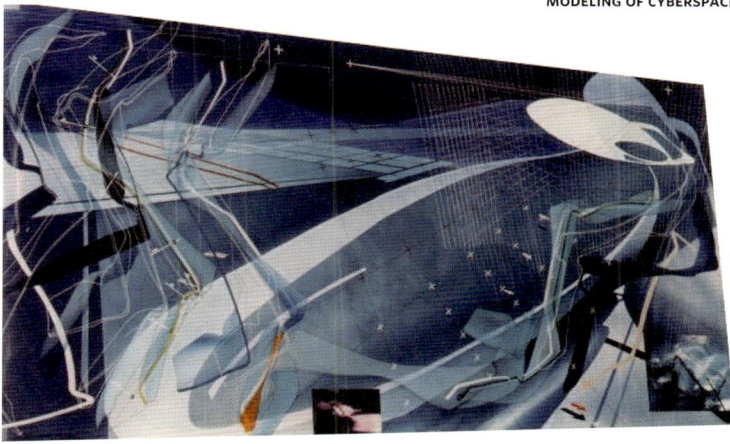

Archulus posits a new landscape at the point where the local river meets the sea at Aldeburgh. The river is separated from the sea by a small spit of land which is prone to the rapid erosion; one day soon the spit will suddenly disappear and the coastline will flood. The proposition creates a series of almost cocked and loaded pieces that suddenly at the moment and point of breach explode into action. This landscape creates a new surface for plant and animal colonisation. The flood shields resonate in the territory of harpooning landscapes linked to each other through an enchanted tectonic loom embroidering and weaving spaces. Through the transitional territories of user - reader - space, Archulus explicitly injects and infuses its own agendas. The movement of waves, tidal imbalances and currents shape the ever changing Archulus profile. The relationship at the neck of the spit, between the water depth in the river and the sea will act as the source of power.

    The coastline of Aldeburgh is a delicately balanced interface with the dynamic behavior of natural and man-made processes. These processes operate over a range of time scales, from short-term fluctuations to long term changes over thousands of years. Archulus was the consequence of recurring floods on the east coast of Suffolk, which loses more than a metre a year for the last 400 years.

    A model is constructed between the environment and the drawing by using a set of derived values from the landscape (temperature, humidity. salinity levels, etc) of the real world and from data and processes of the virtual world, also from numerous techniques of capturing the real and casting it into the virtual. The model is fed time-based data through which the form becomes animate, the architecture vacillating.

Haewon Shin + Liu Yuyang
Chichi Earthquake Memorial
Competition Project_2nd Prize
Taiwan
2004

빈 땅, 남아 있는 벽, 숨겨진 단층선, 소생하는 나무와 같이 눈에 보이고 또한 보이지 않는 땅에 남겨진 흔적에서 영감을 받은 프로젝트다. 이것은 땅, 캐노피, 캠퍼스라는 세 가지 상호의존적 요소로 구성된다.

남아 있는 벽과 땅은 음악을 위한 새로운 곳으로 바뀔 것이다. 이것은 음악을 통해 치료를 위한 피난처가 되는 기념물의 보이는 지형으로 만들어진다. 한스 샤론의 베를린필하모닉 콘서트홀에서 영감을 받아, 대지에 자연스럽게 흐르는 듯한 필하모닉 클러스트를 만들었다. 숨겨진 단층은 가장 위험한 상황을 대지에 부여한다. 지진 때문에 주요 구조물이 알아볼 수 없을 정도로 파괴되었지만, 나무는 땅의 냉정한 폭력을 이겨내고 가장 소생을 잘하는 존재임을 보여주었다. 캐노피 시스템은 인간의 근본적인 거주 욕망을 상기시킨다. 햇빛, 비와 같은 특히 자연재해를 대비하고 기본 안식처를 제공하는 지붕이 필요함을 상기시킨다.

이 마스터플랜은 새로운 정체성을 효과적으로 탄생시킬 수 있는 '캠퍼스'라는 개념을 제안한다. 이 지역이 기념비적인 땅으로 정착되는 동안, '캠퍼스'는 도시에너지를 이 지역으로 끌어들일 수 있도록 차오툰의 경제적인 조직과 Chunghsin Hsintun의 주거조직과 연결될 것이다. 이 방법을 통해 이 지역을 타이완의 새로운 문화 메카로 부활시킬 수 있다고 확신한다. 주변 계획지역은 연관 프로그램으로 개발된다. 음악보존소, 콘서트홀, 체험극장, 음악도서관, 레코딩스튜디오 그리고 기숙사, 게스트하우스, 외국에서 온 음악 애호가와 방문객을 위한 고급 호텔을 포함하는 주거 지역 등이 프로그램에 포함된다.

LANDSCAPE
CANOPY
GROUND
TOPOGRAPHY
RESUSCITATION
INTERCONNECT
EVENT
SHELF
CODE
VIRTUALITY

캐노피 시스템 THE CANOPY OF THE TREE

치치 대지진 추모공원
현상설계 2등작

신혜원 + 유양 리우

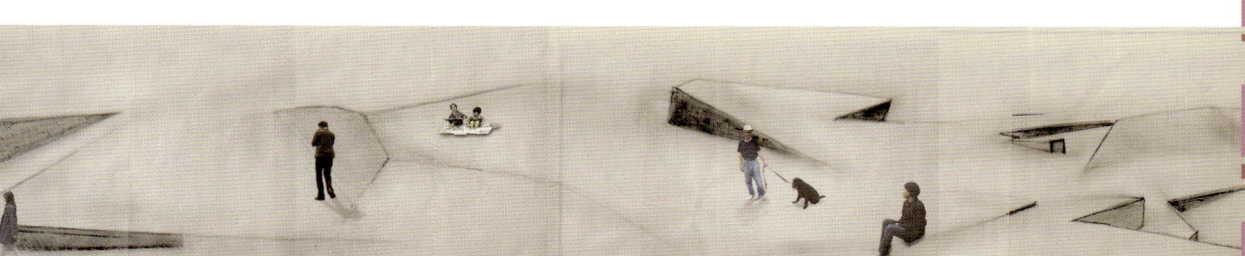

기념물로서 지형 MEMORIAL TOPOGRAPHY _ INTERIOR PERSPECTIVE

The project draws inspiration from both the visible and invisible remains of the site: empty plateaus, retaining walls, hidden fault line, resilient trees. The project consists of three inter-dependent elements: the ground, the canopy, and the campus.

The ground The retaining walls and empty plateaus are transformed to provide a new ground for music. It is through music-the invisible landscape, that the visible topography of the memorial becomes a refuge for healing. Inspired by Hans Scharoun's Berlin Philharmonic Concert Hall, the result is a choreographed cluster of philharmonic plateaus. The hidden fault line presents the most dangerous condition on site. It is also, at the same time, the most didactic representation of the hidden force of Nature. The fact that the Chelongpu Fault Line runs through the middle of the site suggests a new path of pilgrimage to emerge directly over, and on axis with, the fault line.

The Canopy After the earthquake, while the majority of the man-made structures collapsed beyond recognition, trees proved to be the most resilient of all structures against the Earth's unsympathetic violence. The canopy of the tree also recalls the most primordial desire of human habitation, i.e., the need for a roof which provides the basic sheltering against sun and rain, particularly in times of natural disasters.

The Campus The master plan would adopt the notion of a "campus", effectively giving birth to a new identity. This new "campus" will link between the commercial fabric of Tsaotun and the residential fabric of Chunghsin Hsintsun, pulling the urban energy from both districts together, while establishing the memorial site as the central focus. By further removing the main roundabout and reconfiguring the streets, the "campus" also gives an additional 30% of usable open space back to the community. The master plan strategy is conceived with an ambition to revitalize the gentrified provincial headquarters into a new cultural mecca for the whole of Taiwan and beyond. Related programs to be developed around the planned area would include music conservatory, concert hall, experimental theater, music library, recording studio, as well as residential components such as student dormitories, guest houses, and even a high-end boutique hotel catering to music aficionados and other visitors traveling from outside.

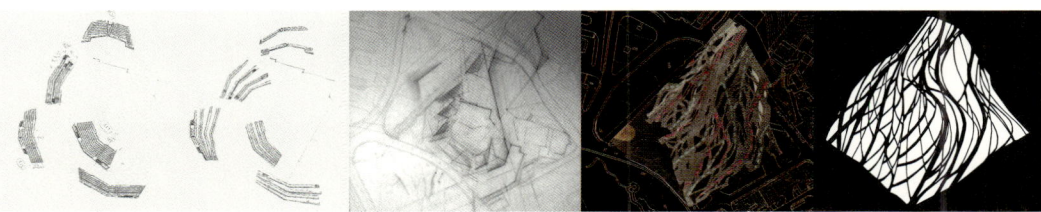

숨겨진 단층의 흔적 CHOREOGRAPHED CLUSTER OF HIDDEN PLATEAUS

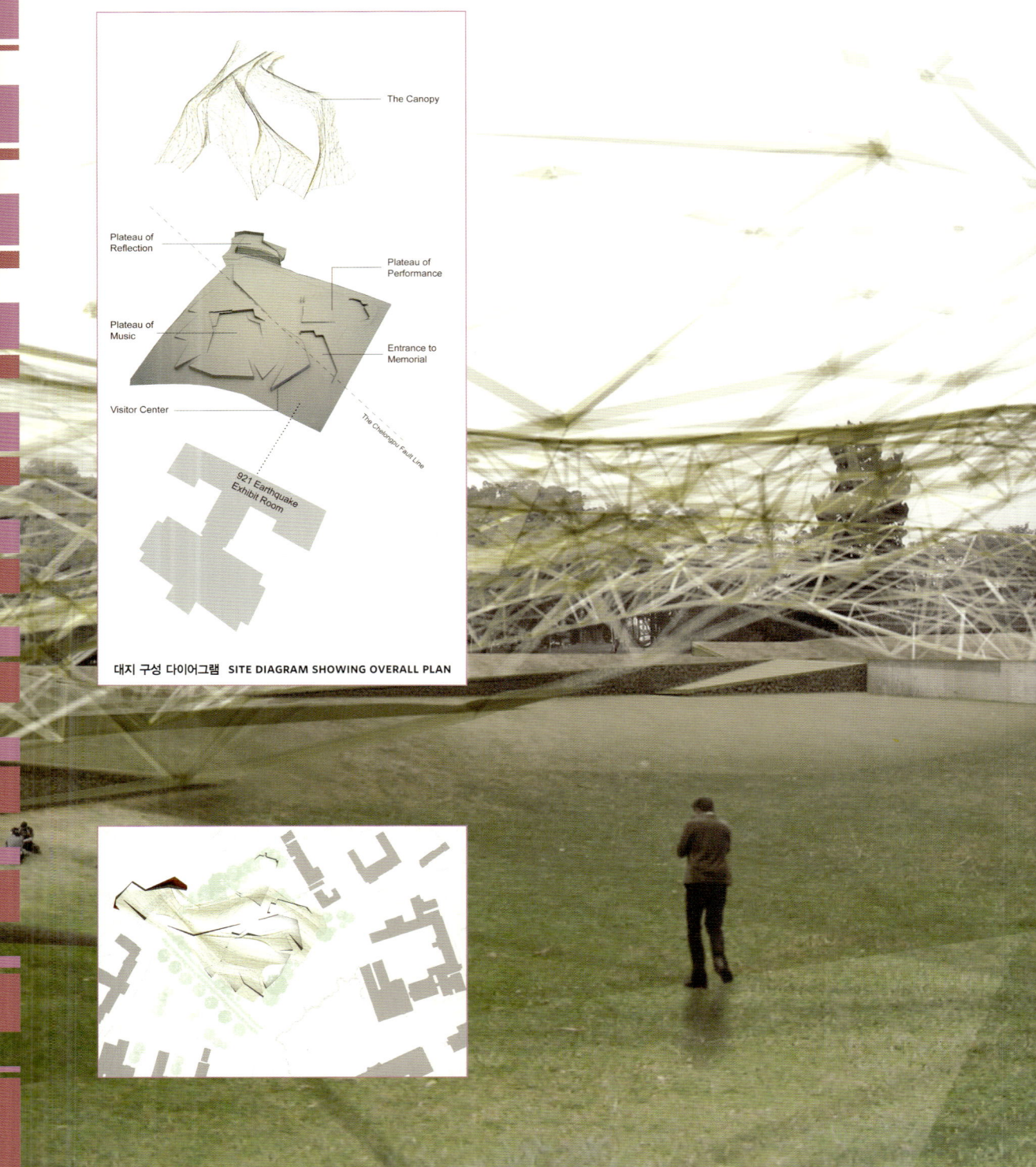

치치 대지진 추모공원
현상설계 2등작

신혜원 + 유양 리우

체험극장 VIEW OF EXPERIMENTAL THEATRE

진입공간 ENTRANCE SPACE

캐노피 하부 콘서트홀 VIEW OF CONCERT HALL UNDER CANOPY

Haewon Shin
Lokaldesign, Korea
Catalog City - Going to Buy an Apartment
Venice Biennale 2006

1962년부터 1964년까지 지어진 한국 최초의 아파트단지에는 642가구가 있었다. 오늘날, 한국 인구의 45%에 해당하는 1,800만 명의 사람들이 아파트에 살고 있다. 아파트를 구입할 수 있는 시장은 크게 두 가지다. 하나는 소위 '중고 아파트'를 사고팔기 위해 항상 열려 있는 시장이다. 다른 하나는 한국의 독특한 시스템으로, 새로 지어진 아파트를 추첨을 통해 구입하는 '분양'이다.

분양은 아파트가 지어지기 전에 이루어지는데, 보통 입주하기 약 2~3년 전에 분양한다. 이미 지어져 있는 집을 사고파는 것이 세계 주택시장의 표준이 되어왔음에도 불구하고, 한국은 수요와 물량의 현저한 차이로 인해, '분양'은 단순히 거래가 아닌 추첨에 의해 이루어진다. 단순히 아파트를 구입하는 것이 아니고 분양받을 수 있는 권리를 얻기 위해 다른 이들과 경쟁하는 것이다.

새 아파트에 대한 높은 수요는 개발업자들의 치열한 경쟁을 가져왔고, 아파트를 브랜드화하여 마치 음료수나 청바지 같은 다른 소비제품처럼 판매하는 독특한 마케팅 장르를 만들어냈다. 한국의 아파트 마케팅에는 TV, 라디오, 인쇄 광고와 함께 카탈로그와 같이 일회적인 건축의 하위문화가 포함된다. 아파트의 인테리어 견본을 실물 크기로 만들어놓은 유닛과 이것들을 전시해놓고 거래하는 '모델하우스'는 그 자체가 하나의 건축적 현상이다. 서울에만 수백 개의 모델하우스가 곳곳에 있다. 각각은 수십 개의 아파트 디자인으로 채워져 있어 아파트 청약을 원하는 이들이 보러오기를 기다리고 있다. 견본 유닛과 그것이 들어가 있는 판매 장소이자 당첨을 기대하는 주부들이 한번 훑어보는 용도로 디자인된 실제와 가상의 건축인 모델하우스를 포함하는 이러한 일회적인 구조가 '카탈로그 시티'의 주제이다.

Temporary

Imagination

Fabrication

Displayed Architecture

Competition

Marketing

Prototypes

카탈로그 시티 CATALOG CITY

카탈로그 시티
2006 베니스비엔날레 한국관

신혜원
로컬디자인

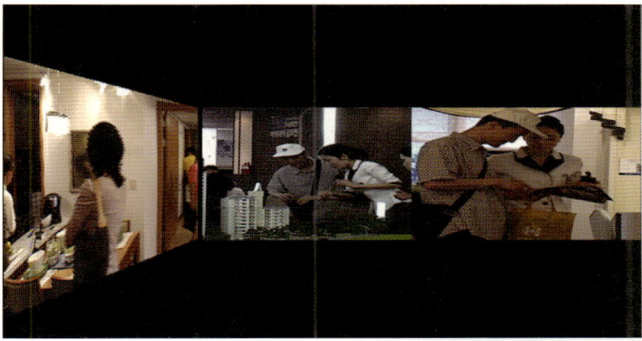

Built between 1962 and 1964, the first apartment complex in Korea contained only 642 units. Today, forty-five percent of the population, twenty million people, lives in apartments, and the market for buying these is divided in two: the open market which deals with all "second hand" units, no matter how recently built, and Bunyang, meaning literally "sale by the drawing of lots", a unique system through which all truly new units are sold in Korea.

Each Bunyang sale is held prior to construction of the apartments - roughly two to three years before move-in date. Though pre-build sales have become a standard of housing markets around the world, due to the extreme discrepancy between the quantity of desire for a home and the capacity to (ful)-fill it in Korea Bunyang is not merely a market, but a Lottery. Participants in Bunyang don't simply purchase an apartment they enter a competition in which the prize is the RIGHT TO BUY.

The extreme demand for new apartments has lead to fierce competition amongst Korean developers who have created their own unique genre of marketing in which apartments are "Branded" and sold like soft drinks, jeans and other consumer items. Along with television, radio and print ads, the marketing of Korean apartments involves a whole sub-culture of "temporary" architectures that function like catalogs. Both the "Show Flats", mocked up prototypes of the would-be apartment interiors, and the marketing complexes or "Model Houses", in which these are displayed, have become architectural phenomena in their own right. At any time, literally hundreds of Model Houses dot Seoul alone, each filled with potentially dozens of different apartment designs ready for aspirants to inspect. This temporary architecture comprising both the Show Flats and their sales venues, the Model Houses, this actual-virtual architecture designed only for the fleeting glance of gambling homemakers, is the subject of Catalog City.

Chanjoog Kim
_System Lab, Korea
Eagon Window Image Shop
Seoul, Korea
2004

시스템 창호회사의 제품 전시장으로 기획된 이 프로젝트는, 공장에 적층되어 있는 창호 프레임들이 바로 회사 이미지를 재현하는 데 사용되었다. 시스템을 구체화하기 위해, 연속되지만 약간 차이를 가지며 변화하는 프레임 단면의 프로파일이 수평적으로 적층되었다. 각각의 단면 프레임들은 공조나 구조, 전기와 같은 서로 다른 기능을 담당하게 된다. 다양한 빌트인가구들은 프레임간 횡력을 잡아주기 위한 구조 역할도 담당하게 된다.

유닛과 프레임 UNITS & FRAMES

이건창호 쇼룸

김찬중
시스템 랩

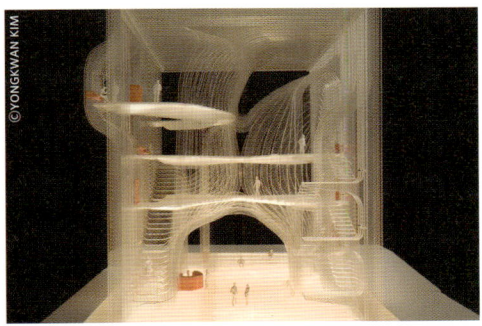

MODEL

This project was initiated for the product show room of system window company, Eagon. Stacked window frames in factory are the project's main system idea as a representation of company's image. In order to realize the system, continuous but slightly different sectional profiles are stacked horizontally. Each sectional frame delivers different functions such as HVAC, structure, and electricity. Also various built-in furniture pieces are applied for resisting against lateral force between frames.

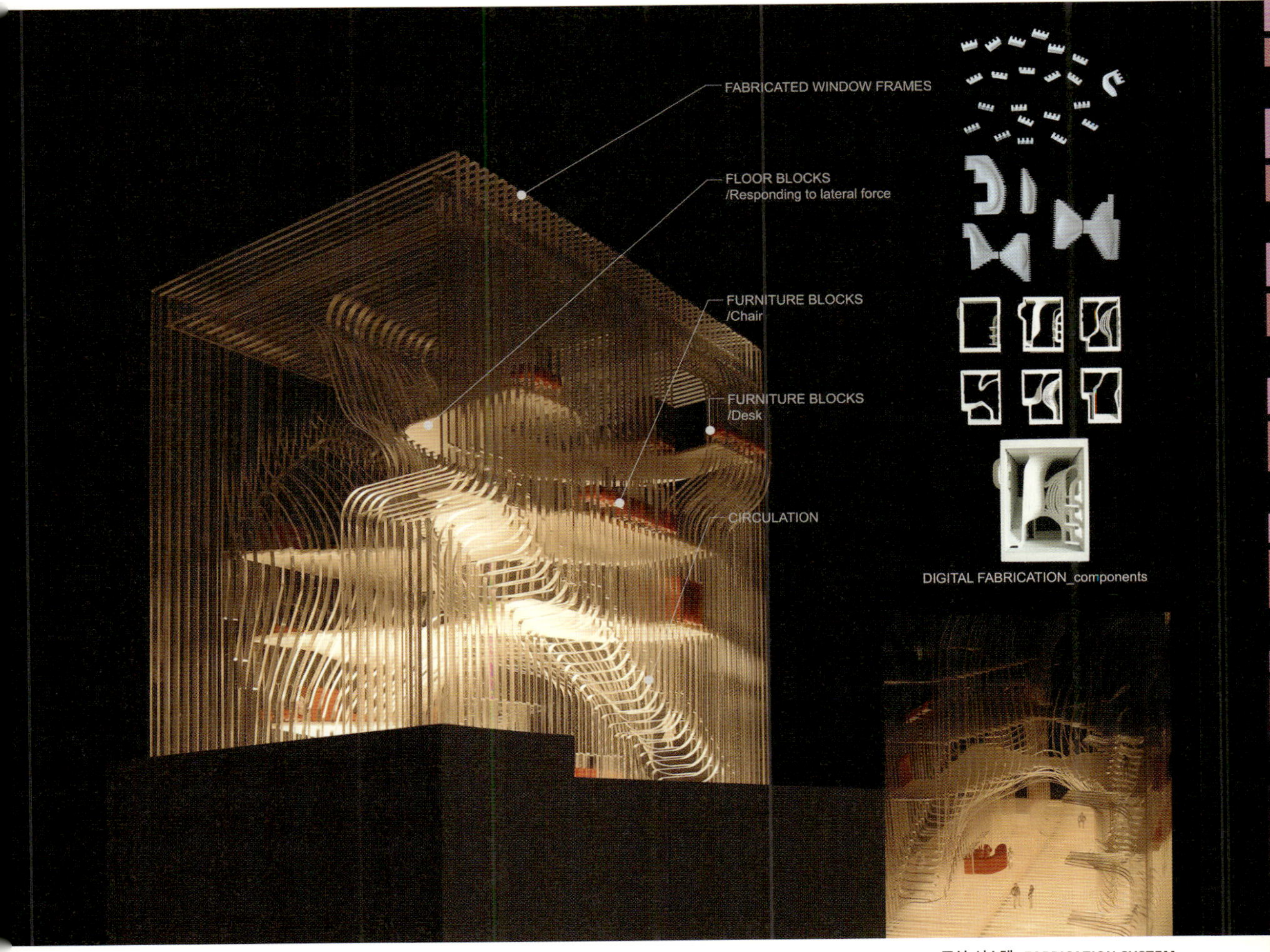

- FABRICATED WINDOW FRAMES
- FLOOR BLOCKS /Responding to lateral force
- FURNITURE BLOCKS /Chair
- FURNITURE BLOCKS /Desk
- CIRCULATION

DIGITAL FABRICATION_components

구성 시스템 FABRICATION SYSTEM

Chanjoog Kim
_System Lab, Korea
The Last House-House for the Dead
Seoul, Korea
2006

유휴 토지가 부족하여 사람들은 더 이상 개인 묘지를 가질 수 없으며, 장묘시설도 거대한 고밀도 수납 공간으로 변모해야 할 상황이다. 새로운 적재 방식을 사용하여 기존과는 다른 유형의 고층 도심형 고밀도 납골시설을 제안한다. 원시 생명체인 방상충의 구조체계에서 착안한 18개의 3차원 구조 블록은 떠 있는 플랫폼으로 결속되면서 구조적 안정성을 획득하고 동시에 이용자에게 사색과 경건함, 도시 경관을 즐길 수 있는 장소를 제공하고 있다. 개인 이동통신기기를 이용한 원거리 디지털 헌화방식과 태양에너지를 생산해내는 납골함은 이러한 시설물이 더 이상 혐오시설이 아니라 도시의 기반시설임을 인식하는 기회가 될 것이다.

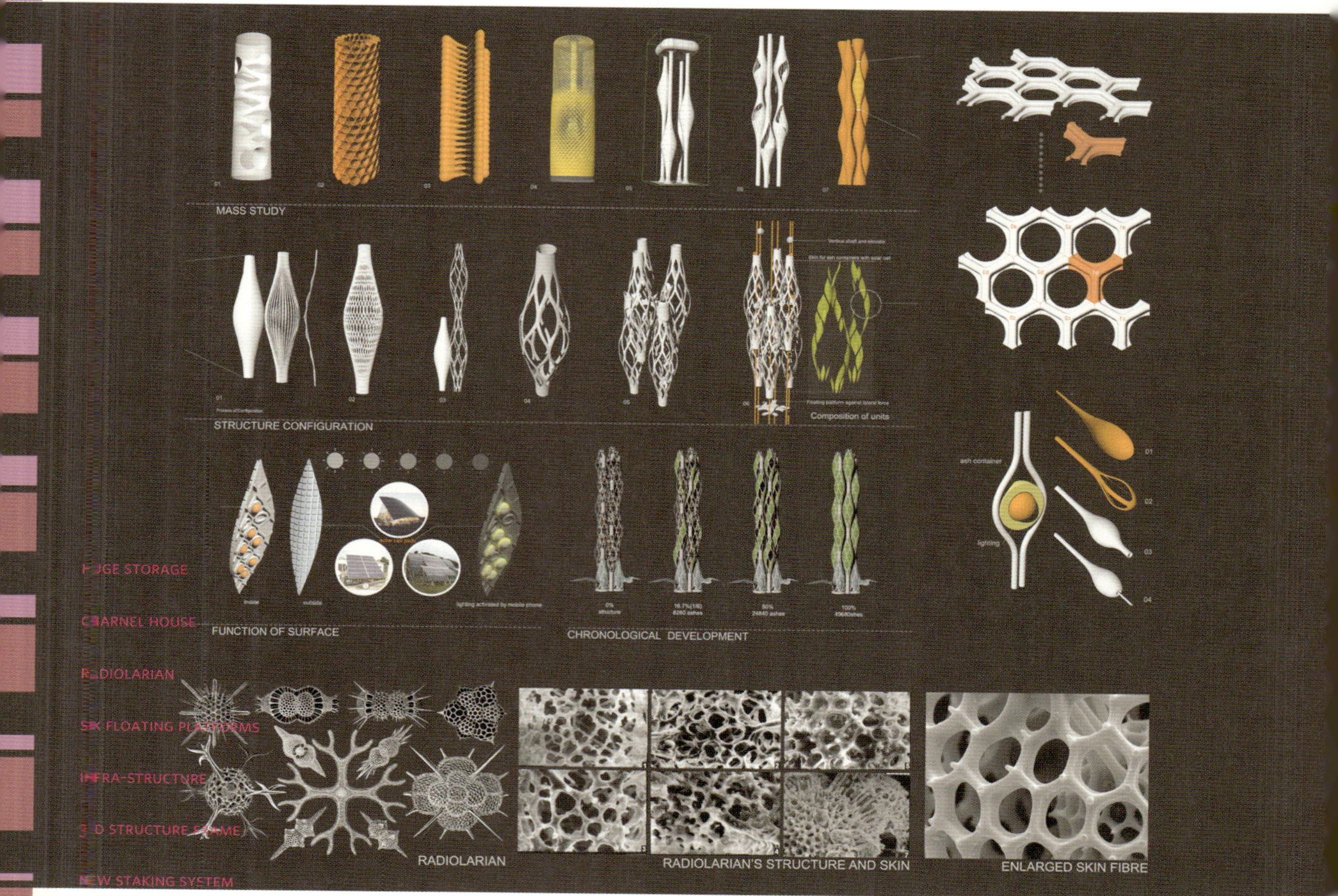

방상충 구조 RADIOLARIAN STRUCTURE

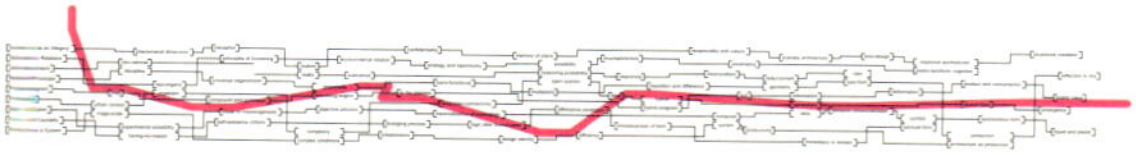

The lack of land doesn't allow people to have personal graveyard anymore, which was deeply associated with the meaning of death in Korea. Like residential situation such as huge apartment complex in Seoul, the function of graveyard is also required to be transformed into huge storage space with extremely high density. This project aims to develop different typology of columbarium (charnel house) by generating new stacking system. Started from the study on structure of radiolarian, eighteen 3-D structural frame blocks are bundled by six floating platforms where people can speculate, worship, and enjoy the urban scenery. By applying solar cell to the surface of ash container, columbarium is not the place of death anymore but the new urban infra-structure generating clean energy for the city.

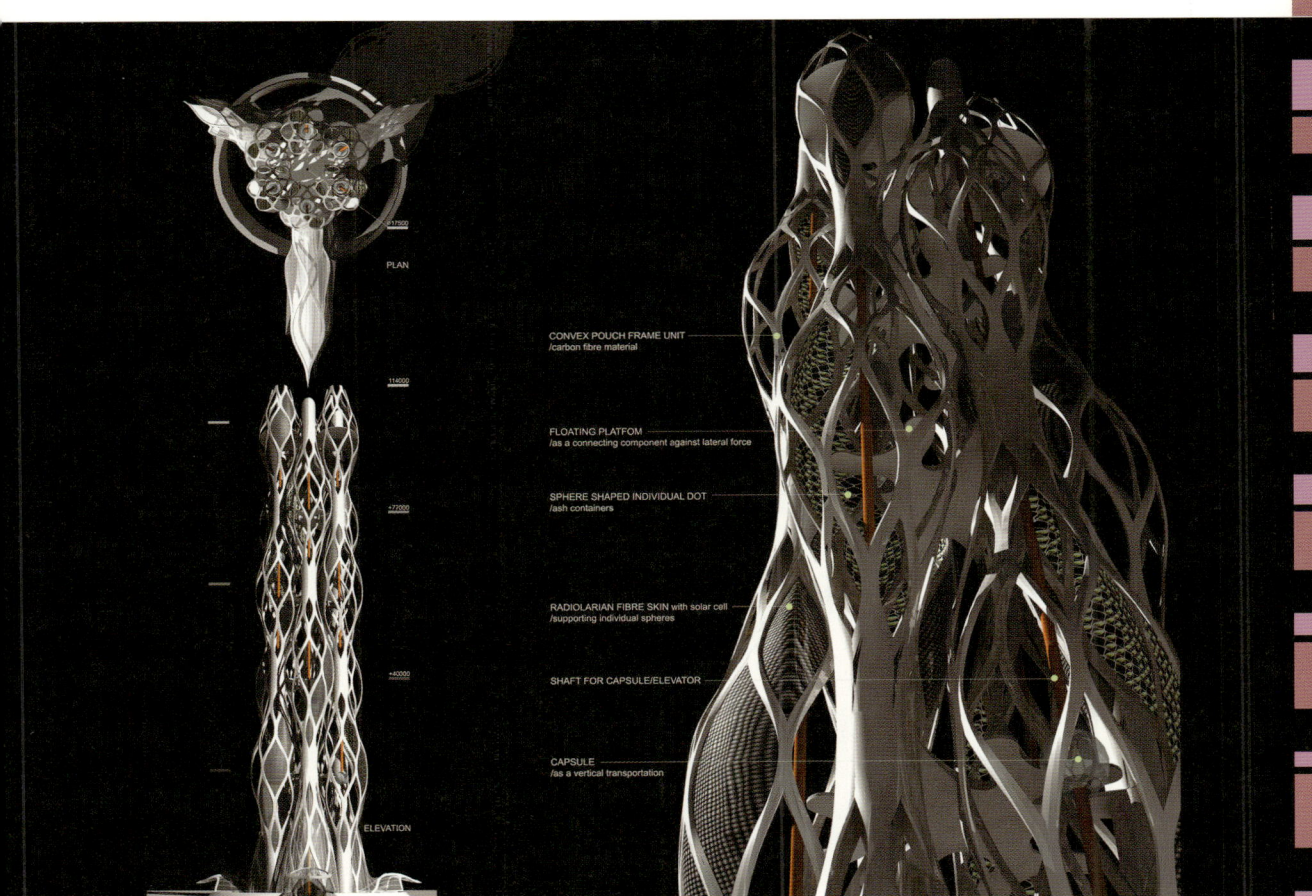

표피와 입면 SKIN & ELEVATION

Sooin Yang
The Living Architecture Lab, U.S.A. / Korea
Better, Cheaper, Faster

양질의 건축과 실리적인 개발이 결국 동전의 양면과 같다면? 우리는 새로운 컴퓨터 기반 제작기술이, '좋은' 건축가와 부동산개발업자 사이에 연결점을 만들고 양자를 위해 전례없는 기회를 제공한다고 가상했다. 이익만 내세우던 분야에서 개념을 개발하고, 개념만 있던 분야에서 이익을 생성한다는 것이다. 이 프로젝트에서는 비용이 저렴하고 경량의 분리와 조립이 쉬운 구조 시스템을 개발했다. CNC 밀링머신로 제작된 4분의 1인치의 합판 슬레이트를 사용해서 실제 규모의 작은 건물용 구조 모형을 만들었다.

높이(피트): 10 구조물의 평방피트: 100 무게(파운드): 120 평방피트별 소요 달러: 9 밀링시간: 6 조립시간: 1 비전문가 작업자 수: 2 작업자별 필요한 툴: 3 합판 장수: 10 강철 패스너: 632 합판 웨지: 52 합판 슬레이트: 144

최종 시연 FULL-SCALE DEMONSTRATION

적재와 운반의 용이성
STACKABLE AND EASY TRANSPORT

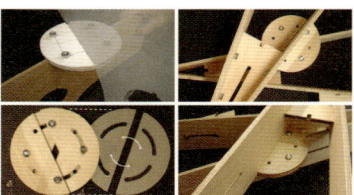

모듈의 신속한 전개를 위한 실험
EXPERIMENTS ON HOW TO KEEP THE MODULE EXPANDED

COMPUTER-BASED FABRICATION

ARCHITECTS AND DEVELOPERS

FULL-SCALE DEMONSTRATION

FLEXIBLE AND WEAK STRUCTURES

EFFICIENCY

접합부 실험 CORNER JOINT EXPERIMENT

더 나은, 더 싸게, 더 빠르게   양수인
THE LIVING ARCHITECTURE LAB

수직적 전개방식 실험 EXPERIMENTS ON HOW TO EXTEND VERTICALLY

충분한 적재하중을 지탱하기 위한 실험
EXPERIMENTS ON HOW TO HANDLE SUFFICIENT LOAD

What if good architecture and bottom-line development were the same thing? Our hypothesis is that new computer-based fabrication techniques can offer a link between "good" architects and real estate developers, creating unprecedented opportunities for both, developing concept where there once was mainly profit and developing profit where there once was mainly concept. For this project, we developed a low-cost, lightweight framing system that is collapsible and easy to assemble. Using slats of one-quarter-inch plywood fabricated with a computer numerically controlled (CNC) milling machine, we created a full-scale demonstration of a frame for a small building.

Height in Feet: 10   Square Feet of Structure: 100   Weight in Pounds: 120   Dollars per Square Foot: 9   Hours to Mill: 6   Hours to Assemble: 1   Non-specialists as Workers: 2   Tools Required per Worker: 3   Sheets of Plywood: 10   Steel Fasteners: 632   Plywood Wedges: 52   Plywood Slats: 144

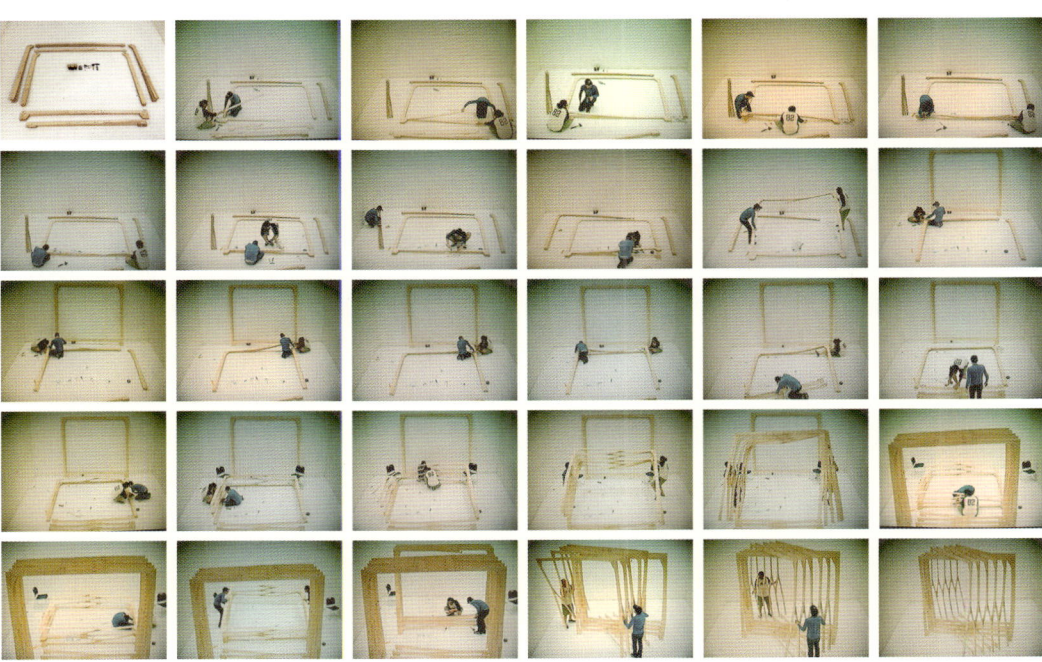

조립 과정 RAPID CONSTRUCTION

Sooin Yang
The Living Architecture Lab, U.S.A. / Korea
Living Glass

건축이 자신에게 반응한다면? 건축 요소가 그 환경에 반응해서 움직인다는 전제에서 연구하고 디자인하여 실제 규모의 동적인 표면을 구축했다. 우리가 사용한 시스템은 일련의 센서를 통해 입력 데이터 및 정보를 수집하고, 마이크로컨트롤러를 통해 이를 처리하며, 휴먼스케일 표면의 국부적이고 전반적인 움직임을 유발한다. 다른 동적 벽면 프로젝트와는 달리 우리가 만든 표면은 얇고 가볍고 투명해서 아무런 동력이나 기계부품을 포함하지 않는다. 운동 시스템은 표면 자체에 들어 있다. 형상기억합금(SMA) 와이어를, 특히 캐스트 실리콘에 다이널로니 플렉시놀Dynalloy Flexinol을 내장시켰다. 전기자극으로 수축하고 gill이 표면으로 파고들어가 개폐시킨다. 이 플래시리서치 프로젝트에서는 표면 소재, 표면 처리, 입력센서, 전기회로를 실험과 다수의 프로토타입을 통해 연구하였다.

운동: 모터나 기계를 사용하지 않음 소음: 없음 표면 두께: 1/16" 표면 투명도: 투명 매스: 초경량 미적: 유기적 구성요소: 교환, 업그레이드 가능

What if architecture responded to you? Beginning with the premise that architectural elements might move in response to their environments, we conducted research on, designed, and built a full-scale kinetic surface. Our system collects input through an array of sensors, processes the input through microcontrollers, and triggers local and global movement of a human-scale surface. In contrast to other kinetic wall projects, our surface is thin, lightweight, and transparent, with no motors or mechanical parts. Our system of movement is contained within the surface itself: we embed shape memory alloy (SMA) wires? "specifically, Dynalloy Flexinol?" in cast silicone, and they contract due to electrical stimulus, causing gills cut into the surface to open and close. Over the course of this Flash Research project, we investigated surface materials, surface treatments, input sensors, and electrical circuits through experiments and multiple prototypes.

Movement: Nonmotorized, Nonmechanical Sound: None Surface Thickness: 1/16" Surface Opacity: Transparent Mass: Extremely Lightweight Aesthetic: Organic Components: Swappable, Upgradable

KINETIC SURFACE
KINETIC WALL
SYSTEM OF MOVEMENT
EXPERIMENTS
PROTOTYPES
ENVIRONMENTAL CONTROL
SIGNALING

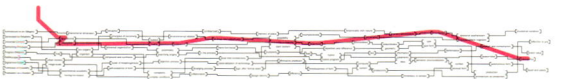
표피의 건축적 적용 POTENTIAL DEPLOYMENT AS BUILDING SKIN

Living Glass

양수인
The Living Architecture Lab

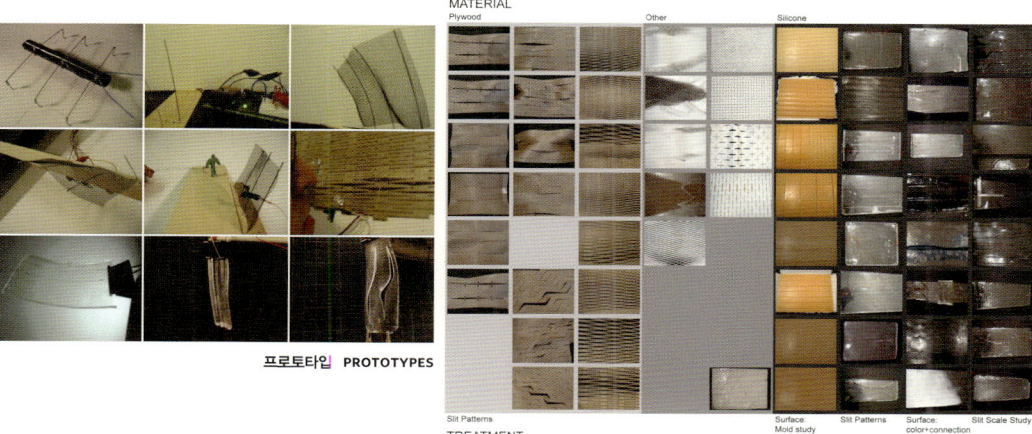

프로토타입 PROTOTYPES

MATERIAL EXPERIMENTS

INPUT: presence sensors
PROCESSING: BASIC stamp II microcoltroller
OUTPUT: gills with Flexinol® wires

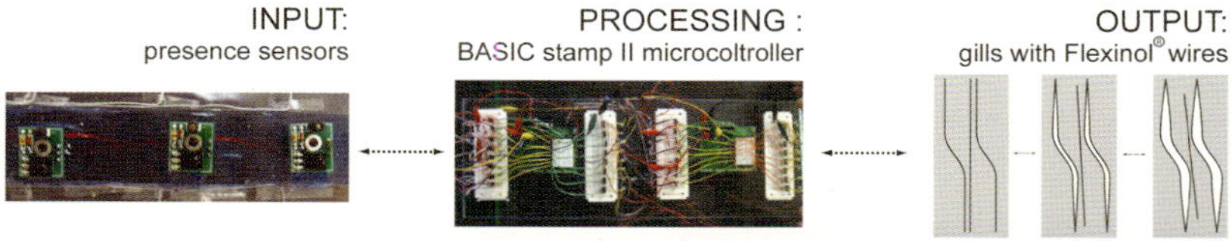

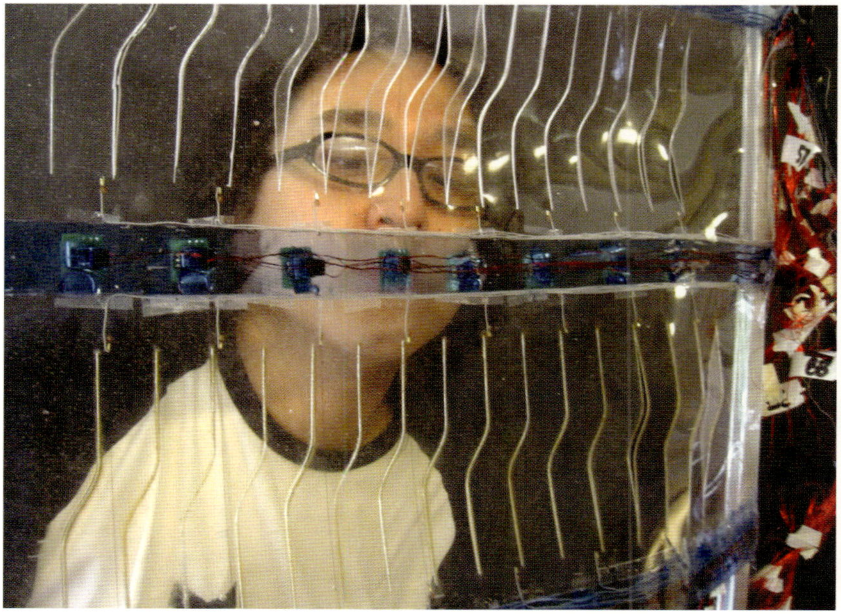

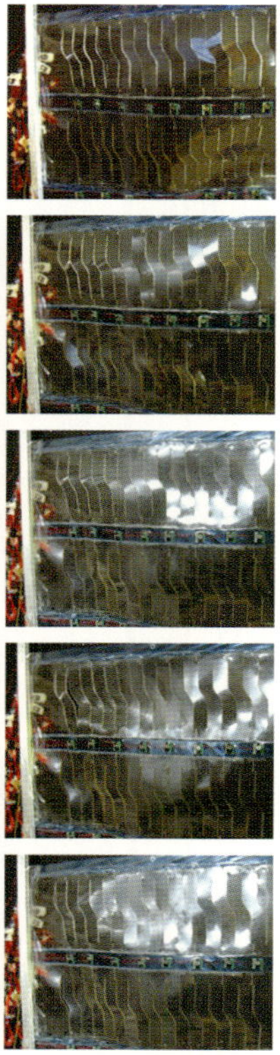

최종 시연 및 시스템 다이어그램 FULL-SCALE DEMONSTRATION AND SYSTEM DIAGRAM

Kihong Kim
Architect KYOSKS, Korea
Residential Complex Plan
Catalunya, Spain

도시를 구성하고 있는 여러 보이지 않는 요소들, 그리고 이를 기반으로 해서 살아가는 우리의 삶을 보다 적극적으로 구체화할 수 있다면…. 물리적, 비물리적인 기반시설들이 건물의 구조와 합쳐지면서 우리에게 보이지 않던 공간까지 보조해줄 수 있다면…. 새로운 주거환경시스템의 제안으로, 도시와 단지에 존재하는 물리적 네트워크를 주거 구성요소로 적극적으로 도입, 구조와 기능 그리고 미적으로 활용함으로써 내외부적으로 가변적인 주거공간을 연출한다. 프로젝트와 환경을 연결하는 실제 프로젝트의 특성은 아래 세 가지다.

1. The Inscription (spatio-temporal): 시공간에서의 현실 환경과 인식 과정의 등록
2. The Indexing: 환경의 성격 규정을 위한 신경조직체(네트워크 그룹)들의 색인 과정을 조정하는 측도
3. The Coulisse: 보이지 않는, 하지만 알 수 있는 공간 인지에 관한 새로운 방법의 은유적 표현이다. 결국 색인이라는 방법과 마찬가지로 인간의 인지에 관한 지적 기술을 이용하는 공간 구성방법 중 하나이다.

입면 투시도 부분 1 ELEVATION PERSPECTIVE 1

입면 투시도 부분 2 ELEVATION PERSPECTIVE 2

실내 공간 투시도 INTERIOR PERSPECTIVE

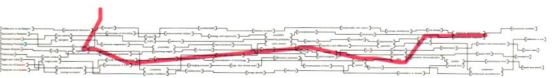

카탈루냐 주거단지 계획   김기홍 / 키오스크

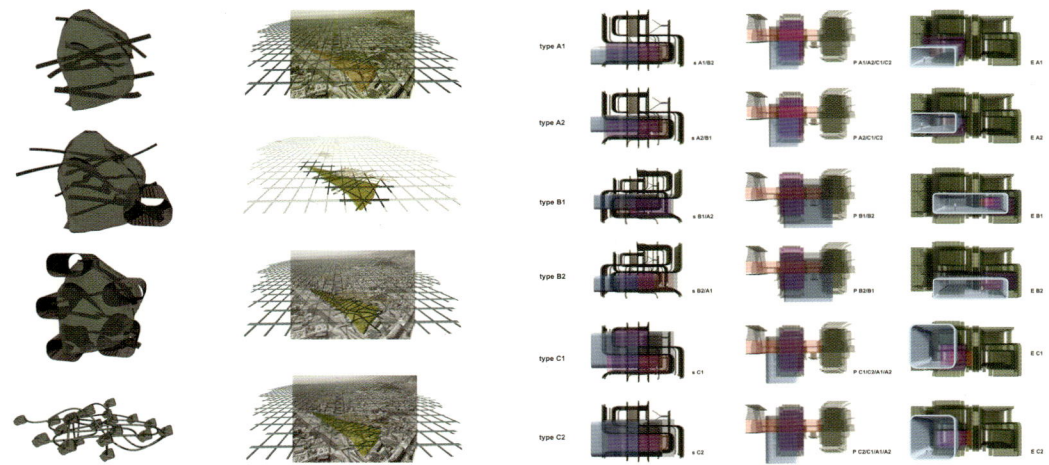

DIAGRAM OF NETWORKS AND COMPOSITIONS   COMPOSITION OF TYPES

THE COULISSE + INTERPLEXUS

If we can actualize more positively invisible components of the cities and our lives on them. If physical and non-physical infrastructures can be integrated into the structures of buildings, and they can support even invisible spaces in results. As a proposal of a new living environment system, we present a flexible living space both in and out, which introduces actively physical networks in the city and complex as a component of living system and utilize them as structures and functions both practically and aesthetically.

Three practical characters between the project and the environment

1. The inscription (spatio-temporal): Registration of the real spatio-temporal environment and perception process
2. The indexing: Standard controlling the indexing process of neural system (network groups) for identifying the environment's characters
3. The Coulisse: The coulisse is a metaphor of a new method to perceive a space which is invisible, but can perceive. Finally, like the indexing, it is one of the environmental construction methods using intellectual technologies.

Kihong Kim
Architect KYOSKS, Korea
Chichi Earthquake Memorial Competition Project_Finalist
Taiwan
2004

가상은 가능성의 항시적 관계를 의미한다. 하지만 가상의 현실화는 잠재성의 변형을 말하는 것이다. 이 현실화는 또 다른 가상을 만든다. 다시 말해 가상은 변형의 연속성 위에 존재한다. 현실화는 가상화의 유동성 속에서 예기치 않은 또 다른 가상을 생성하게 된다. 공간은 시간의 흐름 위에 각각 잠재적 현실성을 중첩시킨다. 이 중첩의 연속은 공간의 잠재적 현실성이 주는 영향을 우리에게 인도한다. 잠재적 가상의 현실화는 현존하는 우리가 어우러지는 공간에 각각의 기억의 파편을 통해 형태들을 형성한다.

이 현실화는 공간과 인간의 행태를 연결하는 고리로 재탄생되고 현실화의 교차 속에 새로운 사건들이 생겨난다. 사건들은 항상 동질적이진 않지만, 이 사건을 통해 기억의 혼돈이 일어나고, 동시에 이 순간의 공간을 형성한다. 매순간 우리의 존재를 우리 주변환경을 통해 인식하게 되면서, 주변 공간들은 각각 현실화의 중첩에 의해 우리에게 다가선다.

그리드와 자연 지형 사이의 경관
LANDSCAPE BETWEEN GRID AND NATURAL TOPOGRAPHY

PERSPECTIVE VIEW

치치 대지진 추모공원
현상설계 최종참가작

김기홍
키오스크

The virtual defines a consistent relationship of possibility. But, the actualization of the virtual defines a transformation of potentiality. The actualization produces another virtual. In other words, the virtual exists on the continuity of transformation. The actualization produces another unexpected virtuals in the mobility of the virtualization. The space piles one underlying virtual on the other on the flow of time. The continuity of this piling leads us the affection of virtual reality of space. The actualization of the virtual forms ourselves through the particles of common memories in the existing space where we are living together.

. The actualization is recreated as a chain connecting behaviors of space and human beings. New things happen in the crossing of the actualization. Although these are not always identical, we may confuse our memories and generate present time and space at the same time. Since we perceive ourselves through the environment surrounding us every moment, the space surrounding us is perceived by piling of the actualizations.

The Composition of the Project
Three vertical components
The core points of the environment are represented into three different modes.
- The movement of the subjective memory
- The movement of the objective memory
- The interaction between two layers.

Then, the four horizontal components are connecting the two layers composed of definite grids and non-definite space with natural curve.

Four horizontal element
- metaphor cube of victims
- artificial liquefaction
- vertical nature
- actualization points

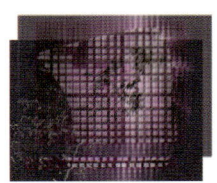

수평 요소와 수직 요소를 연결하는 두 개의 레이어
TWO LAYERS CONNECTING HORIZONTAL AND VERTICAL COMPONENTS

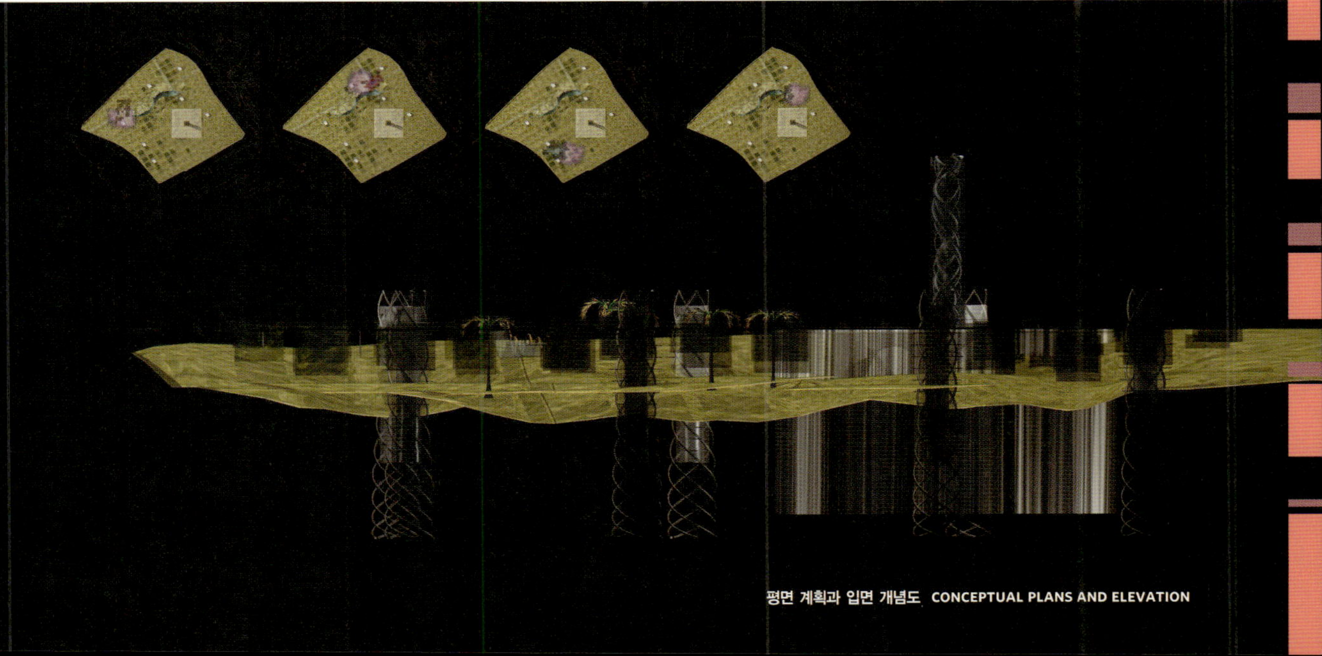

평면 계획과 입면 개념도. CONCEPTUAL PLANS AND ELEVATION

디지털 테크놀로지와 사이버공간

이경훈
국민대학교 건축대학

MATERIALIZATION OF TECHNOLOGY

MODE OF MORPHOGENESIS

SEMANTIC STRUCTURALIZATION

SYMBOL

COMPUTER

DATA

EXPERIMENTAL POSSIBILITY

LIQUID AND PLASTIC

OBJECTIVE PROCESS

AUTONOMOUS FORM

예술은 항상 최선의 테크놀로지를 적용하는 데 적극적이며, 테크놀로지가 물화되고 구체화되는 통로로서 예술은 다시 중요한 의미를 갖는다. 현대문명의 기반을 이루는 디지털 테크놀로지의 출현은 이러한 양자간 교호작용이 가장 선명하고 극적으로 일어나는 교차점으로 성립한다. 이에 대한 관찰과 이해는 현대 예술, 건축 크게는 현대문명을 이해하는 데 중요한 단서를 제공한다.

즉물적 문제의 해결수단으로 도입된 로마의 아치가 출입이나 장식 의도 같은 의미론적 기의가 부착된 아이콘으로 성립되는 과정은 테크놀로지의 적용이 단순한 문제해결 수단을 넘어서서 건축의 개념적 사실에 개입하는 일종의 형태 생성 기제로서 기능해왔음을 의미한다. 고딕의 첨두아치는 아치의 기본 개념에서 출발하고 변형을 거쳐서 종교적 상징을 내포하는 기술과 형태가 결합한 방식의 극명한 예이다. 이러한 경향은 20세기 후반 하이테크건축에 이르는 전 건축사를 관통하는 발전 원동력의 하나였다.

디지털 테크놀로지는 건축에서 기술 적용과 그 결과로 나타나는 중요한 형태와 상징의 생성 그리고 의미론적 구조화 등으로 심화되는 전통의 연장선상에 놓여 있다 할 수 있다. 그러나 건축의 형태 생성 과정에 깊숙이 관여하는 디지털 테크놀로지는 이제까지 즉물적인 문제 해결책로 테크놀로지와 그에 따른 형태, 또 그에 의미가 부착되는 과정과 구별되는 전개 양상을 보인다. 이와 같은 기술 발전에 따른 '디지털건축' 이라 불릴 만한 경향이 분명 성립하고 있다. 이는 다시 디지털 테크놀로지의 개입 정도에 따라 세 유형으로 분류할 수 있다.

첫째, 디지털 테크놀로지 - 구체적으로는 컴퓨터의 건축 - 에 대한 개입을 최소화하는 입장이다. 컴퓨터의 정확한 제도와 사실적인 모사 능력이 건축의 생성 과정에 보조적으로 사용되는 경우이다. 여기서 컴퓨터는 비가시적 개념과 과정을 시각화하는 최적의 도구로 이해되며, 건축사에 지속되어온 드로잉 수단으로 의미를 이어간다. 둘째, 컴퓨터가 적극적으로 개입되어 새로운 형태 생성의 필수 도구로 사용되는 경우이다. 이전까지 만들 수 없었고, 때로는 상상할 수 없던 형태를 구축하고 이를 구체적인 건물 형태로 실현하는 과정에서 컴퓨터는 보조 도구가 아니라 객관적이고 논리적인 형태 생성의 중심요소가 된다. 론 콤, 폴 프라이스너 등이

Digital Technology and Cyberspace

Kyunghoon Lee
School of Architecture, Kookmin University

Art has always been keen to adopt newly available technology, being an important mean that technology is materialized. The emergence of digital technology, shaping current civilization, represents a dramatic intersection with art and technology. Examining and trying to understand this provides us significant view towards art, architecture and, more broadly, on-going cultural foundation.

The adoption of technology in architecture, functions as a morphogenesis in the realm of conceptual sphere as well as a practical solution. For example, Roman arch has been introduced to resolve practical problem and became to be signified as "entrance" or "ornament". The pointed-arch in Gothic, developed from archand modified to imply religious symbolic meaning, is another example of the combination of technology and architectural form. Throughout the history, including Hi-tech architecture in the twentieth century, technology in architecture has been a locomotive in the process of formal development in architecture.

Digital technology's involvement in architecture parallels that of traditional technology in architecture that has created symbolic meanings and their structure. It is, however, shows a different trajectory than traditional one, since digital technology involves more in the process of morphogenesis. It is certain that there are emerging tendencies that are enthusiastic to adopt new digital technology the digital architecture. The digital architecture can be categorized into three different types by the degree of the involvement of digital technology.

The first type is an attitude that utilizedigital technology, but as a drafting tool, and minimize its involvement in architecture. The use of computer is limited to assist to visualize intangible ideas by its ability to represent hyper-realistically, continues the idea of "drawing" in the history of architecture.

Secondly, there is a tendency that computer is involved in the process of form generation. Computer becomes the main component of morphogenesis that creates the forms that was not possible to imagine and to construct otherwise, rather than secondary tool to assist. Lonn Comb and Paul Preissner are the examples for this category and those architects who utilize computer to visualize data for their basis of architectural form.

이러한 범주의 건축가이다. 데이터를 형태 생성의 주요 요소로 삼는 일군의 건축가도 여기에 해당된다. 마지막으로 현재의 디지털 기술이 전혀 새로운 공간, 즉 사이버공간을 형성한다고 보고 이러한 공간환경에서 건축에 집중하는 경우이다. 앞의 경향들이 물리적 실현을 전제로 구체적인 건축에 집중하는 반면, 사이버공간의 건축가들은 건축의 순수한 실험 가능성에 주목한다.

디지털 테크놀로지가 현대문명 전반에 침투해 있는 상황에서 엄밀하고 독립된 의미의 디지털건축은 두 번째와 세 번째 경우로 의미가 한정된다. 세 번째 경향은 사이버건축, 정확하게는 사이버공간의 건축으로 구분하는 것이 타당하다고 본다. 디지털건축과 사이버건축을 비교하고 분석하는 것은 건축과 테크놀로지, 건축의 자기동일성 같은 후기산업사회 건축에 관련된 다양한 이슈를 검토하는 유의미한 출발이 될 것이다.

디지털건축

디지털건축에서는 컴퓨터 테크놀로지의 형태 생성 능력은 인정하지만, 직관이 결핍된 기계 형태로 존재하기 때문에 도구로 활용한다는 입장으로 요약할 수 있다. 즉 컴퓨터는 중요한 부분을 차지하기는 하지만 주어진 초기값이나 매개변수를 변형하고 이를 형태화하는 것으로 역할이 한정된다. 컴퓨터에 의해 변형되는 형태들은 수학적 표현으로서 가치있으며, 이는 유클리드 기하학에 머물러 있던 건축에 수학의 새로운 성과를 반영하는 수단으로 의미를 갖는다. 건축에서 형식적 순수함과 자율성에 대한 관심은 건축 고유의 필수요건을 위협하지 않고 유지하는 범위에서 이루어진다. 디지털건축은 물활론적이며 유기적 모델, 즉 정적인 건축에 동역학의 특성을 반영하는 방법으로 건축을 더욱 발전시킬 수 있다고 믿는다. 이는 건축이라는 예술 형식 고유의 필수조건에 대한 인식과 그 초월에 도달할 수 없는 욕망이 건축의 진보를 가져왔다는 형식주의의 견해와 맥락을 같이한다.

디지털건축에서는 일반적으로 건축에 결핍된 요소들 – 예를 들면, 시간이나 생물학적 성장 유형, 각종 데이터 – 이 보다 직접적으로 개입한다. 각종 힘이나 성장 과정, 알고리즘 같은 비가시적 요소를 형태에 반영하기 위해 유동적이며 가소적인 객체를 동원하는데, 대표적인 것이 블럽blob이다. 블럽은 일종의 위상학적 형태로 주변 환경에 의해 무한 변형 가능한 형태 생성의 기초단위다. 건축가의 역할은 구체적인 형태 생산보다는 블럽의 초기값과 변수 범위를 조정하는 것으로 대체된다. 객관적 프로세스와 수학적 알고리즘을 건축에 도입하려는 시도는 이전에도 있었으나, 건축가의 의지가 개입하는 정도에 따라 차이를 보인다. 전통적인 건축 방식이 이러한 비가시적 이미지를 차용하는 등 우회적이었던 것에 반해, 디지

Finally, there is a group of architects who concentrate on the new possibility in the architecture in cyberspace, resulted by current digital technology. While the former categories deal with the architecture in "real" space and its physical realization, the architects in cyberspace observe pure aspects of architecture in totally new spatial environment, cyberspace.

It is apparent that the first category includes most of arch tects where digital technology is ubiquitous in ordinary life, the definition digital architecture, therefore, has to be narrowed into last two categories. For the convenience of the discussion, the second category is to be called as "digital architecture", and the last one is to be distinguished as cyberarchitecture(the architecture in cyberspace). The comparative analysis on two categories, is believed to be a significant method to examine the subjects on the architecture in late-industrial society, such as self-identity of architecture, technology and its own argument as well.

Digital Architecture

Digital architecture agrees on the ability of computer technology in form-making, but limits its use as a tool since computer exists in the form of machine which lacks human intuition. The computer plays an important role in creating and modifying architectural forms by transforming data and parameters. Resulting forms are considered to convey meaning reflecting new mathematical achievement into architecture which remains in Euclidian geometry.

Its exploration on formal purity and autonomy in architecture is performed as long as the essential elements in architecture are preserved. It is believed that animistic and organic model of digital architecture is reflecting dynamic quality to develop static architecture. This attitude shares the view with the formalist that the unattainable desire for transcend the inherent condition of discipline of architecture.

It is ordinary that digital architects introduce the elements absenting in architecture - for example, time, forces, biological growth pattern and data - more directly. Objects are also introduced to reflect invisible elements into an architectural form, flexible

It is common that digital architects introduce the elements absenting in traditional architecture - for examples, time, forces, biological growth pattern - more directly. The idea of "blob" is introduced to reflect invisible elements into an architectural form. Blob is a topological form that can be transformed and deformed without limit. The role of architects, therefore, is replaced to adjust perimeters and variables of blob rather than actual production of forms. These attempts to introduce

털건축은 다양한 요소의 흔적을 시각화하고 이러한 요소의 조합이 형태를 직접적으로 결정한다. 컴퓨터가 만들어내는 극사실적 가상성은 삼차원 실체를 실제로 축조하는 대신에 드로잉이라는 이차원 수단을 통해 생각 과정에 생겨나는 존재론적인 것이며, 건축에 고유하게 부착된 특성으로 건축과 가상성의 범위를 확장한다.

사이버공간에서의 건축

사이버공간에서의 건축(이하 사이버건축)은 디지털건축의 일부이지만 사이버공간이라는 새로운 개념에 대한 태도 차이로 구별된다. 사이버건축에서 의미하는 사이버공간은 현실공간을 모사하는 것에서 출발하였으나 완벽하게 자율성을 획득한 새로운 대지환경이다. 디지털건축은 실현을 위한 매개체 또는 과정으로만 의미를 갖지만, 사이버건축은 사이버공간에서만 구축되는 독립된 예술형식으로 간주된다. 그 안에서 건축은 현실건축과 마찬가지로 대지의 특성을 반영한다. 대부분의 사이버 건축가들은 컴퓨터의 형태 생성 능력에 주목한다는 점에서 디지털건축과 맥락을 같이하나, 현실건축으로의 전환보다는 사이버공간이라는 새로운 매체의 가능성에 주목하는 것이다. 따라서 중력으로 대표되는 물리법칙 등 건축을 제약하는 구속이 제거되는 상황에서 이제까지 건축에서 불가능했던 개념의 표현이 가능하다는 입장이다.

실현을 전제하지 않는다는 점에서 사이버건축은 아방가르드건축의 전통에서 그 선례를 모색하고 이어가려고 노력한다. 건축에서 아방가르드의 전통이 특별한 것은 아방가르드가 문학이나 다른 예술 형식과 달리 구체적인 실현 통로를 갖지 못하고 초기 개념 단계에 머무는 페이퍼 아키텍처로 이어져왔기 때문이다. 문학이 활자와 인쇄라는 일종의 비물질적 기제에 의해 생산되는 예술 형식이고, 회화 또한 순수한 이상의 실천이 가능한 물적 수단이 있다는 사실은 각 예술 형식에서 아방가르드가 그 지적인 급진주의에도 불구하고 매우 개인적 차원에서 실현될 수 있는 조건을 가졌음을 의미한다. 건축은 실현에 필요한 물적 특성뿐만 아니라 사회·경제적 조건에 의해 종속된 양상으로 발전되어왔고, '실현'의 문제는 다른 예술 형식의 아방가르드와는 다른 의미를 갖는다. 이런 맥락에서 사이버건축은 아방가르드가 모색해온 실험정신이 실현 통로를 갖게 됨을 의미한다. 동시에 실현 문제로부터 자유로워진 건축은 순수 자율 형식으로 새로운 모색과 성찰을 시작한다.

사이버공간은 전 과정을 컴퓨터 연산체계에 의존하고, 여기에서 형태의 생성은 건축의 존재론적 조건을 유지하는 범위에서 새로운 기제에 의해 이루어지게 된다. 따라서 사이버건축은 자연공간에서와는 달리 인공적·비실제적 발생 구조를 가지며, 환원적이며 형이상학적 접근으로부터 출발한다. 즉 실제적인 문제해결

objective and mathematician process to architecture have been pervasive, it varies by the degree of thee involvement of the architect' intention. While it has been more indirect means to adopt in architecture, digital architecture distinguish itself by direct process of visualizing the trace of elements and the combination of the elements. Its hyper-real virtuality by computer technology is conceived as ontological in the process of designing three dimensional reality in the form of drawing which inherently connected to architecture, expanding the notion of architecture and drawing

The Architecture in Cyberspace

The architecture in cyberspace (to be referred as cyberarchitecture hereafter) is a part of digital technology, but has a different notion of cyberspace from that of digital architecture. The cyberspace has started with imitating real space and has been developed to achieve autonomy as a totally new environment. While digital architecture utilize cyberspace as a mean or a process to realize, cyberarchitecture is conceived as an independent discipline, reflecting its own site condition: cyberspace.

Cyberarchitect admits computer's form generating capability as digital architects, but focusing more on the possibility of cyberspace as a new media of communication than on transferring to the architecture in real space. Architecture, therefore, is able to express the concepts under the conditions without restraints such as gravity and other laws of physics.

Cyberarchitecture finds its precedent in the tradition of avant-garde architecture and continues it. It is noticeable that the avant-garde in architecture has been remained in paper unlike other forms of art, since its lacks a mean of realization otherwise. Literature is generated through immaterial mechanism, and painting is also possessive physical tool to commit pure ideal, which means avant-garde in other disciplines have their own physical means, possible in a personal level despite of their intellectual radicalism. Architecture has developed under the influences of the problem of "realization", as a result it has been subordinate socio-economic conditions. In this context, it means that architecture has new way of incorporating various experiments and liberate itself from the problems to explore as a pure form of autonomy.

The entire process of cyberspace depends on the logic system of computer, and the form generating process is through new process under the limit of preserving the ontological condition of architecture. Unlike the architecture in real space, that in cyberspace follows an artificial and unrealistic genesis, and starts with reductive and

과정에서 다양한 양식의 관점을 수용하는 기존 건축과 달리 이상과 개념이 앞서고, 개념 자체가 형태로 표현되며, 의미를 갖는 표현의 특성이 우월적으로 나타난다고 볼 수 있다.

디지털건축과 사이버공간

디지털건축과 사이버건축은 현실건축과 비교하여 그 특성과 의미를 점검할 수 있다. 양자간 차이는, 구체적으로 형태 생성 방식의 차이는 주체로서 건축가가 개입하는 정도에 따라 일차적으로 구별된다.

전자의 경우 변수와 알고리즘에 조정된 조작을 통해 형태를 생성하되 여전히 건축가의 역할을 유효하고 주요한 생성인자로 파악한다. 반면 후자의 경우에는 보다 적극적으로 컴퓨터의 형태 생성 능력을 인정하고 활용하는 경향으로 나타난다. 즉 디지털건축에서 사이버공간은 이미 주어진 선험적 공간이기보다는 현실공간과 간접적으로 연결되고 경험할 수 있는 대상이다. 타자로서 현실공간의 규범은 사이버공간을 통제하는 동일한 규범으로 재현되며, 그 목적이 현실공간의 수행 목적에 종속되는 목적지향적 공간으로 파악된다. 그러나 사이버건축에서 다루는 사이버공간은 본질적으로 초월적이며 형식도 자율적이다. 따라서 사이버건축은 자율체계에 의한 형태 생성 논리를 발견하는 것에 주목한다. 반면, 디지털건축의 관심은 현실건축에 적용 가능한 형태의 실험에 집중하고 있다.

디지털건축은 보다 실체적이다. 규범에 충실한 현실공간에 머물며 건축의 도구적 한계를 극복하기 위한 증강현실augmented reality에 가까운 개념으로, 실재적 요구에서 출발한 이유로 사변적이며, 신비한 도구와 프로세스 적용으로 연금술적 공간이라 할 만하다. 반면 사이버건축은 인간이 경험하고 인식하는 현실공간과 유사한 실체가 현실 이외에 존재한다고 믿는다는 점에서 애니미즘적이다.

건축에 컴퓨터의 개입이 본격화된 1990년대 이후 디지털건축과 사이버공간에 대한 담론이 풍부하게 생산되었다. 하지만 이는 디지털건축에 대한 관심의 증거이긴 해도 아직 확립되지 않은 영역임을 반증하는 것이라 볼 수도 있다. 일치되지 않은 다양한 견해가 엄존하고 그에 따라 디지털건축의 경향을 범주화하고 각 범주의 인식, 특히 사이버공간에 대한 인식 차이를 논하였다. 그러나 각 범주와 인식 차이가 고정적이거나 단절된 것은 아닐 것이다. 디지털건축과 사이버건축의 끊임없는 갈등과 소통이 사이버공간의 건축 담론과 창발적 실천을 유지하는 커다란 에너지임을 부인할 수 없을 것이다.

metaphysical approach. In other words, while conventional architecture derives from problem solving process, ideas and concepts proceed and are expressed as forms in cyberspace.

Digital Architecture and Cyberspace

The differences between digital architecture and cyberarchitecture can be clarified by the comparison of their attitudes towards respective relationship with the architecture in real space. The former generates the form through the process of adjusting parameters algorism, maintaining architect's role as a subject, while the latter recognizes and utilizes computer as a form-generator.

The cyberspace in digital architecture is not a priori, or given space, but a space connected to real space and experiential. The norms in real space as an "other" re-appears in cyberspace as same manner and meaning, and its purpose is subordinate that of real space. The cyberspace of cyberarchitecture is, however, transcendent in its nature and autonomous in its formal structure. While cyberarchitecture is concerned in finding a logic of morphogenesis, digital architecture focuses the experiment of the forms that can be adopted in the architecture in real space.

Digital architecture is more realistic and remains in normative real space, and its notion of cyberspace is closer to the idea of "augmented reality" to overcome the constraint of the discipline of architecture. Starting with practical problems and solutions, it is alchemic with exquisite tools and processes. Cyberarchitecture, on the contrary, believes in a space similar to real space that can be experienced and recognized and thus animistic.

Many of the discussions on digital technology and cyberspace have been made since 90's when computer began to involve in architecture seriously, and demonstrate on-going discovery on new territory of architecture. Various views on cyberspace exist, so do that on digital architecture. Continuous discord and communication between digital architecture and cyberarchitecture are the energy in creating architectural discourses and commitments.

Paul Preissner
Qua'Virarch, U.S.A.
Jeongok Prehistory Museum
Competition Project_2nd Prize
Gyeonggi-do, Korea
2006

전곡선사박물관은 유적이 보존된 장소와 주변 지형이 이어지는 거대한 구조물이다. 온화하고 자연적인 지형에서 강인하고 이질적인 물질(건물)이 부지 위로 모습을 드러내도록 계획하였다. 설계자료 조사과정에서 형성된 구조적인 틀을 바꾸어, 서로 연결된 개방형 전시공간을 설계안으로 제시했다. 모든 전시공간은 오직 이동공간에 의해서만 구분된다. 공간이 구분되어도 다른 공간에 대한 가시성은 여전히 확보될 것이다. 이렇게 공간을 구성하면 박물관 어느 곳에서든 소장품 전체를 감상할 수 있다. 이러한 구성은 박물관에 소장된 역사 유물의 중요성을 더욱 부각시킬 것이다.

이번 계획에서 초점을 맞춘 부분은 박물관이 시설물 이용자를 위한 직접적인 공간일 뿐만 아니라, 한국 문화의 전반적인 발전을 위한 문화교육사업을 증진시킬 수 있는 방편이 되어야 한다는 점이다. 전곡선사박물관은 발굴 및 역사학습의 공감대를 형성하는 새로운 중심지가 될 것이다.

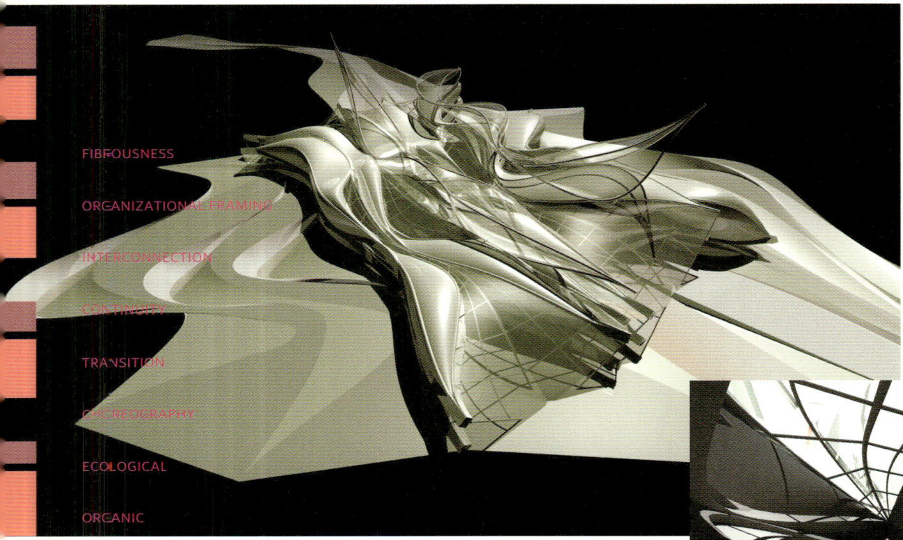

FIBROUSNESS
ORGANIZATIONAL FRAMING
INTERCONNECTION
CONTINUITY
TRANSITION
CHOREOGRAPHY
ECOLOGICAL
ORGANIC
CONTINUATION OF LAND SCAPE
MONSTEROUS
METAMORPHOSIS
NATURAL SKYLIGHTS

실내 공간 투시도 INTERIOR PERSPECTIVE

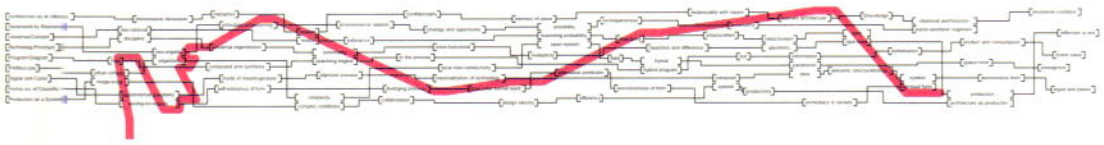

전곡선사박물관
현상설계 2등작

PAUL PREISSNER
QUA'VIRARCH

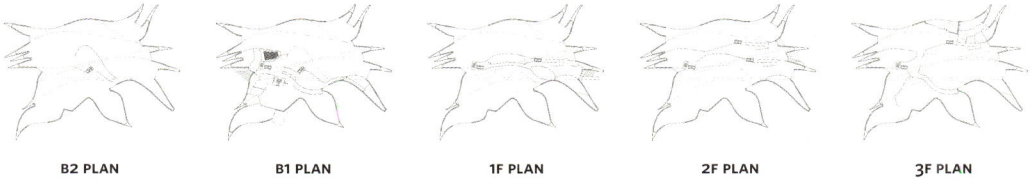

B2 PLAN   B1 PLAN   1F PLAN   2F PLAN   3F PLAN

The Jeongok Prehistory Museum is envisaged as an extension of the surrounding terrain and archival sites, and as a fibrous volume that emerges from itself, expanding onto the site, ferociously changing from a soft and natural surface (terrain) into a strong and foreign material (building). Using a policy to promote cultural and educational activities, our proposal performs as a central character in the play, not only for the users of the facility, but also for the entire culture of Korea; becoming a new center for the discovery and empathetic learning of history.

    The concept for museums of cultural prehistory, particularly those that organizational framing in the process of research, have changed. Our design proposes an open, interconnected terracing of exhibition spaces. All exhibition rooms are separated by the platform circulation, while visibility is constantly maintained. This new form of curatorial framing allows the entire contents of the museum to be constantly appreciated from every position, reinforcing the magnitude of historical artifacts contained within.

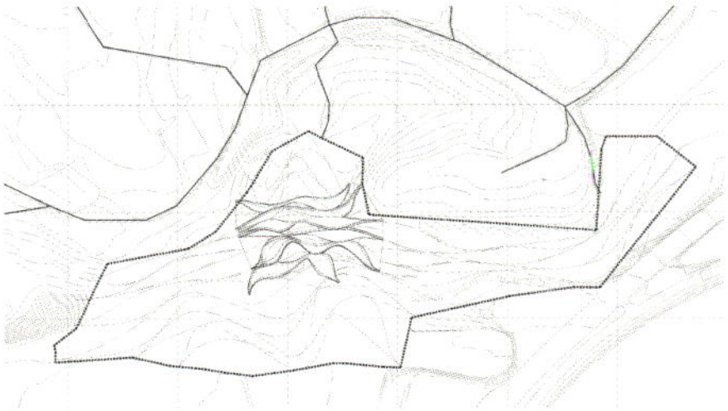

SITE PLAN

지형적 배치 개념을 보여주는 모델링 MODELING TO SHOW TOPOGRAPHICAL LAYOUT CONCEPT

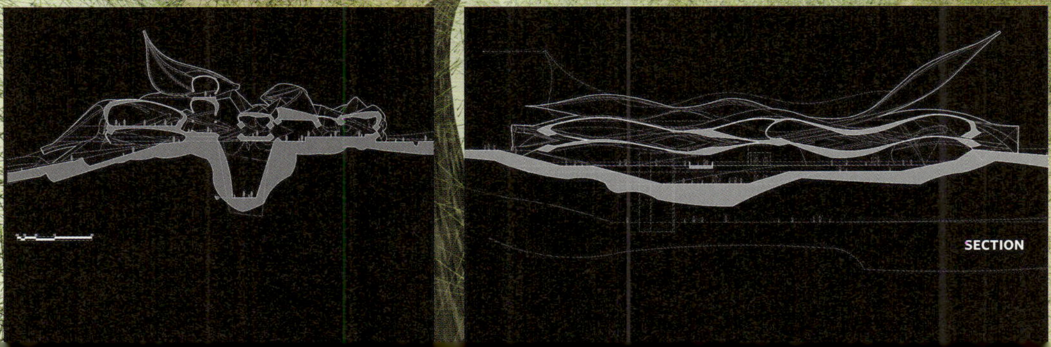

SECTION

Paul Preissner
Qua'Virarch, U.S.A.
Haarun, South Bank Planning Development
Cape Town, South Africa
2006

사우스뱅크 개발 계획은 남아프리카에 새로운 예술문화센터를 세우고자 하는 진취적인 제안이다. 농촌 지역을 세계적 수준의 새로운 문화공동체로 개발하고 탈바꿈시키며 비즈니스 구역, 고급 주거지, 콘서트홀, 박물관, 극장 등 문화시설과 2010년 월드컵축구경기 시설과 호텔 등 레저 프로그램을 갖추게 된다. 해당 부지는 주요 고속도로, 철도와 연계되는 중요한 인프라 연결의 합류점에 위치한다. 우리 제안은 문화와 교육 활동을 촉진하는 정책을 포함하기 때문에 시설 사용자뿐 아니라 남아프리카 전체 문화를 위해 중요한 요소로 작용한다. 이곳은 역사를 발견하고 학습을 위한 새로운 센터가 될 것이다.

The Southbank Site urban organization is an aggressive proposal for a new artistic and cultural center within South Africa. The plan is the new development and transformation of an agricultural site into a new an world class cultural community, complete with a central business district, high-end residential development, cultural facilities such as concert halls, museums, and theatres, and leisure programs including a proposed facility to accommodate FIFA World Cup 2010 matches and tourist hotels. The site lies at the confluence of several important infrastructure links, including the major highway and rail link.

Using a policy to promote cultural and educational activities, our proposal performs as a central character in the play, not only for the users of the facility, but also for the entire culture of South Africa; becoming a new Center for the discovery and empathetic learning of history.

URBAN FABRIC
VISUAL FLUIDITY
INTERACTIVITY BETWEEN PEOPLE
PUBLIC SPACE
FIBROUS GEOMETRY
RENEWABLE SOURCE
SAVING RESOURCE
COMMUNICATION
FLUID TOPOGRAPHY
MOVEMENT DIAGRAM
ZERO IN THE ENERGY REQUIREMENT

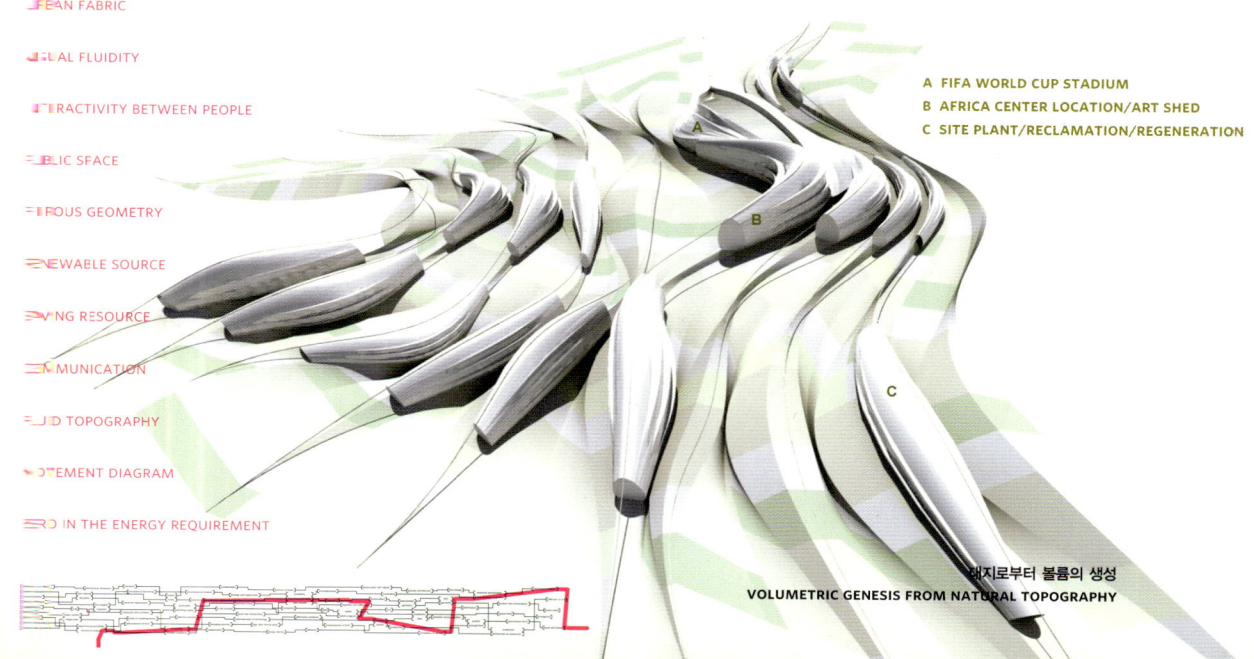

A FIFA WORLD CUP STADIUM
B AFRICA CENTER LOCATION/ART SHED
C SITE PLANT/RECLAMATION/REGENERATION

대지로부터 볼륨의 생성
VOLUMETRIC GENESIS FROM NATURAL TOPOGRAPHY

사우스뱅크 개발 계획

PAUL PREISSNER
QUA'VIRARCH

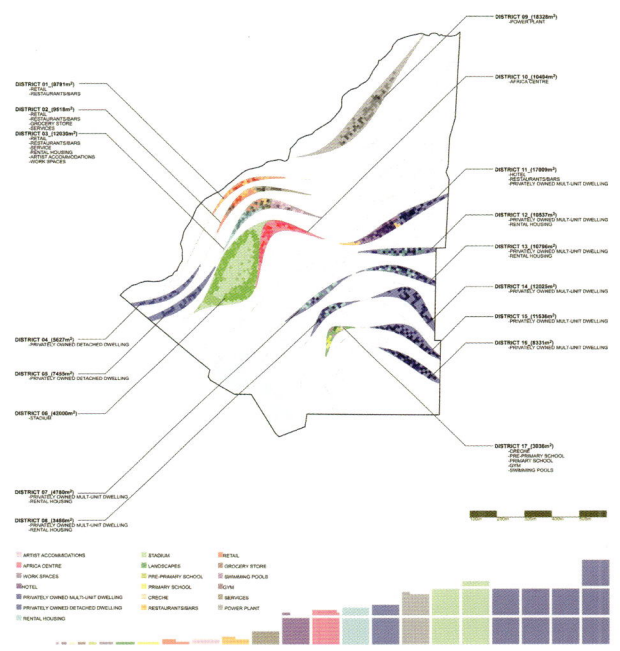

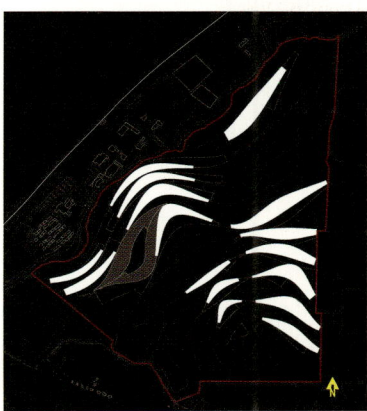

프로그램 배치 DISTRIBUTION OF PROGRAMS

배치 개념도 SITE DRAWING

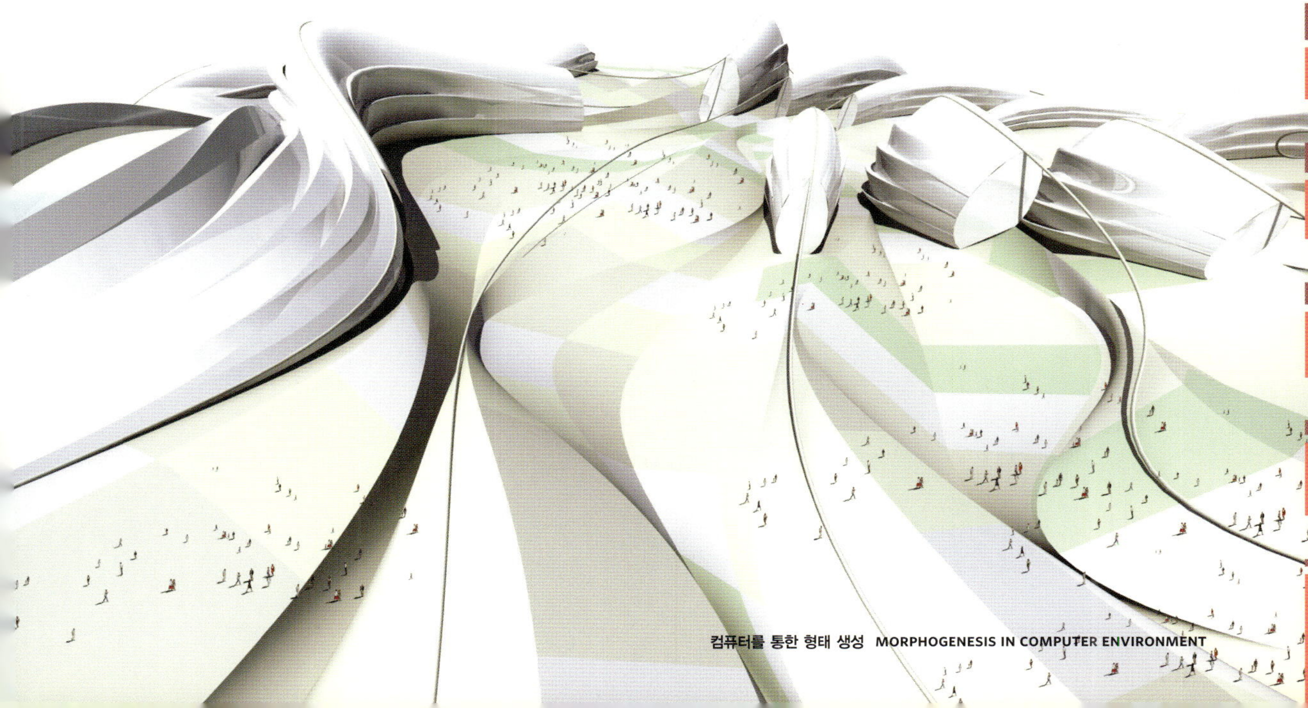

컴퓨터를 통한 형태 생성 MORPHOGENESIS IN COMPUTER ENVIRONMENT

Lonn Combs + Rona Easton
Easton+Combs Architects, U.S.A.
Jeongok Prehistory Museum
Competition Project_3rd Prize
Gyeonggi-do, Korea
2006

이 설계안에는 두 개의 벡터가 교차하고 있다. 뚜렷하게 구분되는 두 가지 인식, 즉 가치와 시간 사이의 개념적 관계가 구축된다. 하나의 벡터는 마음에서 출발해서 현재와 미래의 관계를 보여주는 수직선을 이룬다. 다른 하나의 벡터는 시간과 역사 그리고 미래에 대해 더 근본적이고 직관적인 생각을 보여주면서 지면에서 하늘로 비스듬하게 연결된다.

이러한 개념상의 교차점이 건축 구성과 소장품 전시, 보관의 전략적 핵심이 된다. 구석기 시대 유물 수백 점은 건물을 대각선으로 가로지르며 지면과 하늘을 연결하듯이 배치된 일련의 전시실에 수장될 것이다. 작은 전시공간의 운집 형태로 이루어진 전시실들은 그 중심까지 자연광이 들어오고, 통풍이 잘되는 구조로 계획되었다. 갤러리 아래로는 공원의 지면이 내려다보인다. 이러한 전략을 통해 건물의 거주 목적을 잠정적으로 제한함으로써, 자칫 손상되기 쉬운 부지의 생태환경에 미치는 영향력을 줄이고 박물관 부지와 선사공원 사이의 인접성을 보장했다.

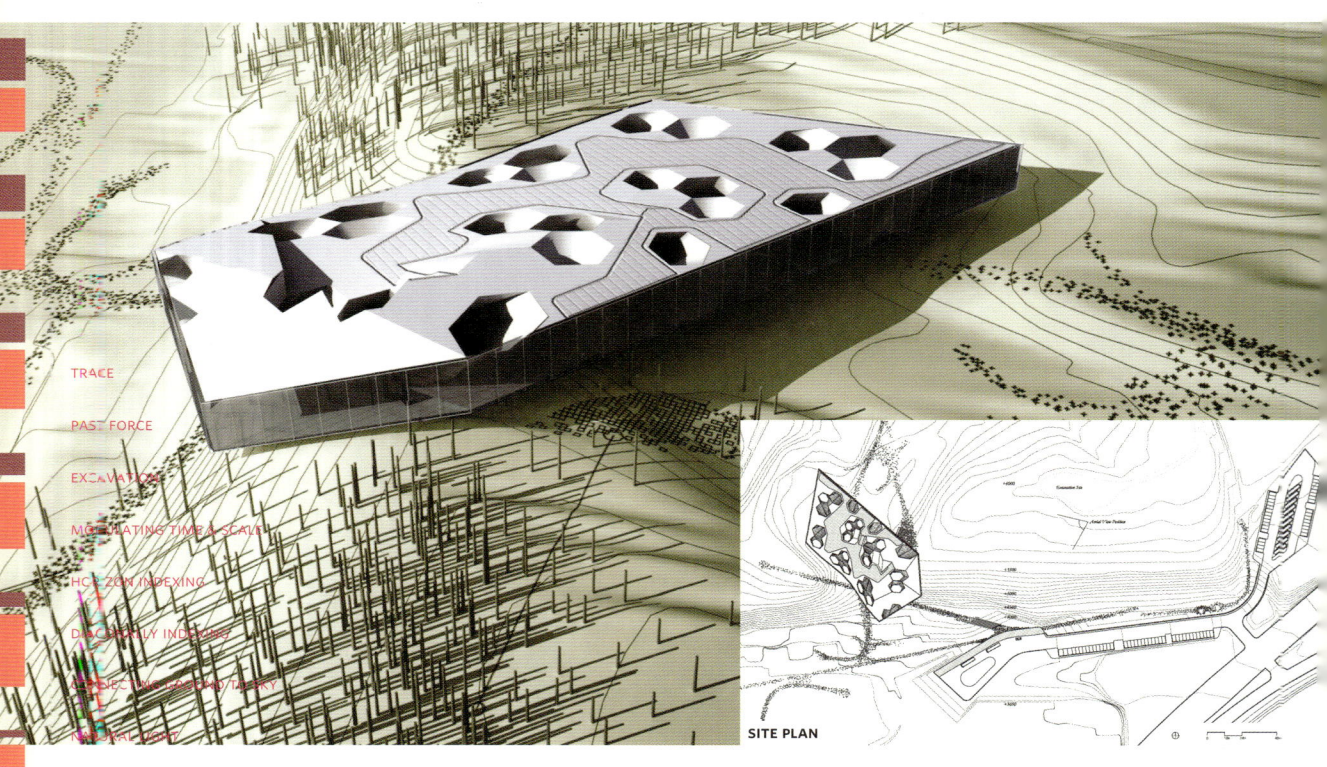

SITE PLAN

Lonn Combs + Rona Easton
Easton+Combs Architects

전곡선사박물관
현상설계 3등작

인접 공원 입구측에서의 조망 VIEW OF CONTIGUOUS PARK LEVEL AT ENTRANCE

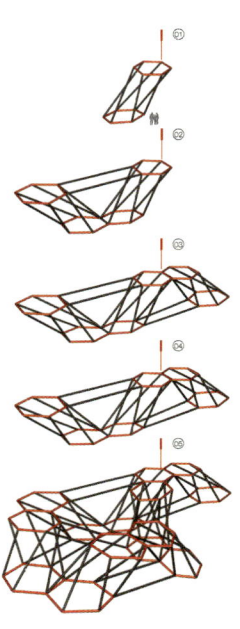

비틀린 형태의 구조 유닛을
구성하는 전략
TORQUED STRUCTURAL UNIT
COMPOSITE STRATEGY

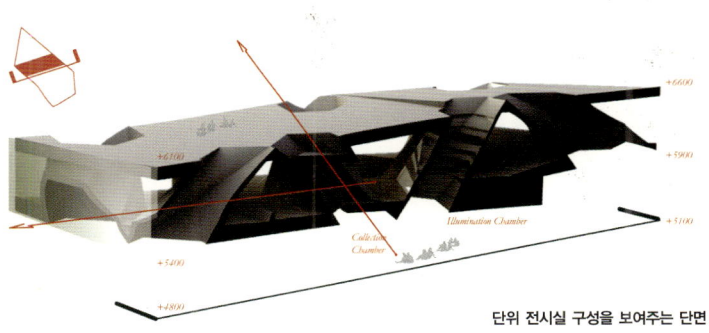

단위 전시실 구성을 보여주는 단면 모형
SECTION THROUGH EXHIBITION SHOWING COLLECTIONS CHAMBERS

Two vectors of consciousness intersect in this proposal, establishing a conceptual relationship between two distinct ideas about scale and time. One vector travels from the mind to the horizon, indexing the relationship of present time to the future. A second vector travels diagonally from the earth to the heavens, indexing a more primal and instinctual idea about time, history and the future.

It is this conceptual intersection that drives the organizational strategy of the architecture and the curatorial strategy for the collection. The several hundred artifacts of Acheulean and proto-Acheulean hand axe technology are housed in a series of suspended galleries cut diagonally through the building, connecting ground to sky. These galleries reside in a larger constellation of aerated chambers that allow natural light to travel through the main exhibition hall and reach the park-surface below. This strategy expresses the building as a suspended, inhabitable surface, thus reducing the scale of the footprint in the fragile ecology of the site and allowing the contiguity of park space between the site of the museum and the prehistory park.

Lonn Combs + Rona Easton
Easton+Combs Architects, U.S.A.
New National Library Czech Republic
Czech Republic

체코 신국립도서관은 500만 권 이상의 도서를 소장하게 되며, 이는 도서관의 진화와 더불어 현 체코 사회와의 관계에 기여하는 기회가 될 것이다. 소장품 규모 자체와 그 유지와 조직에 소요되는 상당한 작업이 일반대중이 접근하기 어려울 것으로 예상되었다. 이러한 이슈를 개념화하는 데서 기관의 새로운 패러다임이 등장했다. 얇은 층의 통기 전략을 통해 도서관과 대중의 관계를 구상하면서, 도서관 서비스 운영과 소장품 부분의 매스를 이용해 공적 공간을 묶는 새로운 공간 유형학이 발생했다. 이를 통해 직원과 공적 운영에서 접근이 용이한 풍부한 매트릭스를 생성하고, 대형 도서관 기능에서 요구되는 프로그램 분리를 엄격하게 유지한다.

LONN COMBS + RONA EASTON
EASTON+COMBS ARCHITECTS

체코 신국립도서관

The housing of the 5 million plus volumes of the New Czech National Library constitutes a unique opportunity in modern history to contribute to the evolution of the library and its relationship to contemporary Czech society. The shear scale and magnitude of the collection coupled with the substantial commitment to its maintenance and organization challenges the notion of accessibility with respect to the society at large.

It is in the conceptualization of this question that a new paradigm of the institution emerges. In framing the library's relationship to the public through a strategy of laminar aeration a new spatial typology emerges which binds the public spaces through the mass of the library's service operations and collections areas to produce a rich matrix of adjacency in staff and public operations whilst not compromising the strict programmatic separations required by the function of the large institutional library.

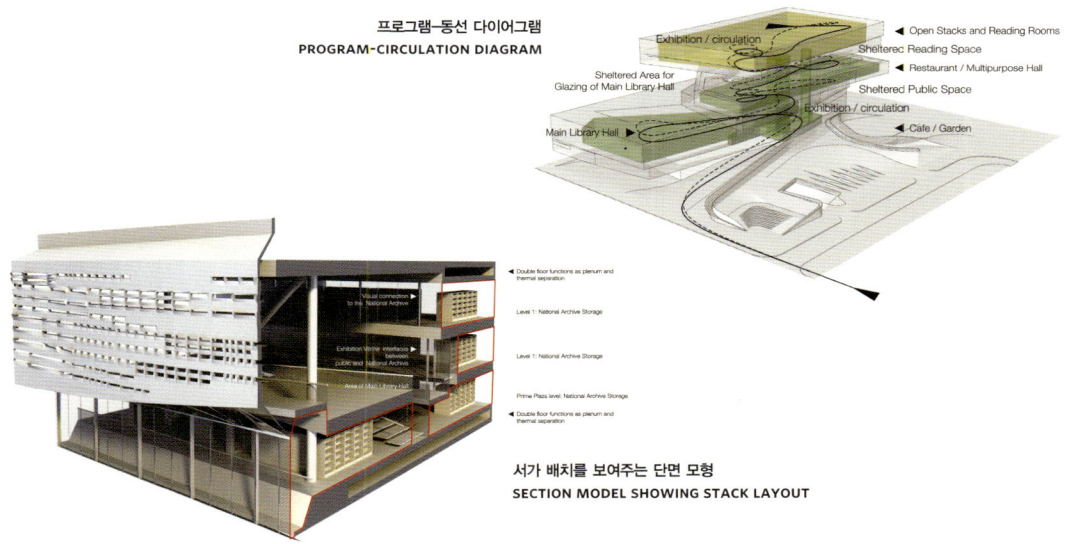

프로그램-동선 다이어그램
PROGRAM-CIRCULATION DIAGRAM

서가 배치를 보여주는 단면 모형
SECTION MODEL SHOWING STACK LAYOUT

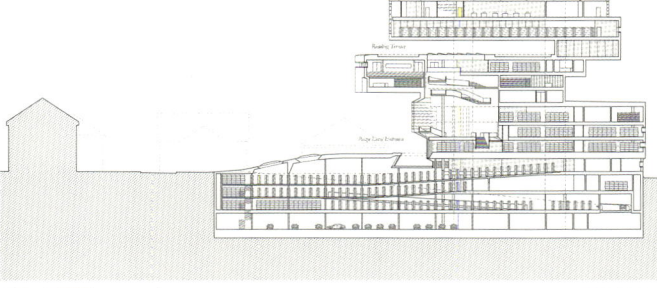

SECTION

내부 공간 투시도 INTERIOR PERSPECTIVE

Marcos Novak
UCLA, U.S.A. / Venezuela
Allo Series

그가 신종이라고 부르는 일련의 디지털 프로젝트는 건축이나 오브제 또는 미술로 정의되지 않는다. 다분히 구조적이지만 물상의 공식을 떠나고, 음악을 동반하며 그래픽으로 전개되지만 비디오아트로 불리기를 거부한다. 그의 작품은 원천적으로 비디오로 표현되어야 하기에 이 뜻을 인쇄매체로 전달하는 데는 한계가 있다. 그러나 그의 작업 과정에는 자연으로부터의 발상과 물리적 모형으로 확인하는 작업이 포함된다. 그가 말하는 'Allo'는 사전적으로는 의미가 없다. 이전에 없던 새로운 종의 탄생이기 때문이다. 기존 단어로는 명명할 수 없는 수사에 물리 또는 생물학적 은유가 더해질 뿐이다. 그래서 그의 작업을 굳이 건축이라고 하지 않는다. 미술이라고도 하지 않는다. – 박길룡

The "new type" of digital project as he calls it, is not a new type in the sense of architecture, objet or even art. It is quite structural but free of physical limitations, is accompanied by music and graphics but refuses to be called video art. As his work must be expressed in video, putting it into words is inherently and grossly limited. But his work process also includes the naturally occurring and verification through physical models. Allo is not in the dictionary. It indicates a new species heretofore non-existent. Physical or biological metaphor is added on a description that cannot be articulated using existing vocabulary. This is why his work is not called architecture. Nor is it called art.

- Text by Kilyong Park

MUTANT

MECHANISM

NATURE AND CULTURE

SIMULATE PROCESS

NATURE'S OPERATION

SIMULATIONS

BIOLOGY OF EVOLUTION

ALLOBIOLOGY OF INNOVATION

PRODUCTION OF THE ALIEN

RAPID PROTOTYPING

TECHNIQUES AND TECHNOLOGIES

ALGORITHMS

GENETIC CODE

FORMAL AND ORGANIZATIONAL PRINCIPLES

DUAL READING

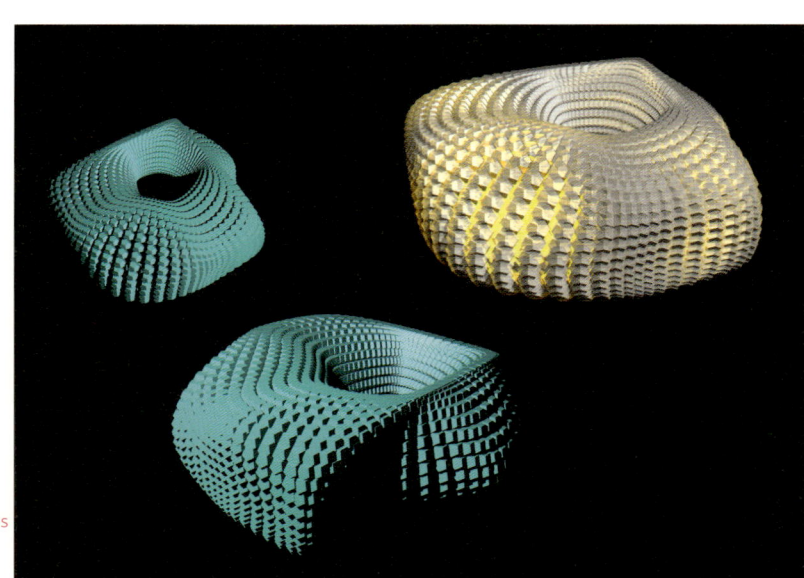

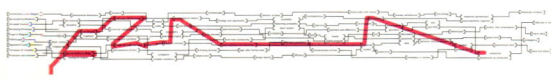

Allo Series — Marcos Novak
UCLA

Allo Cortex Shell

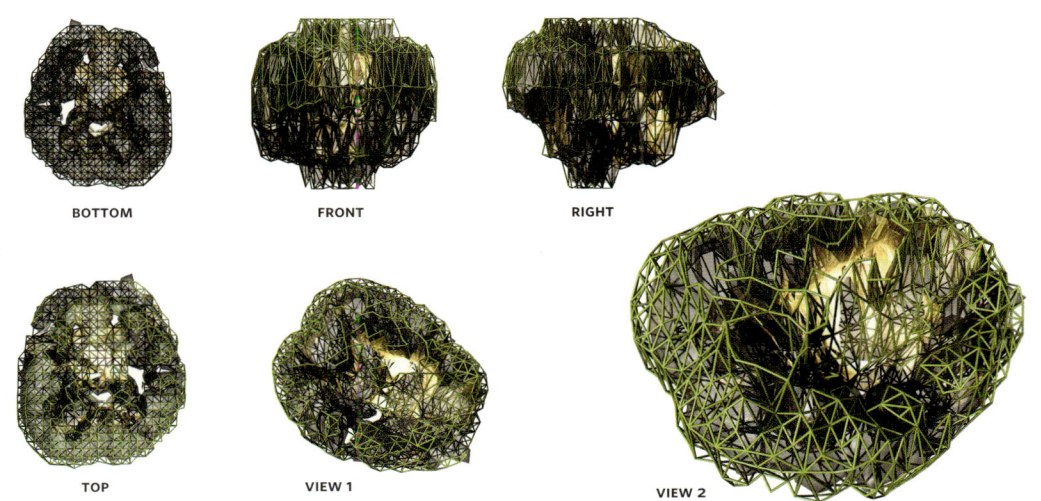

Allo Cortex Structure

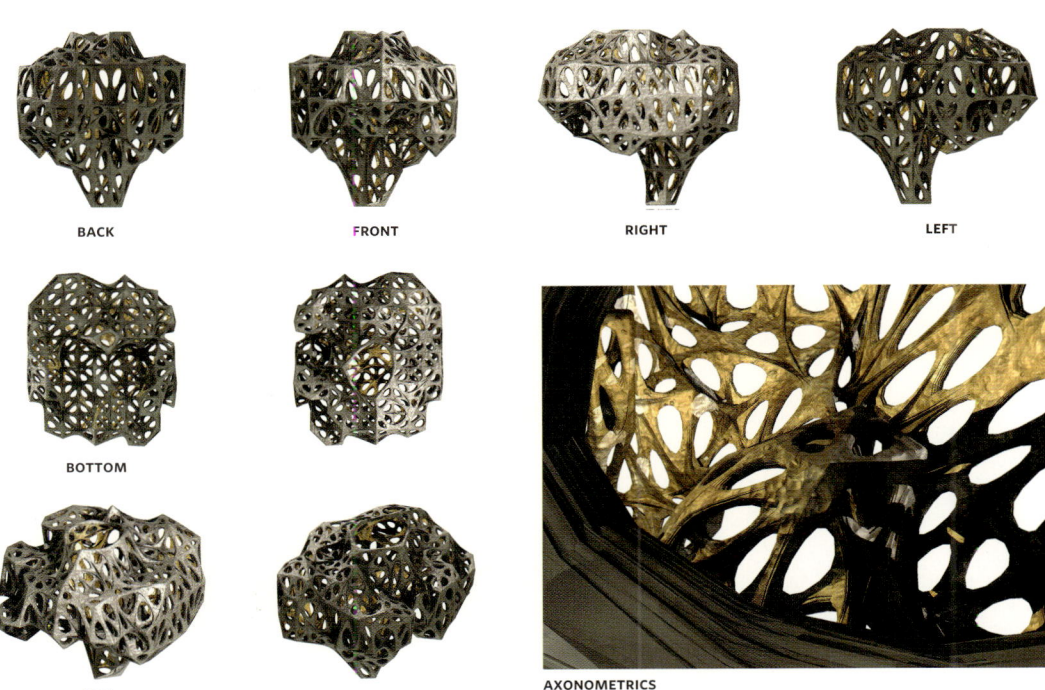

Allo Atoms

ALLO ATOM의 모형

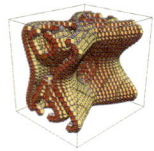
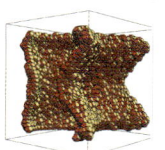
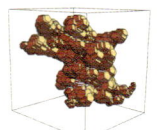
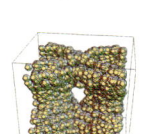

ALLO ATOM의 기하

Allo Neuro

Allo Series — Marcos Novak

Allo Genhome

MAK Z Renders

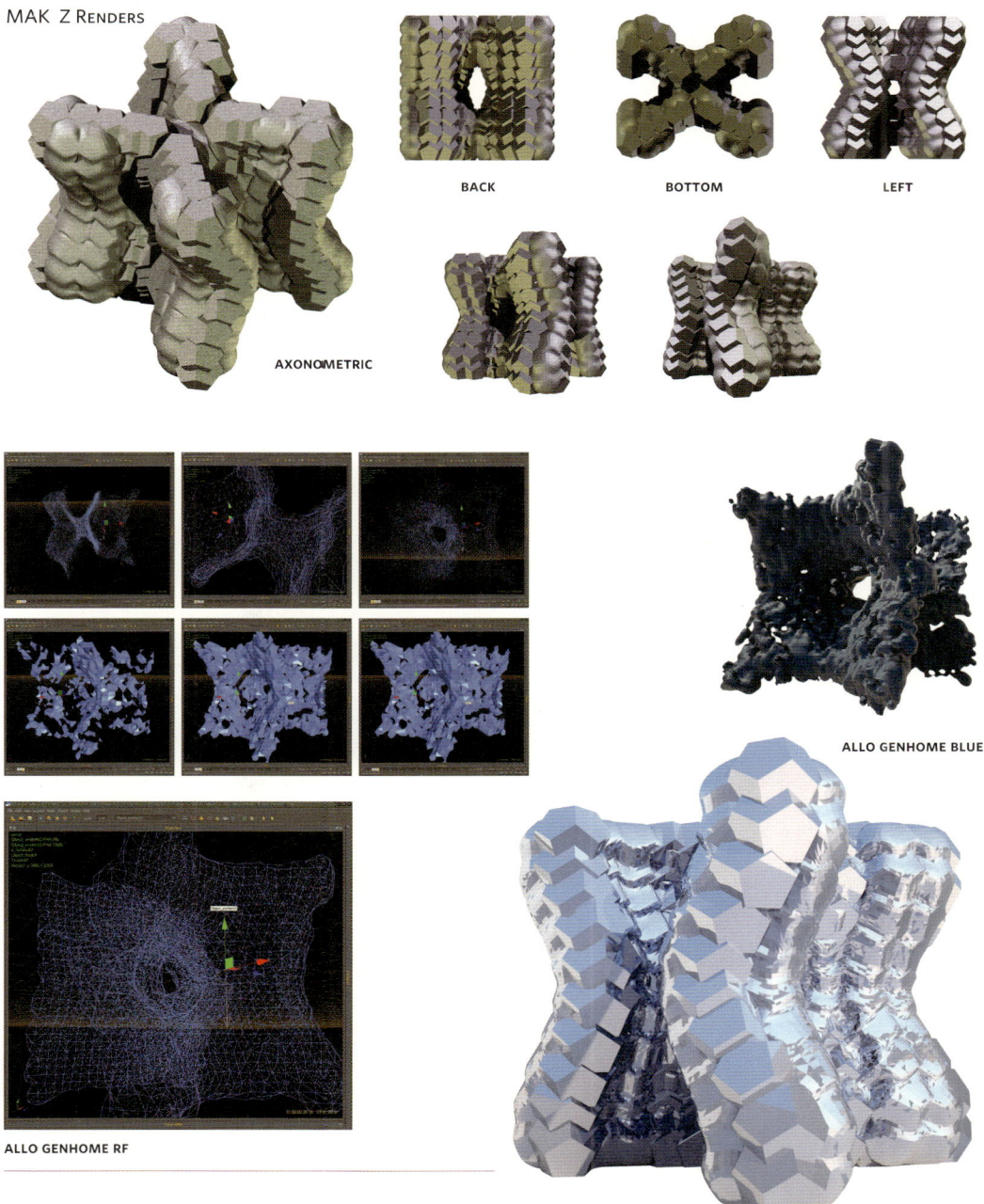

AXONOMETRIC BACK BOTTOM LEFT

ALLO GENHOME BLUE

ALLO GENHOME RF

Junsung Kim + Hailim Suh

hANd Architecture + Architecture Studio Himma, Korea

House of Open Books

Paju, Gyeonggi-do, Korea

2005

파주출판단지에 위치한 서가유형의 이 박스형 건물은, 초기에는 번역센터와 사옥 두 동으로 계획되었으나 그중 사옥동이 먼저 지어졌다. 이웃 건물에 내재된 한강 쪽 사이공간, 심학산 쪽 원경을 건물 내부로 끌어들이는 요소인 두 동 사이 보이드 공간과 '번역' 이라는 프로그램이 지닌 결코 만날 수 없는 수평의 선들이 계획의 밑그림이 되었다. 한강부터 심학산까지 이어지는 각 동은 인식과 움직임을 지속적으로 제공하도록 접힌 벽들로 구성되어 있다. 그 벽들은 각기 다른 인공 지형을 건물 표피나 내부에 끌어들임으로써 주변과 연계되고 있다.

번역센터동 중앙을 통과하는 접힌 반투명 유리스크린은 내외부의 빛과 움직임을 접힘으로 조율하며, 사옥동에서 접힌 콘크리트는 외피를 이루는 구조벽체가 되어 움직임과 경계 사이를 조율한다. 투명과 불투명 두 이질적 물성이 지니는 대조적 경험은 사이공간들에 건축적 긴장을 주고 있다.

Located in Paju Book City, type of the building was given namely "Bookshelf Type", of which the enclosed bar spaces are to be layered with voids in between. From the Han River on one side to the Simhak Mountain on the other, each bar building is organized through folded walls to produce continuous perception and circulation. Yet two bar buildings contain different kinds of artificial landscape to produce connection; one with folded translucent glass screen located in the center of the building where light penetrates as one moves through, and the other with the concrete folded wall wrapping the edges of the building as one negotiates the movement along the boundary.

VOID IN BETWEEN

FOLDED WALL

CONTINUOUS PERCEPTION

CONTINUOUS CIRCULATION

LIGHT

ARTIFICIAL LANDSCAPE

서가유형의 매스 BOOKSHELF TYPE MASS

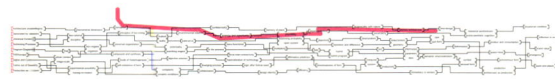

열린책들 사옥   김준성 + 서혜림
에이치엔드 + 힘마건축

틈새를 갖는 접힌 콘크리트 파사드   SPLIT AND FOLDED CONCRETE FACADE

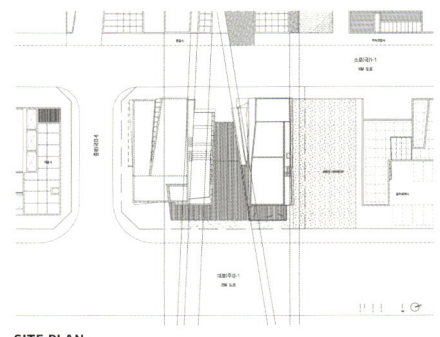

SITE PLAN

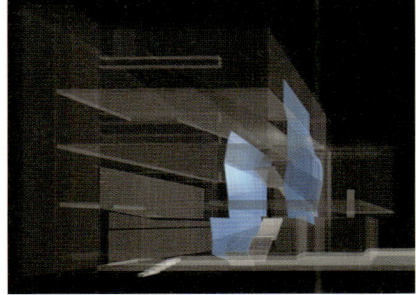

계단을 통한 연속적인 지각과 동선
CONTINUOUS PERCEPTION AND CIRCULATION THROUGH THE STAIRS

복도에서의 외부 조망
OUTSIDE VIEW FROM CORRIDOR

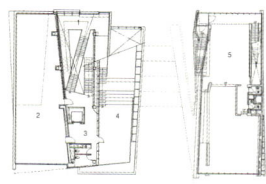

3F PLAN

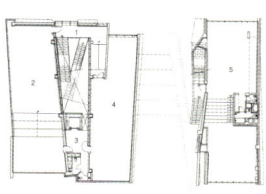

2F PLAN

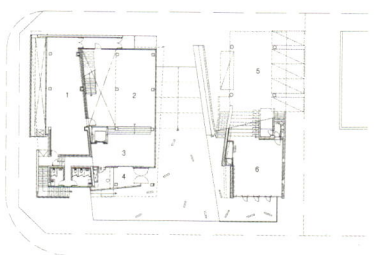

1F PLAN

열린책들 사옥

김준성 + 서혜림
에이치엔드 + 힘마건축

공간과 빛의 상호관입 INTER-PENETRATION OF SPACE AND LIGHT

내부의 공간조직을 보여주는 유리 입면 GLAZED FACADE SHOWING INTERIOR SPACES

Halim Suh + Junsung Kim
Architecture Studio Himma + hANd Architecture, Korea
Borim Publishing House and Marionette Theater
Paju, Gyeonggi-do, Korea
2004-2005

서가유형의 이 건물은 파주출판단지에서 심학산과 한강을 시각적으로 그리고 공간적으로 열리도록 연결하는 기본 틀에 따라 나열된 수직적 바 형태 매스(어린이 책 출판을 위한 사무공간)와 수평적인 인형극장 매스가 교차하여 위치한다. 2개 바 형태의 사무공간 건물은 이중 표피로 유리 커튼월과 펀칭 메탈 표면으로 구성되어 외부에서는 불투명한 단일 박스로 읽힌다. 하지만 내부에서는 시야가 풍경에 열려 있다.

풍경에 열린 시야는 내부에서 보이는 눈높이에 따라, 또한 심학산의 산세와 한강 물결을 연결하듯 펀칭 메탈 패널의 홀 크기가 변이되면서 수평적으로 물결친다. 반면 인형극장의 매스는 줄에 매달린 인형들의 움직임처럼 수직적으로 변이되는 프레임으로 구성되어 있다. 인형 다리가 바닥에 닿을 듯 떠 있는 것처럼, 극장의 매스는 외부의 그라운드 레벨에서 아이들 눈높이에서만 극장 하부 놀이 공간이 보이도록 살짝 들려 있다.

As "Bookshelf type" in Paju Book City, the building is organized with two vertical bar buildings providing office space for children's books publishing house which intersect with horizontal marionette theater. While the vertical volumes contain the multiple curved horizon lines embedded on the flat surface of the perforated metal panels, which seems invisible from the outside, yet the view towards outside is clearly perceived from the inside. The marionette theater's undulating structure has been generated from the animated fluctuating gravity lines based on movement. As the puppets' feet float from the floor, the theater is slightly lifted from the ground at the eye level of children.

DOUBLE SKIN

INTERSECT

UNDULATING STRUCTURE

ANIMATED FLUCTUATING GRAVITY

LINE

SLIGHTLY LIFTED

SKIN VARIATION

떠 있는 인형극장의 매스 FLOATING MASS OF MARIONETTE THEATER

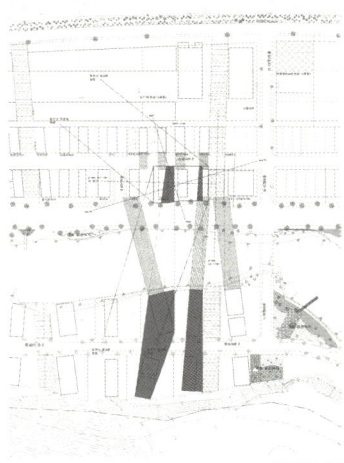
SITE PLAN

표면 스터디 모델 SURFACE MODELING

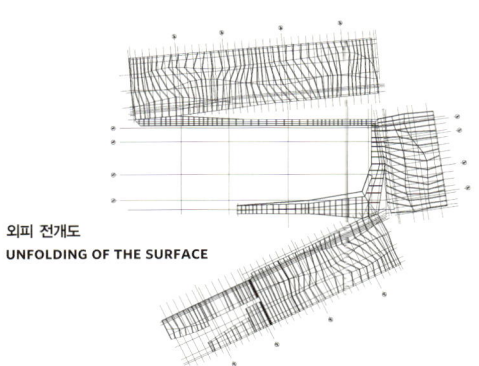
외피 전개도
UNFOLDING OF THE SURFACE

인형극장의 물결치는 구조를 위한 스터디
STUDY ABOUT UNDULATING STRUCTURE OF MARIONETTE THEATER

대조를 보여주는 입면 상세 CONTRAST IN ELEVATION DETAIL

Hailim Suh + Nader Tehrani
Architecture Studio Himma, Korea + Office dA, U.S.A.
Obzee Fashion
Seoul, Korea
2006

NEW ICON WITHIN THE CITY

SLIVER OF SPACE

PARAMETER

INTERACTIVE FUNCTION

FOLD THE PLANNING

VERTICAL LINKS

UREANISTIC IDEA

FAIR UP THE LOBBY

EXTENDING

PUBLIC BALCONY

SHARE

UREAN FABRIC

올림픽공원을 마주하고 있는 이 프로젝트는 한 회사가 자사 브랜드로 소유한 다양한 패션라인의 새로운 본사이다. 확장되는 스튜디오공간들과 이벤트공간으로 구성된 이 건물은 옆 대지의 기존 사옥과 서로 공유하는 사이공간을 두어 빛과 바람을 통과시키며 서로간의 시각적 커뮤니케이션을 추구한다. 이러한 개념은 의상디자인의 용어를 빌어 'Pleating'이라는 시스템을 도입하여 건물의 매스가 위로 올라가면서 접혀지고 변형됨으로써 다양한 빛을 드리우며 적극적인 유기적 관계를 의도한다. 단순히 올림픽공원을 마주하는 채워진 건물이 아니라 건물 사이에서 형성되는 도시의 '틈'공간을 적극적으로 디자인하여 도시의 풍경을 활성화한다.

이러한 도시적 개념은 건물의 로비공간에서 패션쇼, 이벤트 등을 개최하여 공공공간으로 접목시켜 런웨이와 경사진 공연공간들은 도시로 향해 연장되며, 거리의 활기를 확장시킨다. 건축을 구축하는 주개념의 하나는 의상디자인에서 쓰이는 테크닉과 효과를 건축적 용어로 번역하는 것이었는데, 이것은 구조와 스킨의 관계 설정이 중요했다. 건축의 구조와 스킨의 근본적인 차이를 수용하지만 시그램빌딩의 순수주의 접근이 아니고, 동시에 자유의여신상처럼 구조와 의장 사이의 이분법적 관계도 아닌, 구조와 표피가 서로 자기유사적인 기하학적 시스템을 지니며 관계하도록 구축되었다.

도로를 향한 주 입면 MAIN FACADE

오브제 사옥　　　서혜림 + **Nader Tehrani**
힘마건축 + Office dA

Facing the vast Olympic Park in Seoul, the Obzee Project will serve as the new headquarters for the various fashion lines that the company owns under its larger brand. The company currently operates out of a neighboring building, but it now seeks to expand its studio and public event spaces. Wedged between two existing buildings, the site will share a party wall on one side, but will remain apart from the current headquarters, leaving a sliver of space folded between them for better lighting, air circulation and visual communication. As the massing of building "turns the corner" into the common alley, this carving of the mass at the top rotates the building in a contra-posta fashion permitting more light while acknowledging that the building not only faces the park but also its sister building as well.

The basic strategy is to fold the planning efforts into section, creating vertical links between programs to foster the best working relationships. The lobby is paired up with the main public space of the building: a theater and an accompanying runway, where fashion shows, events, and trial runs can be held. This gives Obzee the opportunity to publicize its programs and events while extending the life of the city sidewalk into its domain. The organization of the main trunk of the tower is to create separate and autonomous studios, but to create connections between each design studio and their respective sewing/tailoring shops by way of double height shared spaces where easy communication can occur.

Having researched the art of tailoring - its techniques and its effects - our main goal was to translate it to the architectural medium in such a way that it formed the basis of an integrative strategy for the building, bringing the structure, cladding, mechanical and iconographic agendas into and organic relationship. Accepting the fundamental difference between structure and skin, we sought to create a symbiotic relationship between the two, using geometry as a vehicle. we designed a self-similar geometric system that establishes a direct correspondence between structure and skin, though at different scales. In essence, the inside of the building is like inhabiting a truss, yet the skin has a variety of functions that could gauge its geometry in relationship to its desired performance at any given instance.

All perforated panels are created from ruled surfaces, and thus developable for fabrication. In this way, no geometry is extraneous, and nor is any panel cut in inattentive ways. Thus, the skin is developed from the same morphological code as the structure, though pleated, folded, and fenestrated to respond to local circumstances - all within a strict language that is guided by the discipline of geometry.

표면 스터디 모델 SURFACE MODELING

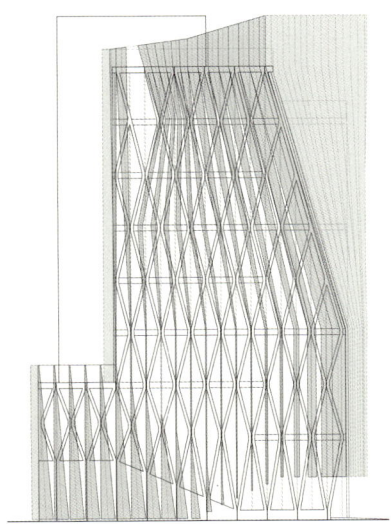

구조화된 입면을 위한 연구 STUDIES FOR STRUCTURED FACADE

오브제 사옥

서혜림 + Nader Tehrani
힘마건축 + Office dA

다양한 조직의 내부 공간들 DIVERSITY OF INTERIOR SPACES

Why Not?
형태의 인과율에 던지는 물음

늘길범
□□대학교 건축대학

> ...
> 얼굴 잠깐이야
> 맘 이쁜게 최고지
> 어른들은 항상
> 그렇게들 말하지
> 그 말만 철썩 믿고
> 마음만 가꿨더니 아이고 ♬
> ...
>
> 럼블 피쉬, 'I GO' 가사의 일절

일찍이 19세기 중반, 파리의 에콜 데 보자르Ecole des Beaux-Arts 교수였던 비올레 르 뒤Viollet-le-Duc 선생께서 말씀하셨다. ". . .눈에 보이는 형태와 같은 순수하게 예술적인 문제들은 우리의 '지배적인 원리들'에 비하면 부수적인 조건일 뿐이다." 오래 지나지 않아 시카고의 루이스 설리번Louis Sullivan 소장은 "형태는 기능을 따른다"는 일성一聲으로 역사에 이름을 올렸고, 이미 "형태를 발명해내는 것이 건축의 과업은 아님"을 깨달았던 베를린의 미스 반 데 로에Mies van der Rohe가 바로 그 건축의 진실에 관해 고민할 때, 모스크바에서는 OSA(현대건축가동맹)의 모제이 긴즈부르크Moisei Ginzburg 부의장이 "형태는 미지의 변수 X의 함수"라는 결론에 도달했다. 그리고 마침내 70년이 흘러 MVRDV는 이렇게까지 선언하기에 이른다. "주어진 프로그램에 대한 다이어그램적 해석과 재배치가 건물의 형태를 생성한다."

이견은 없는 것처럼 보인다. 형태의 유희에 지나지 않는다고 비판받았던 소위 해체deconstruction의 건축조차도 명시적으로 형태의 자존을 공표한 바는 없다. 건축의 전역사에서 보기 드물게 만장일치의 합의가 이루어졌던 바가 있다면 바로 형태 자체의 개념적 가치를 소거하고 그것을 종속변수의 위치로 강등시키려 했던 지난 200여 년의 부단했던 노고라 해도 무방할 듯하다. 이렇듯 건축 혹은 건축론의 역사는 형태가 그 자체로서 자족적인 개념의 경지에 오를 수 없음을 공론

SELF-EXISTENCE OF FORM

DIVERGING PROCESS

AFFIRMATIVE PREDICTION

INNOXIOUSNESS OF FORM

IMMEDIACY TO SENSES

SENSUAL FORM

LOGIC AFTER FORMAL SEED

HAVING-NO-REASON

Why Not?
a Question to the Law of Causality about Form

Ilburm Bong
School of Architecture, Kookmin University

> . . .
> Don't see the features
> See the hearts
> Daddy always prays
> I, trusting him, prune
> my heart only then. . . OOPS ♬
> . . .
>
> from the song "I Go" by Rumble Fish

Early in the mid 19th Century, Professor Viollet-le-Duc in École des Beaux-Arts once said, ". . . purely artistic questions of symmetry and apparent form are only secondary conditions in the presence of our dominant principles." Not long after then Louis Sullivan in Chicago raised his name to the Architecture Hall of Fame by the very notorious remark: Form Follows Function! When Mies van der Rohe in Berlin, who had already realized that the task of architecture was not to invent form, was struggling to find the way to the architectural truth, at Moscow, Moisei Ginzburg, the vice-chairman of OSA(the Union of Contemporary Architects) came to the conclusion that Form is a Function X. And finally 70 years from then on, MVRDV declared, "The diagrammatic interpretation and redistribution of the given program has generated the form of the building."

There seems not to be any different opinions. Even the deconstructivists, who were condemned for so called the play of pure form, has never elucidated the self-existence of form. It is through the ceaseless efforts of the past 200 years to devaluate the conceptual independence of form itself and to drag it down to the dependent variable that the whole history of architecture has rarely reached to unanimity. In this way the history of architecture or architectural discourse seems to be a consistent enterprise or the enormous conspiracy to make it public opinion that forms for forms' sake cannot be ascended to the level of self-sufficient concept. And this historical cruelties to form has not yet come to an end with us, who are always asking "WHY?" like a conditioned reflex whenever seeing designed forms of architecture.

화하기 위한 일관된 기획 또는 거대한 음모처럼 보인다. 그리고 그 '형태의 잔혹사'는 디자인된 건축의 형태를 보기만 하면 조건반사와도 같이 어김없이 "WHY?"를 묻게 되는 지금의 우리에 이르러서도 아직 끝나지 않았다.

건축의 설명과 비평이 "WHY?... BECAUSE..."의 구조에 의탁하기 십상인 이유는 아마도 그것이 실제로 지어지기 전, '앞으로 이렇게 하리라'는 계획만을 대상으로 평가와 판단, 투자가 이루어져야 하는 (현실적으로 불가피하기는 하나) 다소 기형적이라고도 할 수 있는 과정을 통해 생산되기 때문일 것이다. 그러므로 "WHY?"의 물음과 그에 대한 답변들이라는 인과율에 따른 사고체계는 이처럼 실체가 존재하지 않는 상태에서 거의 유일하게 설득력을 갖는 도구가 될 수밖에 없다고도 할 수 있을 것이다. 그런 이유로 간혹 우리는 지어지고 사용되는 증거가 도래했을 때, 혼란에 빠지거나 당위화의 논리를 새로 찾거나 아니면 애써 외면하는 것 외에 달리 탈출구를 갖지 못하는 '논리적인 건축가/비평가'의 딱한 처지를 목도하게 되는 것이 아닌가...

이 글의 문제 제기는 지극히 단순하다. "이유 없는 것은 없다"는 인과율에 기반한 사고의 준거 또는 '통제율'이 "이유 없는 것은 있어서는 아니 된다"는 금기의 억압 또는 '배제율'로 변질되어가고 있는 것은 아닌가를, 아니 이미 배제율이 되어 건축가들의 발등을 찍고 있는 것은 아닌가 묻고 싶은 것이다. 그리하여 우리가 던지는 단 하나의 물음이 이와 같아진다.

"WHY NOT?"

수학적 참/거짓의 문제가 50:50의 균형이 아니라 '오직 하나 : 그밖의 모든 것' 사이의 문제이듯, "WHY?"와 "WHY NOT?"의 문제 역시 동등한 이항대립은 아니다. "WHY?"의 물음을 따라 이루어지는 일의적一意的 프로세스는 오로지 단 하나의 명시적 해답이라는 참을 위해 암묵적으로 또는 우회적으로 해답이 될 수도 있었을 모든 거짓을 말살시킨다. 그것은 더 이상 분기와 비약, 전환과 융합을 허용하지 않는다. 다시 말해 일의적 연쇄구조를 따라 진행되는 인과율을 배제율로서 고집하게 될 때, 그 결과는 다음의 세 가지 중 하나가 될 수밖에 없다는 것이다. 결과로서의 형태가 무미건조할지라도 인과율에의 부합 하나에 만족하고 말거나, 실제

The reason why the explanation and criticism of architecture are apt to rely on the catechism of "WHY? . . . BECAUSE . . ." is that architecture is produced through the (unavoidable in reality though but) abnormal procedures in which almost all the evaluation and judgement are done before its realization in accordance with the project of what will be. Therefore the way of thinking depending on causality between the question of "WHY?" and its answer cannot help but become the only convincing way before construction of the real. We often see, as a result, the unpleasant "logical architects/critics" facing the advent of the empirical evidences constructed and used, who are just confused, searching for the new logic of justification, or intentionally overlooking those evidences.

So I would like to raise a very simple question. I would doubt whether the reference system or control rule based on the law of causality that there is nothing without reason becomes increasingly the rule of exclusion or suppression that there must be nothing without reason; whether it has already been petrified into the rule of exclusion and shackled the freedom of architects. Hence my simple question is:

"WHY NOT?"

As the True/False problem in mathematics is not that of alternative equilibrium but that of "between only one (true) and all the other (falses)," so the problem of "WHY? or WHY NOT?" is not the problem between the two equivalent alternatives. The univocal process always referring to an answer to the question of "WHY?" obliterates all the Fs that could be the Ts in implicit or suggestive ways for the sake of only one explicit T. It does not allow any divergence, leap, conversion, nor crossbreed. In other words, when you adhere - in architectural design process - to the law of causality as the law of exclusion based on the univocal chains of logic, the result has a difficulty in escaping from one of these three: satisfying only with the coherency to the law of causality in spite of boredom of resultant form, retroactively making up the "process" in accordance with the evident result as if it has perfectly followed the univocal line of logic, or lamenting the sterility of not reaching the stage of embodiment with paranoiac virginity of logic.

Architectural design is not so much reasoning as incantation. It is to make what there was not; to say the magic words having a premonition of what there will be. For that reason, the process of architectural design is not that of convergence but that of divergence. In other words, it is the process of paradox in which adjudication precedes an evidence; interpretation precedes a dream. We can, I suggest, break the

로 일의적 연쇄구조를 따르지 않았던 과정을 그 과정의 결과에 맞추어 소급적으로 그러했던 것처럼 포장하거나, 그도 아니면 논리의 순결주의에 매여 구체화의 단계에까지는 이르지 못하는 불모를 애석해하고 말거나. . .

건축은 추론이라기보다는 차라리 주술呪術이다. 있지 않았던 무엇인가를 있게 만드는 일, 이러이러한 모습으로 있게 되리라 직감하고 예언하는 주술이다. 그런 이유로 그 과정은 수렴적인 것이 아니라 발산적인 것이며 물증에 앞서 유무죄가 선고되는, 꿈에 앞서 해몽이 선행하는 역설적인 과정이기도 하다. 이 위험한 과정 속에서 누군가 거쳐 가지 않았던 새 길을 찾는 일, 그것이 곧 "WHY NOT?"으로 "WHY?"를 대체하는 간단한 사고의 전환으로부터 시작될 수 있다는 것이 이 글의 주장인 바. . . 발상을 바꾸어 "WHY NOT?"을 묻는 순간, 우리는 단 하나의 참이라 '믿어 왔던' 것을 제외한 그 외의 모든 가능성의 보고寶庫를 손에 쥐게 된다. 그리고 이것은 "WHY NOT?"의 응답이 "WHY?"의 응답과 같은 방식으로는 이루어질 수 없음에서 그 타당성을 얻는다.

만일 우리가 "WHY NOT?"의 물음에 "합당한 이유가 없기 때문"이라고 답한다면 그것은 질문에 답을 한 것이 아니라 질문 자체를 회피하는 것이라는 사실을 인정하는 데서 "WHY NOT?"의 방법론이 갖는 함의를 이해해야 한다. 풀어 이야기하자면, 우리가 만일 "왜 안 되는가?"의 물음에 "타당한 이유가 없기 때문에 안 된다"고 답한다면 그것은 곧 "왜 안 되는가?"의 이유를 찾은 것이 아니라 "왜 안 되는가?"라는 물음 자체를 부정함으로써 (그렇게 부정된) 물음에 답할 필요가 없음을 이야기한 것이 되고, 그것은 곧 " '왜 되는가?'가 아니기 때문에 안 된다"는 동어반복의 폐쇄회로에 빠지게 됨을 의미하는 것이다. 이와 같은 논리라면 우리는 "왜 되는가?"의 모든 물음에 대해서도 마찬가지로 " '왜 안 되는가?'에 답할 수 없기 때문에 된다"는 형식논리상의 우회로를 해답으로서 제시할 수 있게 된다. 그렇게 되면 답하지 못할 "WHY?"는 존재하지 않게 되고, 인과율은 급기야 배제율조차 아닌 절대율의 경지로 승격되고 만다.

모든 참인 경우를 일일이 귀납적으로 증명하는 것이 불가능하다는 현실의 제약으로부터 파생된 반증법, '가정이 거짓이라는 전제가 거짓임을 보여 결론이 참일 수밖에 없음을 입증하는' 반증의 방법과 마찬가지로 "WHY NOT?"에 정확히 답함으로써 "WHY?"의 답이 될 수 없는 대안들을 가려내는 소거법적 방법으로 쓰이지만 않는다면, "WHY NOT?"의 물음은 인과율의 제2항을 찾아내기 위한 방법

fresh ground and find a new way nobody could go in these contradictory conditions by the simple switching of conception to the way in which we will substitute "WHY NOT?" for 'WHY?"; we will be able to acquire a treasury of all the possibilities except only one that we have believed as true as soon as we switch to ask "WHY NOT?" And it is possible because the way of answering to "WHY NOT?" is different from that of answering to "WHY?"

The first step to understand the implication in methodology of asking "WHY NOT?" must be to acknowledge that it is not answering but detouring the question if you answered "No" to the question of "WHY NOT?" by saying that there was not the proper reason. Put in other way, if you made an negative answer to the question "WHY NOT?" by saying that it is because there you could not find the reason, i.e. proper answer to "WHY?", your answer would not be direct response but negation of the question itself. By doing so, it should be clear, you just say no more than that you need not to response to the negated question and you are now trapped into the loop of detoured tautology: WHY NOT? NO because NOT WHY! If we accepted the logical consequence like this, we would be able to make affirmative answers to all the question of "WHY?" through the evading formal logic by saying that WHY? YES because NOT WHY NOT! Then there is no question "WHY?" to which we cannot answer; the law of causality will be even the absolute rule, not to say the rule of exclusion.

Questioning "WHY NOT?" should be the way not to make a consequence of the second term from the law of causality but to define the first term in the logical sequence to develop the design, if only - like the method of *reductio ad absurdum* which verifies the consequence from the wrongly assumed premise based on the law of non-contradiction, which is derived from the fact that it is actually impossible to verify all the true cases one by one - it were not used as the tool for erasing the alternatives that could not be the answer to "WHY?" by means of negative answering easily to the question of "WHY NOT?". In other words, it is the meaning of the conscious questioning "WHY NOT?" to convert the architects' profession of imagining resultant forms from the past-oriented explanation (there comes this decision because of this reason or that logic) to the future-oriented affirmative prediction (there will be such a good things as a result of this form becoming). Form as the first term stemming from the question "WHY NOT?", therefore, is not any more univocal, dependent, nor convergent. Subversion of the logical order as such will be the very simple but effective way to rid architectural design of the blind spots, which has been blinkered by the law of causality.

이 아닌 논리의 출발점이 될 제1항을 규정하는 방법이 되어야 한다. 달리 말하면 결과로서 드러날 형태를 상상하는 건축가의 작업을 "이러이러한 이유와 논리적 과정이 있었기 때문에 그 결과로 이와 같은 결정이 이루어졌다"는 과거지향의 설명으로 끝맺는 것이 아니라 "이와 같은 것이 생겨남으로써 그 결과로 앞으로 이러이러한 좋은 일들이 벌어질 것이다"의 예언적 긍정으로 전환하는 일이 "WHY NOT?"을 묻는 의의라는 것이다. 그러므로 "WHY NOT?"으로 출발하는 제1항으로서의 형태는 더 이상 일의적이지 않으며 종속적이거나 수렴적이지 않다. 이와 같은 논리상의 선후관계에 대한 전복은 인과율의 눈가리개로 인해 볼 수 없었던 건축 디자인의 사각死角을 밝혀보기 위한 간단하면서도 효과적인 하나의 방법이다.

그렇다면 인과율의 철가면을 벗겨주어야 할 억울한 유배자로서, 한국 건축의 현재가 내면에 숨기고 있을 새로움의 광맥을 짚어보기 위한 탐침의 하나로서, 왜 이 글은 유독 '형태'에 주목하는가? 그것은, 우선 이처럼 장구한 억압의 역사를 지녔음에도 건축가들의 혈관 속에는 늘씬한 미녀의 각선미를 (혹은 근육질 장정의 팔뚝을) 훔쳐보듯 형태를 힐끗거리게 만드는 특수한 종류의 감각수용체가 녹아 흐르고 있음을 부인할 수 없기 때문이기도 하지만, 그보다 더욱 진솔한 이유는 바로 형태가 다른 어떤 건축의 요인들보다도 그것의 실질적 효용이라는 측면에서 '무해' 하다는 단순한 사실이다. 역설적으로 말하자면 형태는 '중요하지 않기 때문에' 그만큼 자유로울 수 있고 자유의 여지가 큰 만큼 중요하다는 것이다. 건물의 '생김새'가 조금 이상하다 한들, 조금 낯설고 생경스럽다 한들. . . 그것이 대관절 무슨 문제가 되겠는가?

그런 이유로 지금 이 글이 주목하고 있는 형태라는 것이 비례체계나 기하학 따위의 엄연한 상위의 법칙을 인과율의 제1항으로 내포하고 있는 종류의 미학적 범주에서 바라본 형태가 아님은, 속물적인 고급취향의 심미안을 가질 때에만 비로소 알아볼 수 있는 '아는 만큼만' 보이는 종류의 세련되고 현학적인 형태가 아님은 물론이다. 반복과 증식을 통해 각인되고 다수의 지지에 의해 공인됨으로써 일정 기간 주관적인 검증의 고단함을 면케 해주는 복수複數적인 '스타일'로서의 형태가 아님도 물론이다. 사고를 통해 이해할 필요가 없는, 타인의 판단과 비교해 감상능력의 수준차를 고민할 필요가 없는, 다수의 합의 속에 참여하고 있다는 안도감으로 안일한 면책을 득할 수도 없는, 무엇보다 스스로의 감흥을 인과율을 빌

If then, what is the reason of my arguing that form is the unjustly accused exile from whom we must remove the iron mask; what is the reason of my attention to form as one of the probes for searching the vein of "newness", which could be buried under the surface of Korean architecture? it is not only because I cannot deny that there flows in the blood vessels of architects, in spite of that long history of oppression, the special kind of receptors susceptible to form just as we are keenly susceptible to shapely legs of beautiful girls (or the muscular arms of nice guys) but also because -it is a simple but more straightforward reason- form is less intoxicating than any other constituents of architecture in terms of operative effects. To say in paradoxical way, form can be liberated as much because as it is "unimportant", and at the same time it is as much important as it has enormous room to be free. Let the shape of a building be unfit or ugly, be strange or scandalous; . . . what does it matter?

In these reasons, it is needless to say that now I am arguing about neither the form seen from the traditional aesthetic view which has the inherent dominant rule such as proportion or symmetry as the first term of the law of causality; nor the form too much refined or sophisticated which can be recognized only when we have the snobbish sense of discrimination. It is not the form as a collective style which exempt us from the burden of subjective verification for a certain moment through the acquaintance by repetition and proliferation and the authorization by the approval of majority. Again, I want to see such a form that does not need comprehension through thinking; does not need worry about level of appreciative power in comparison with others'; does not allow the immunity by the relieved feeling that you participate in agreement of majority; and above all does not need concealing the inspiration of yourself by a clear explanation in accordance with the law of causality. In short, I want to see such a never-repeated sensual form that penetrates directly into our senses and make immediately the elaborated chain of causality incompetent.

In terms of the form as this, we can realize that the phrase, "logic of morphogenesis" is a contradiction in itself. For, considering logic as the mode of thinking operation (i.e. the control rule or to a certain extent the rule of exclusion) to persist in the premises by neglecting all the sensed facts that are contradicted to those rationally preset premises, it is clear that the process of diverging, proliferating, evolving, developing to "generate" form contradicts with the univocal line of logic. In other words the forms which, we usually regard without any doubt, are thought as the logic or more precisely the law of causality generates may have been already given from the very starting point of logical process, or at the place of the first term of causality; nevertheless, they were only covered up with the words, which is the

어 '앞뒤가 맞는' 명료한 설명으로 포장해야 할 필요를 갖지 않는... 순식간에 우리의 감각을 파고들어 치밀한 인과의 연쇄들을 무력화시키고 마는 단 한 번의 관능적인 형태들에 우리는 주목하고 싶은 것이다.

그와 같은 종류의 형태를 생각하다 보면 형태 생성의 '논리'라는 말 또한 그 자체로 모순을 내포하고 있음을 깨닫게 된다. 논리란 이성으로써 정한 원칙에 위배되는 감각된 사실들을 모두 배제함으로써 그 원칙을 끝까지 지켜나가는 작동기제, 다시 말해 사고의 통제율이자 이미 일정 정도 배제율임을 생각하면 형태를 '생성'한다는 분기, 확산, 진화, 발전의 과정이 논리의 작용과 충돌함은 자명하기 때문이다. 달리 말하면 논리, 좀 더 정확히는 인과율의 과정이 '낳았다'고 생각되는 형태들은 그 논리의 출발지점 즉 인과율의 제1항에서부터 이미 정해져 있었는지도 모른다는 것이다. 다만 감각과는 다른 수용과 표현의 경로를 사용하는 이종異種의 기호체계 즉 '말'로 포장되어 있었을 뿐... 형태 생성의 '논리'를 거론하는 것은, 그러므로 가장 부정적인 의미에서의 실증주의positivism, 또는 전문가로서의 지적우월감으로 벌거벗은 몸을 가리려는 자기기만으로 전락할 위험과의 불편한 동거일 뿐이다.

사고의 통제율로서 작용하는 '논리'는 형태를 생성시키기 위한 과정이 아니라 (그 이유를 정확히 말할 수 있든 없든) 이미 만들어진 형태가 갖게 될 가능성을 발견하고 해석하고 구조화하는 '형태 이후'의 작업이 가져야 할 작동기제가 될 때 비로소 제자리를 찾는다. 형태의 이전에 작용하는 것은 육감, 직관, 영감, 상상... 아니면 잘 해야 모방, 각색, 윤색, 번안이 아니겠는가! 종속적인 인과율의 제2항이 되는 대신에 독립적인 논리적 전개의 제1항으로 자존하게 될 때, 형태는 진부한 말의 향수에 오염되지 않은 그 순수한 살냄새로 우리의 감각에 직핍直逼하게 될 것이다. 소심한 도덕주의, 미숙한 순결주의, 더 나아가 자기만족과 합리화에 머무는 지적 방기放棄가 될 수도 있는 형태 생성의 '인과율'이 아니라 오히려 아무 이유도 없는 불경한 것들을 감히 만들어 놓은 자로서의 선구자적 뻔뻔함에 주목하는 것이야말로 최소한 지난 10여 년을 옭아매어 온 주지주의 또는 지적 우월감과 배타적 합리주의로부터 건축의 디자인을 해방시키는 첫걸음이 될 것이다.

한국 건축의 현재에서, 이미 그 뻔뻔한 선각자들은 힐끗거리던 시선을 당당히 들어올려 "이유 없는 형태"의, "인과율로 설명되지 않는 느닷없는 형태"의 무르녹은 관능을 똑바로 응시하기 시작했음이 분명해 보인다.

different kind of sign system using different paths of impression and expression from senses. To mention the "logic" of form-generation is, therefore, no more than uncomfortable sharing the same bed with the danger of positivism in its most negative meaning or self-deceit to hide a naked body with the intellectual superiority complex as a professional.

The "logic" operating as the control rule of thinking can be properly placed not at the preliminary stage of form generating but at the stage after form making in which we should probe, interpret, and articulate the possibilities of readily made form (regardless of the exact reason of that form). What if it were not intuition, inspiration, imagination, or in other way, imitation, adaptation, or transformation that operated before generating the seed form? Instead of being the dependent second term in deductive process based on the law of causality, form must be the independent first term in inductive processes; then it will rush on to our senses not with dazzling perfume of the words but with pure smell of its flesh.

Taking notice of those having-no-reason, shameless pioneering conducts, not of the law of causality for morphogenesis that is apt to be polluted with hypocritical moralism, naive virginity, or intellectual negligence satisfying with only self-consolation or justification: it will be the first step to unshackle the architectural design from obsessive intellectualism, intellectual superiority complex of architects, or exclusive rationalism of at least last 10 years.

At the present of architecture in Korea, those shameless pioneers must be already starting to stare at the ripe sensuality of the "form without reason", of the "abrupt form inexplicable by the law of causality".

Hoon Moon
Moon Hoon Design Lab., Korea
New-Type-Body-Architecture 1.0

바이오-에로-테크노-판타지-재생공간

지구상의 모든 생명체에게 배운다. 외강내유인 홍합은 부드러운 쿠션과 같은 공간을 상상하게 한다. 그 가볍고 단단한 껍질은 우리의 영약한 몸과 마음의 보호자이자 그에 대한 시그널이다. 래피드 프로토타입이 건축 스케일에 적용되는 그날부터 우린 더 복잡하고 미묘한 공간에서 일상을 보내리라. 육체의 노-가다, 노-프레임들은 온데간데없고 뉴-간담들이 조립하는 그런 공간에 살면서, 정신의 노-가다를 꿈꾼다. '열려라 참깨' 라고 외치면 바위도 열리며, 거대한 조개 형상의 공간도 열릴 것이다. 전자제품의 공간화 속에 아마도 너바나-방도 곧 개발되리다. 기적을 바라보고, 꿈을 되씹는 그런 공간에서 많은 시간을 보낼지도 모른다.

Bio-Ero-Techno-Fantasy-Recycling-Bin (BETEFARB)

Fresh mussels are great when they are just boiled. The hard crust is a contrast to the softness of the inner mass. They stick to rocks with bio-bonding mechanism stronger than anything we have invented yet. We are slowly learning the secrets... what shall we find within the softness of the inner space? We will find what we imagine. It always has been.

BIO
EROTIC
TECHNO
FANTASY
RECYCLING
SUSTAINABLE
MIRACLE
DREAM

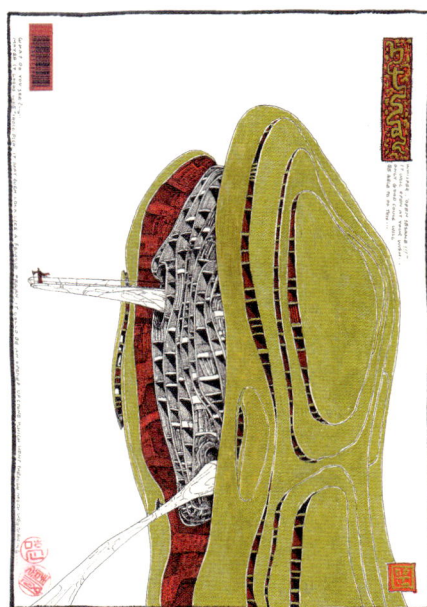

개념 스케치
CONCEPTUAL SKETCH

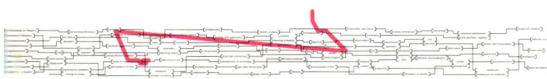

신몸건축 1.0

문훈
문훈건축발전소

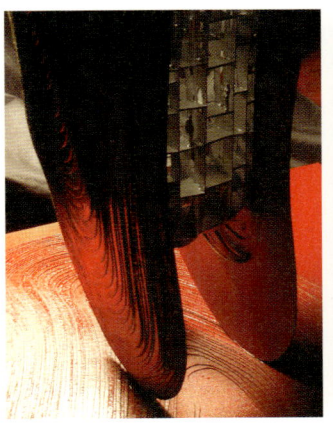
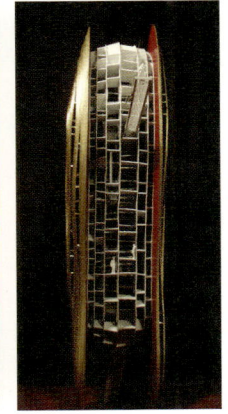
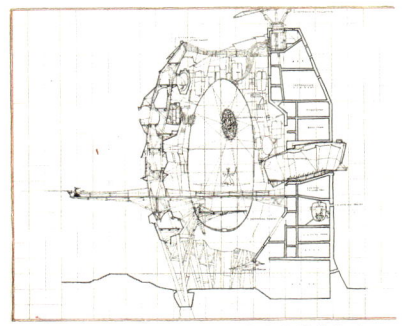

단면 개념도 SECTIONAL CONCEPT

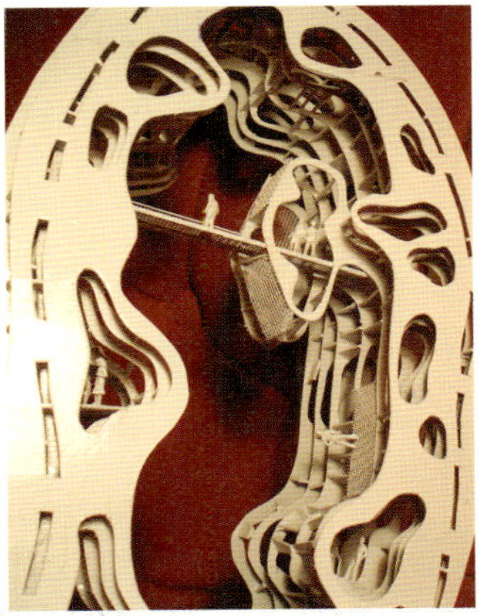

두 겹의 육질 안에 놓인 내부 조직 INNER TISSUE IN-BETWEEN THE TWO OUTER SHELLS

Hoon Moon
Moon Hoon Design Lab., Korea
New-Type-Body-Architecture 2.0

바이오-에로-테크노-판타지-재생공간

미래 건축의 방향 중 하나로 떠오르는 것들. 건축가는 마치 로마의 폐허 혹은 광산의 폐허처럼, 거대하거나 거친 구조체를 설계하고, 거기에 부착되는 다양한 기능의 공간은 히타치, 삼성, 미츠비시, GE, 노키아 혹은 유명한 바이오테크 브랜드들이 제작하는 것이다. 아직까지 우린 여성의 몸을 통해 이 세상으로 나선다. 그 기억나지 않는 첫 공간의 느낌을, 촉각을, 시각을 상상해본다. 여성과 건축의 짬뽕이여. 거대해진 여성의 몸을 닮은 건축을 바라보며, 그 누군가는 반응하고 있다. 비대해진 단순한 시각이여.

Bio-Ero-Techno-Fantasy-Recycling-Bin (BETEFARB)

The one possible way of Architecture in the future comes to my mind in a millionth of a nano-second. The Architects are providing massive structures, all rough and brutal... They remind us of Roman ruins, deserted factories, mining towns, and Lou's Salk Lake Institute without the wooden finishes. Hitach, Toyota, GE, LG, Nokia, Samsung and Bio-companies will design and build intricate and complex pods that will be attached to the brutal but beautiful structures. We will live happily or unhappily ever after. The form, shape, contours of the female body inspires and stimulates many bodies and minds. The clear notion of inanimate existence of the mannequin still moves many minds. What is this? A human will be aroused to Female-Body-Architecture and will try to make love to it. The poor guy in the movie CRYING GAME falls in love with the form of the woman. Within the female body is the primordial space where everyone has been, but unable to recollect anything.

NEW TYPE

TECHNOLOGY

WOMAN'S BODY

FUNCTIONAL

REACTION

SENSE OF TOUCH

SENSE OF SIGHT

BEGINNING OF THE WORLD

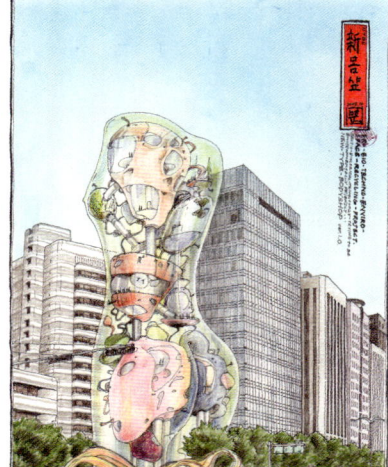

개념 스케치 CONCEPTUAL SKETCH

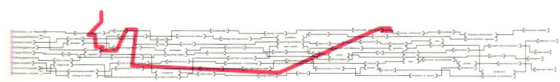

신몸건축 2.0 문훈
문훈건축발전소

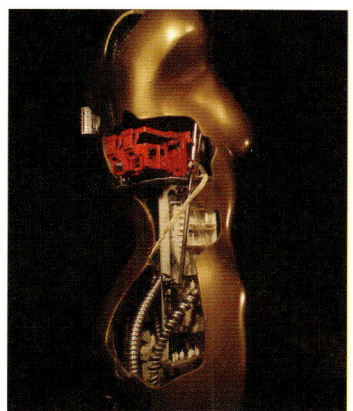

개념 모형 CONCEPTUAL MODEL

Hoon Moon
Moon Hoon Design Lab., Korea
Ssalon de Sson 1.0
Yangpyung, Gyeonggi-do, Korea
2006-2007

수의사인 건축주는 아내와 장인장모, 고양이 3마리, 개 2마리와 함께 살고 있다. 그는 침실 3개가 있는 40여 평의 집을 원했다. 또한 중정과 동물들을 위한 공간을 원했다. 대지가 약 225여 평이어서 공간상으로는 충분했다. 각 방마다 다른 크기와 분위기 그리고 기능의 중정을 계획하였다. 2층에는 독실한 신자인 건축주를 위해 멀리 강을 바라볼 수 있는 방향으로 기도실을 배치하였다.

집의 형태는 땅 모양을 그대로 따르고 있으며, 처음 계획 때 외부 중정들의 연결을 생각하다가 마치 양파껍질 같은, 여러 층의 돌음공간을 생각하게 되었다. 그 공간의 층들이 중앙으로 향하면서 그 기운이 결국 2층 기도공간으로 상승하게 된다. 최외곽의 자갈 깔린 돌음공간은 외부의 다양한 기능공간과 만나고, 그 공간들은 집의 외곽을 에워싸는 툇마루 같은 돌음공간과 만난다. 그 공간은 집의 방, 중정과 만나고 그 공간들은 실내화되는 복도공간과 만나고, 그 공간은 중앙의 중정에 이르게 된다. 또한 중앙의 정원은 물과 만나고 그 물은 화장실을 밑으로 통과하여 집 밖 최외곽의 돌음공간과 다시 만나게 된다. 집의 외곽 툇마루 공간 일부에 동물들을 위한 놀이기구를 계획했고, 서재 앞 중정을 동물들을 위한 공간으로 계획했다.

중정들을 갖는 1층 주거공간과 독특한 형상으로 돌출된 2층 기도실
LOWER LIVING SPACES WITH COURTYARDS AND
UPPER PRAYER ROOM IN PECULIAR SHAPE

LAYER

SHAPE OF THE LAND

AMBULATORY

CONNECT

CENTRIPETAL FORCE

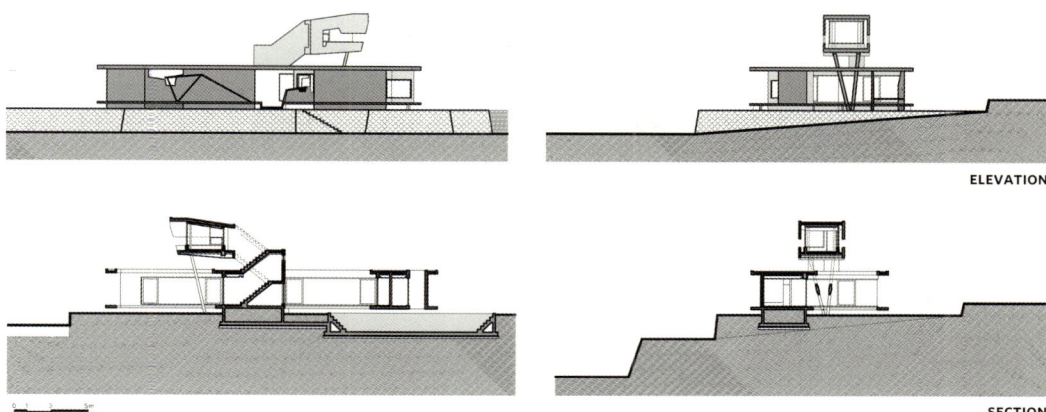

ELEVATION

SECTION

The client is a vet, living with his wife, her parents, 3cats, and 2dogs. He had a very simple program in that he wanted 3bedroom house with a possible atrium: he finds them very attractive. He also wanted to take a good care of his pets, by having a proper space for them. The land is pretty large, compared to urban situations, so I was able to provide a atrium for each room which developed into distinct seven gardens. A prayer room was designed for the faithful christian client, which is elevated high enough to provide a distant river view.

     The form of the house follows the shape of the land. and the initial conception of the house came as many layered space - like onion - that moves towards the center, where the second floor is born which is also a prayer space. The outer ambulatory covered with gravel is met by various functional outdoor space which then is met by an ambulatory of the house which is met by various outdoor atriums and rooms which is met by the corridor of the house, which is met by the central garden which is connected directly to the pool which passes under the bathroom to the outskirts of the site. Part of the ambulatory is equipped with play structure for the pet, and a entire garden is also devoted to them.

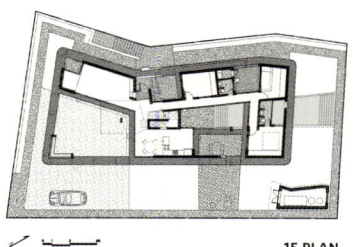

1F PLAN

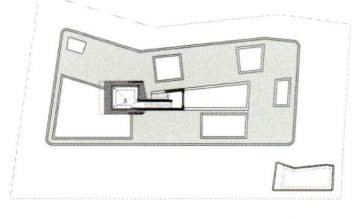

2F PLAN

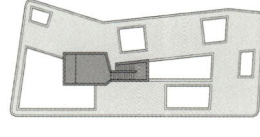

ROOF PLAN

Rem Koolhaas + Ole Scheeren
OMA, the Netherlands + Samoo Architects & Engineers, Korea
Museum of Art, Seoul National University (MoA)
Seoul, Korea
1997-2005

서울대학교 미술관의 디자인은 캠퍼스와 지역커뮤니티간 관계에 주안점을 두고 이들 사이 연결고리로 작용한다. 이러한 연결은 프로젝트의 형태학을 뒷받침하는 작용이다. 그 작용은 최대 건물 외피를 통한 슬라이스로 공동체와 캠퍼스 사이 보행자 연결점을 구성한다. 슬라이스로 생성되는 배회하는 매스는 이동경로와 부지 지형으로 조절된다. 매스는 콘크리트 핵을 포함한 캔틸레버 구조 강철 셀이다.

건물 내 이동은 슬라이스의 연속으로 경로와 나선, 두 갈래로 갈라지며 건물 안으로 진입하면 여러 프로그램으로 연결된다. 프로그램은 전시공간, 교육공간, 도서관, 사무공간 네 영역이다. 교육공간인 강의 홀과 강당은 슬라이스로 형성되는 경사로 지형을 그대로 활용한 계단식으로 계획했다. 도서관은 건물의 중심으로 구조적 핵에 해당한다. 주변과 중앙의 이동경로는 두 개의 나선형 루프를 만들어서 건물 내부 프로그램에 접근하도록 한다. 전시공간은 외피의 주공간으로 최상층에 위치한다. 전시 층은 교육공간으로 확장할 수 있도록 디자인되었다. 교육공간을 다용도로 사용할 수 있는 것은 램프의 이동경로와 물질성에서 표현된다. 마감재는 건물의 구성요소와 기능 요건을 명확히 하는 데 한몫한다. 단일 볼륨은 전략적으로 특정 전망을 향해 있고 그 구조 프레임의 모멘트를 드러낸다.

진입로에서 본 건물 하부의 슬라이스 공간 SLICED SPACE UNDER THE BUILDING MASS, VIEW FROM APPROACH

RELATIONSHIP OF THE CAMPUS TO THE COMMUNITY

LINKAGE

MORPHOLOGY

HOVERING MASS

MONOLITHIC VOLUME

CIRCULATION

PROGRAM / THE EDUCATIONAL SPACES

CULTURE

ARTICULATE

STRUCTURAL FRAMEWORK

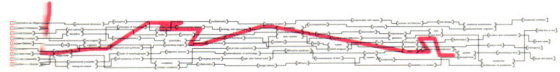

서울대학교 미술관　　OMA + 삼우설계

쇄기형 볼륨 하부의 잘려나간 외부 공간의 조망　VIEWS OF SLICED SPACE UNDER THE WEDGE VOLUME

SITE PLAN

The design for the Seoul National University Museum is driven by the relationship of the campus to the community and serve as a link between them. This linkage is the defining operation behind the project's morphology. The operation is a slice through the maximum building envelope and establishes a pedestrian connection between the community and the campus. The hovering mass generated by this slice is modulated by the circulation path and site topography. This mass is a cantilevered structural steel shell bearing on concrete core.

Circulation through the building is a continuation of the defining slice, internally the path bifurcate and spirals inward. As one enters the building the circulation affords connections to the different programs. There are four basic program areas: Exhibition, Educational, Library and Operations. The educational spaces, the lecture hall and auditorium, benefit from the slope formed by the slice and internally accommodate ramped seating. The library inhabits the center and structural core of the building. Peripheral and central circulation paths create two spiraling loops, which allow program contiguity in the building. The exhibition space being the primary space in envelope is located at the top.

The exhibition level is designed for expansion by allowing its invasion of the educational spaces. This invasion and resultant multi use of the educational spaces for exhibition purposes are articulated by a ramped circulation path and in materiality.

The building is further defined by a selection of finishes and materials, which articulate its compositional elements and functional requirements. The monolithic volume is strategically punctuated towards specific site views, consequently exposing moments of its structural framework.

SOUTH ELEVATION

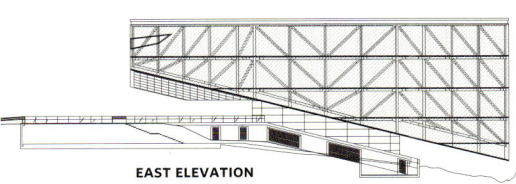

EAST ELEVATION

전시공간에 이르는 수직 동선 VERTICAL CIRCULATION TO THE EXHIBITION SPACE

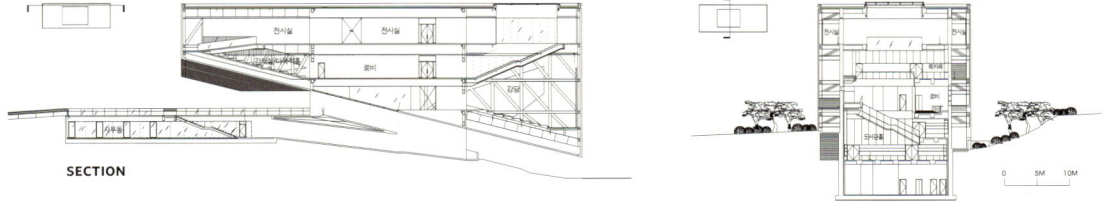

SECTION

서울대학교 미술관   OMA + 삼우설계

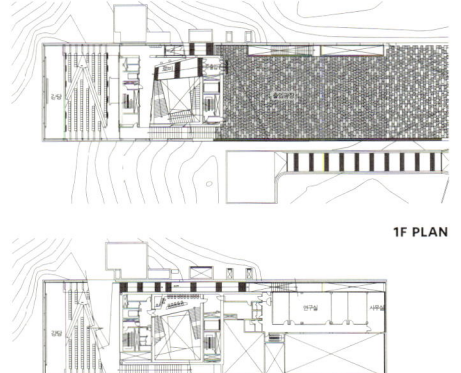

1F PLAN

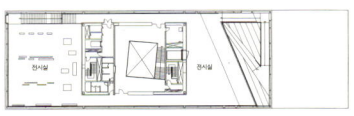

3F PLAN

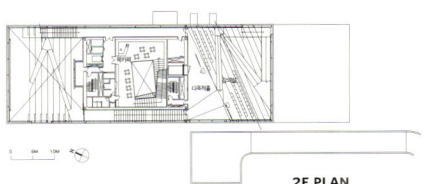

B1 PLAN

2F PLAN

교육공간의 경사진 내부 공간 RAMPED SEATINGS IN EDUCATIONAL SPACES

Rem Koolhaas + Ole Scheeren
OMA, The Netherlands + Samoo Architects & Engineers, Korea
Leeum Samsung Museum of Art
Seoul, Korea
2002-2004

삼성미술관 리움은 OMA가 1997년 설계한 27,000m²의 마스터플랜이다. 도심 주거지역인 한남동에 위치하며 OMA, 마리오 보타, 장 누벨이 설계한 세 개 건물로 이루어진다. OMA가 설계한 건물은 총 13,100m² 면적에 기획전시공간, 미디어 및 사무공간을 포함한다. 세 건물은 로비와 정보 영역에 해당하는 중앙 홀로 모인다.

미술관은 OMA 건물의 램프를 통해 진입해서 중앙 홀로 바로 연결된다. OMA 설계의 주요 특징은 육중한 블랙콘크리트 박스로 입구에서 방문객을 맞는다. 이 박스는 물결치는 지형의 커다란 공간에 매달려서 공간 내 다양한 조명 조건을 만들어낸다. 블랙박스 밑, 그 안 그리고 그 아래로 이동이 이루어진다. 이러한 움직임을 통해 방문객은 건물과 부지, 도시의 역동적인 관계를 경험할 수 있게 된다.

Samsung Museum of Art project is a 27,000m² masterplan designed by OMA in 1997. Located in Hannam-Dong - a residential district near the city centre - the complex comprises three buildings by OMA, Mario Botta and Jean Nouvel. OMA's building covers a gross area of 13,100m² for temporary contemporary exhibitions, media and office spaces. The three buildings converge into a central mixing chamber that forms the lobby and information area.

The museum complex entry is through the OMA building via a ramp leading directly into the mixing chamber. The dominant feature of OMA's design is a massive black concrete box, which confronts the visitor immediately at the entrance. The box is suspended within large excavation in the undulating topography creating varying light conditions within the space. Circulation is conceived around the experience of the black box by descending under it, into it and moving above it. This movement provides the visitor a rich experience of the dynamic relationship between the building, the site and the city.

Central Mixing Chamber

Complex Entry

Massive Black Concrete Box

Excavation

Movement

Rich Experience

Dynamic Relationship

City Centre

Three Buildings

삼성미술관 리움　OMA + 삼우설계

상층부 데크에서 본 건물 전경 VIEW FROM THE UPPER DECK

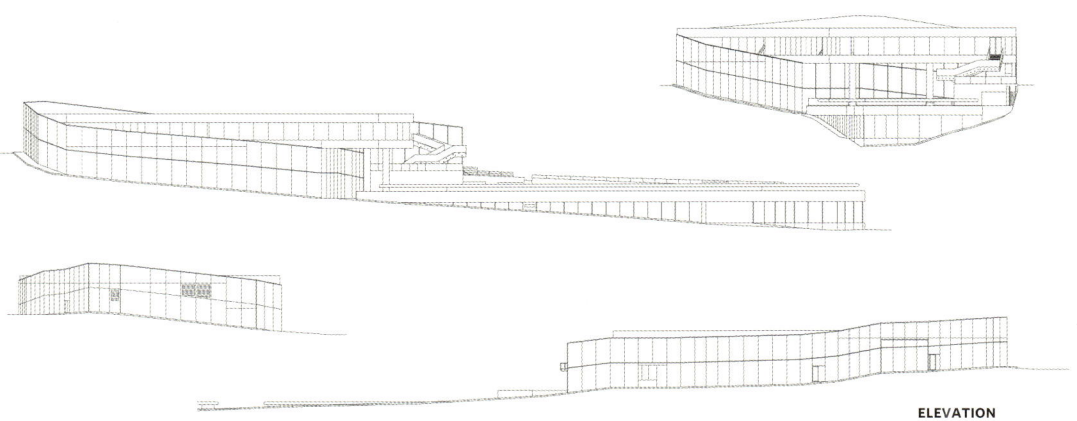

ELEVATION

중앙 홀로 연결되는 램프공간 RAMP DIRECTLY LEADING TO MIXING CHAMBER

표피와 블랙박스가 이루는 사이공간 IN-BETWEEN SPACE OF SURFACE AND BLACK BOX

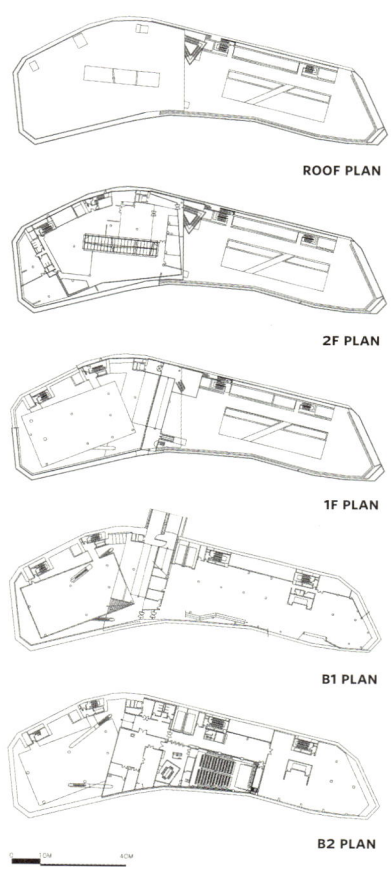

ROOF PLAN

2F PLAN

1F PLAN

B1 PLAN

B2 PLAN

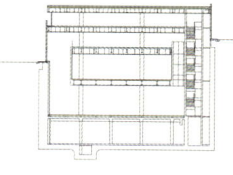

SECTION

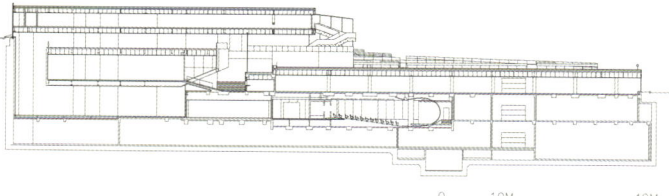

삼성미술관 리움　　OMA + 삼우설계

블랙박스와 그 주변의 틈새공간들
BLACK BOX AND INTERSTITIAL SPACES AROUND IT

Ken Min Sungjin
SKM Architects, Korea
S-Gallery
Seoul, Korea
2001-2002

S 갤러리의 북쪽과 서쪽으로는 고급 주택단지가, 동쪽에는 저소득층 주택들이 밀집되어 있다. 갤러리 주변은 도로 폭이 좁아 자동차 통행이 불가능하고 10~15평의 작은 집들이 모여 있다. 갤러리는 1, 2층 면적이 각 20평이며 연면적이 40평인 작은 건축물로, 주변 환경에서 핵심 역할을 하면서 긍정적인 충격을 의도했다.

갤러리의 외피는 도시적이고 딱딱한 껍질로 쌓여 있지만 내부 공간은 여러 레벨이 서로 연결되어 있다. 기존의 작은 주택 두 채와 새로 지어진 갤러리 세 동이 연계되면서 안뜰이 있으며, 데크가 조그마한 정원과 중정을 연결한다. 밝고 아늑하고 조용한 내부 느낌과 외부의 스케일 없는 단단한 셸, 이 두 요소의 대비를 염두에 둔 것이다. 밖으로 중정을 노출시켜 열린 느낌을 주기보다는 안쪽을 감추고 보호해서 후면 정원과 데크의 고요하고 아늑한 느낌을 살려두었다.

SCALELESS

SILHOUETTE

CONTRAST

COURTYARD

물결치는 스케일 없는 표면 UNDULATING SCALELESS SKIN

S 갤러리   켄 민 성진
SKM 아키텍츠

주입면 MAIN FACADE

옥상정원의 조망 VIEW OF ROOF GARDEN

S-Gallery was designed as two-story building - 20 and 40 pyongs, respectively - to play a core role in the area, shocking it in a positive way. Whether you see S-Gallery on a photo or actually visit it, its exterior will betray its interior space. The skin or a hard shell may look urban, while the inside space consists of various levels interconnected. Two small existing houses and a new gallery building are interlinked with each other only to create courtyards. Small gardens and patios are connected by a deck. Bright, cozy and still inside is well contrasted with the scaleless and hard shell. Patios are neither exposed or open to outside, but the inside is concealed and protected to enhance the quiet and cozy backyard and deck.

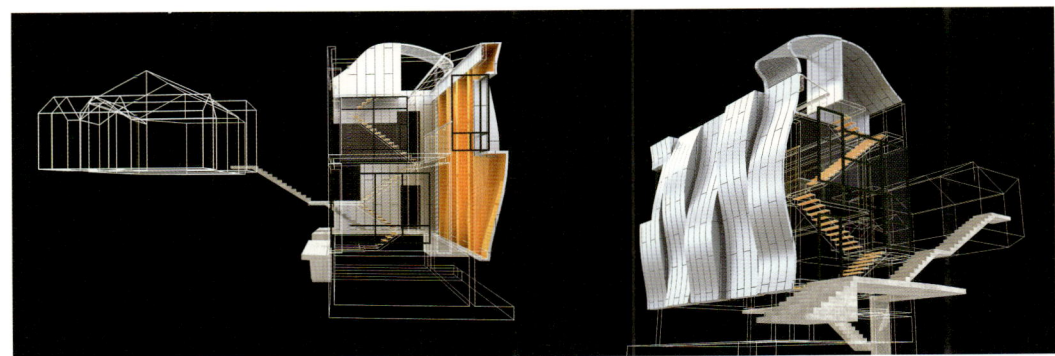

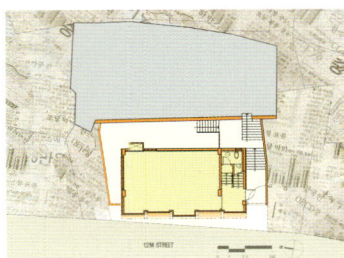
1F PLAN

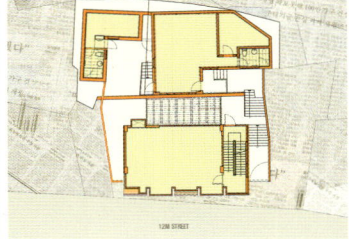
2F PLAN

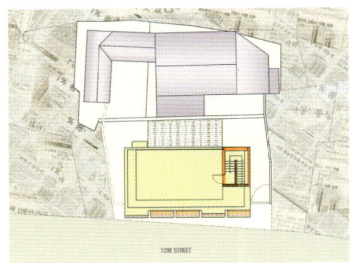
3F PLAN

Ken Min Sungjin
SEM Architects, Korea
Hilton Namhae Golf & Spa Resort
Namhae, Gyeongsangnam-do, Korea
2006

힐튼 리조트는 경남 남해군 덕원리 해안의 반도와 골프장이 만들어질 인공 매립지를 부지로 한다. 골프장을 내려다보면서 바다를 조망할 수 있는 사이트 A와 삼면이 바다로 둘러싸여 있는 사이트 B, 클럽하우스가 위치할 수 있는 능선은 각기 개성 있는 조망과 지형적 특색이 있었으며 이는 건축물 배치에 중요한 요소였다. 기존 호텔 디자인과 차별화된 전망을 중시한 평면과 자연친화적 건축을 지향하며 모던하면서도 자연과 어우러지는 디자인을 하였다. 초기 계획안에 있던 고층 단일 타워형 건축물을 배제하고 3층 이하의 소규모 건축물을 그룹별로 배치하여 쾌적하며 휴먼스케일의 휴양지다운 환경을 만드는 데 주력하였다.

The site addressed at Deongwon-ri, Namhae-gun, Gyeongsangnam-do is a landfill for a seaside golf course. Each part of the site has its own unique view and topography; site A commands a view of the sea, site B looks like an islet surrounded by the sea on its three sides, and the ridge provides for an ideal place for the club house. Such conditions would be taken into consideration as important elements available for layout of buildings. The primary design points were the planes offering panoramic view and program functions modernistic but harmonized with the nature. Thus, small-scale buildings as low as 3 stories or lower were laid out in groups to be more environmental-friendly.

클럽하우스 전경 CLUBHOUSE, PERSPECTIVE

힐튼 남해 골프 & 스파 리조트

켄 민 성진
SKM 아키텍츠

클럽하우스와 호텔을 포함하는 모형
MODEL, CLUBHOUSE AND HOTEL

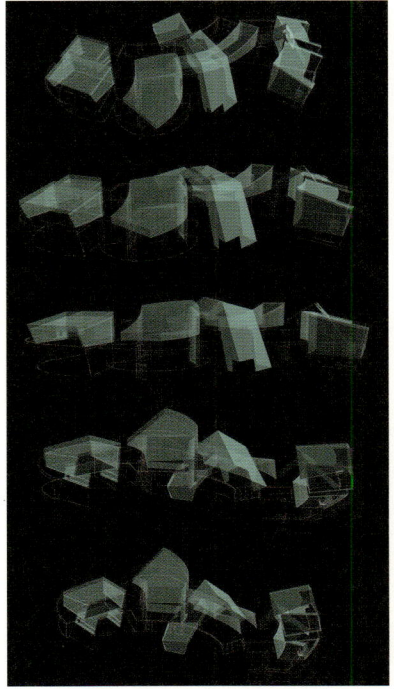
MASS MODELING

극적인 자연경관과의 조우 ENCOUNTERING THE DRAMATIC SURROUNDING NATURE

연속적인 흐름을 볼 수 있는 내부 공간 CONTINUOUS FLOW OF INTERIOR SPACE

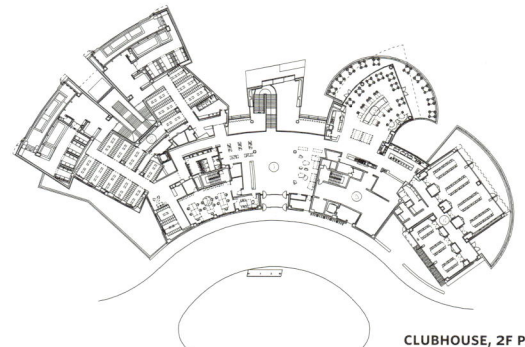
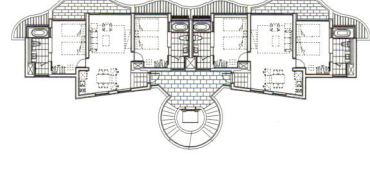

CLUBHOUSE, 2F PLAN   HOTEL PLAN

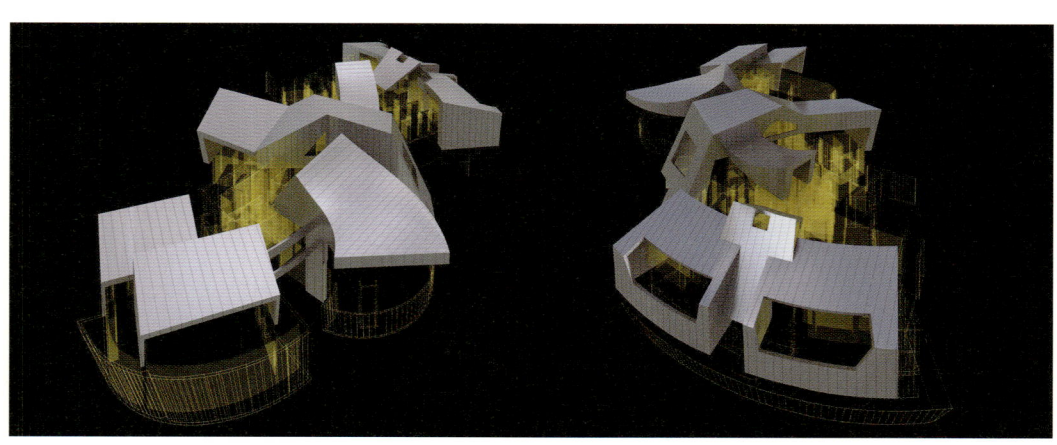

힐튼 남해 골프 & 스파 리조트   켄 민 성진
SKM 아키텍츠

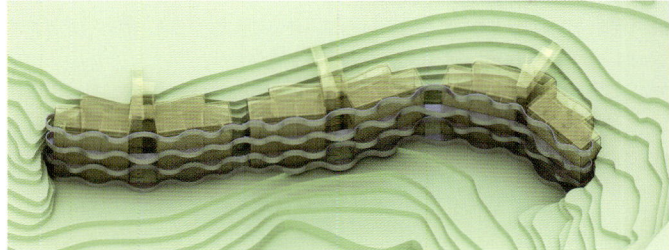

물결치는 호텔 입면 상세
UNDULATING FACADE DETAIL OF HOTEL

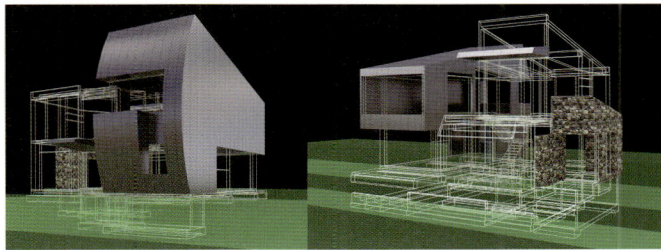

Zaha Hadid
Zaha Hadid Architects, U.K. / Iraq
Ewha Campus Center
Competition Project_Participation
Seoul, Korea
2004

캠퍼스의 중추적인 복합문화공간이 되기 위해 지형의 특징을 살려 자연스러운 공원 같은 공간을 만들고자 하였다. 지상부를 단층으로 분할하여 캠퍼스센터의 지하부를 깊이 드러내는 커다란 협곡 스타일을 제안한다. 부지 북쪽과 남쪽 입구 사이에는 상당한 단차가 있는데, 솟아오른 듯한 불룩한 모습을 띤 단층이 그 높이 차이를 유연하게 연결하여 역동적인 랜드스케이프를 형성한다. 이처럼 그 흐름에 따라 자연스럽게 동선을 끌어가는 방법이 우리의 설계 특징이기도 하다.

기본적으로 공간은 캠퍼스센터의 입구를 공공적인 성격으로 구분하고, 끝으로 갈수록 조용한 행정시설을 비롯한 교육공간으로 구성했다. 센터는 전체 3층으로 0.00, +4.50, +9.00 높이의 지형처럼 세워지며, 주요 관건인 주차공간은 모두 지하 2층에 구성된다. 이는 자연스러운 접근과 일광을 유입하기 위해 표면을 분할하여 조도 깊이에 따라 필요한 공간들을 배치한 것이다. 따라서, 가장 낮은 레벨에는 일광을 비교적 최소로 적용해도 되는 강당과 상점 같은 공공공간을 배치하였다.

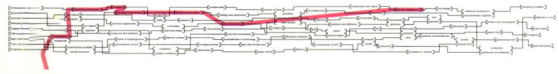
캠퍼스 정문에서 바라본 전경 VIEW FROM THE MAIN ENTRANCE TO THE CAMPUS

이화여대 캠퍼스센터
지명현상설계 참가작

ZAHA HADID

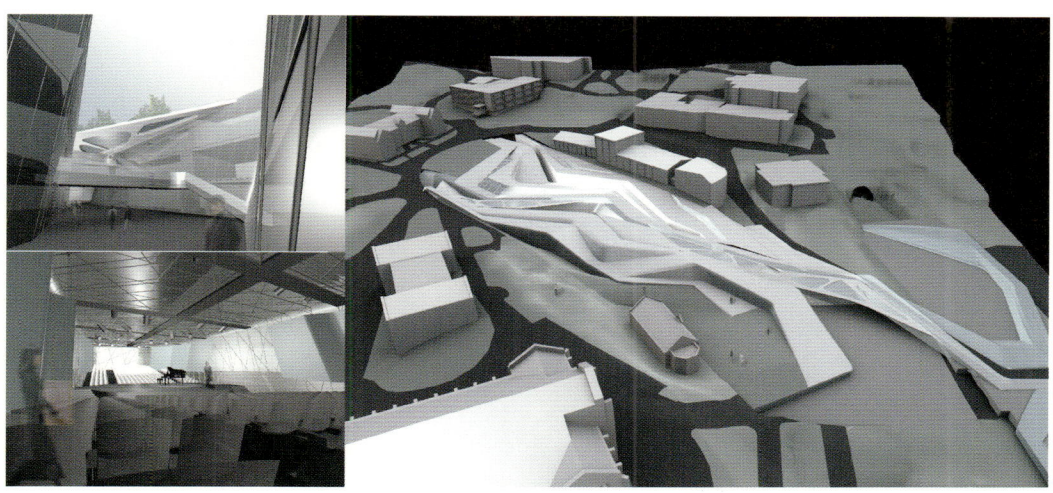

외부의 역동적인 경관으로부터 내부 공간에 이르는 장면들 SCENES FROM DYNAMIC LANDSCAPE TO INTERIOR HALL

The key challenge of the project is to create an urban event space and communication hub that is buried under the ground. Given the animated topography and the informal character of the campus as a whole we decided against a formal geometric plaza design and in favor of a more fluid and landscape-like approach to structure the field. The primary feature is a canyon that splits the ground and reveals the deep interior of the campus centre. The split is articulated as a fluid fault-line along which the two sides lift up and bulge as if under the pressure/energy from below. This primary Canyon is surrounded by a series of secondary crevices. Together they form a dynamic landscape that will be a pleasure to traverse. The facilitation of obvious and easy access, as well as the smooth guidance of all movements within, is the fundamental ethos of our design.

    In principle there is a more public front end versus a more academic and quiet back end. The shops, the theatre, the large reading room and the auditoria oriented towards this front end. The classroom, and the administration spaces are located more in the depth of the back end. There are three levels for the student centre plus two underground levels for all the required parking spaces. However, the three levels (0.00, +4.50, +9.00) build up only as the terrain builds up, so that the volume of the student centre does not create an over ground building but appears only as a surface split open by cuts for access and daylight penetration. The cuts are laid out in such a way as to allow for an efficient packing of classrooms along these linear light-wells. On the main entrance level - the lowest occupied level (0.00) - we allocate those spaces with most public access and with the least requirements for daylight: the various auditoria, the student gallery, the theater and the shops

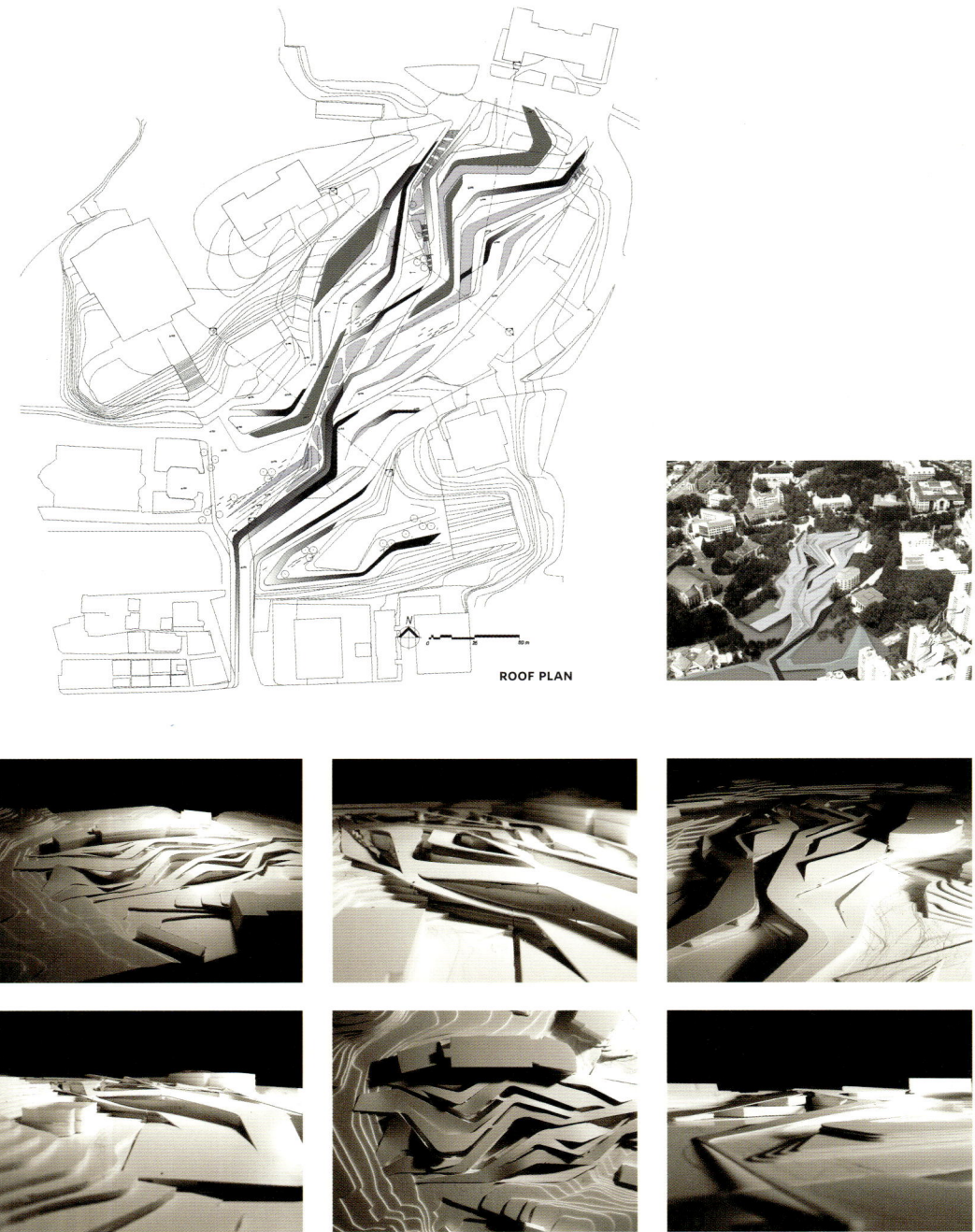

ROOF PLAN

물결치는 형태의 띠들을 보여주는 모형 사진 MODEL VIEW WITH UNDULATING STRIPS

이화여대 캠퍼스센터
지명현상설계 참가작

ZAHA HADID

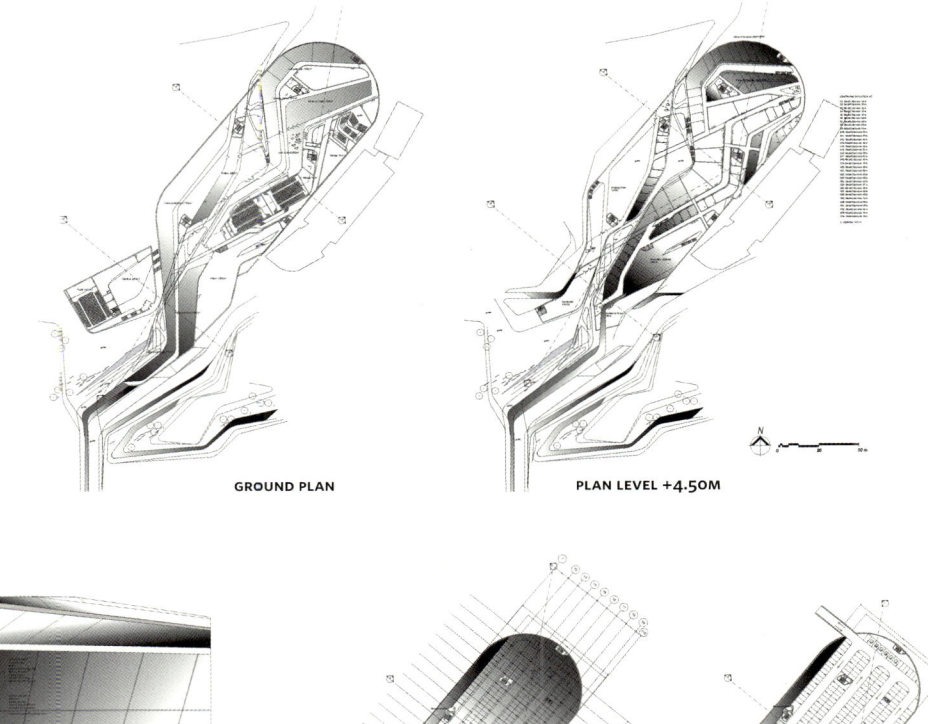

GROUND PLAN

PLAN LEVEL +4.50M

DETAIL SECTION

PLAN LEVEL -7.00M

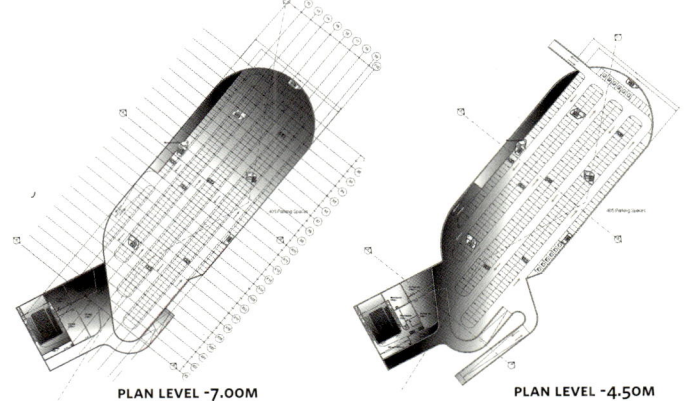

PLAN LEVEL -4.50M

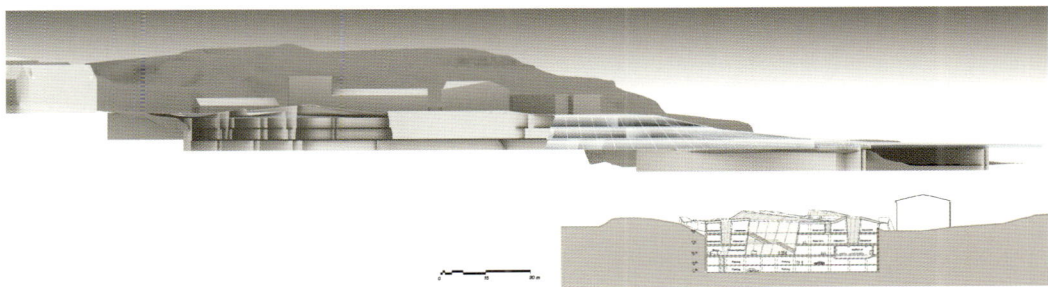

SECTION

Alberto Francini + Andrea Boschetti
Metrogramma, Italy
Asian Culture Complex
Competition Project_2nd Prize
Gwangju, Korea
2005

아시아문화전당은 기존 지역의 기억, 역사적 발견 등과 더불어 전망의 생산적인 문화 프로그램으로 대표되는 광주의 미래를 둘러싸는 커다란 경계벽으로 계획되었다. 문화의 성을 은유하는 새로운 아시아 문화전당은 단순하면서도 복잡한 세계 구축을 보여준다. 움직이는 인프라로 구성되어 어디서나 누구나 쉽게 접근할 수 있다. 높이와 경계 연속성에서 일정하며 표면, 공원을 향한 출구, 기억 면에서 이질적이다.

박람회 공간, 공공 홀, 지하철 역사, 주차장은 지하에 위치한다. 역사적 건물들 사이에서 침투 가능한 유기적인 속성의 공원에는 바위, 광장 그리고 조밀한 지형으로 변한다. 자연과 예술로 가득한 형이상학적 부지는 단순히 중앙공원이 아니라 광주의 명상 숲이 된다. 이곳을 에워싸는 모든 거리는 자연적인 원형극장 내 기하학적 보행로처럼 서로 교차하고 다시 만난다. 공원과 현대적인 거주 벽은 프로젝트의 집합성과 공동체가 갖고 있는 민주 문화적 가치의 중심임을 상징한다. 분수는 기하학적 중심으로 자주 언급되는 균형의 상징이라 할 수 있다.

Our project for the Asian Culture Complex (ACC) is conceived as a big boundary wall which encloses the foundation memories, the historical finds and at the same time the future of Gwangju which is represented by the productive cultural program of ACC. Metaphor of the castle of culture, the new ACC reveals itself as a simple and at the same time complex world-building. It consist of infrastructures in movement and is easily accessible to everybody from everywhere. Regular because of the height and border continuity. Heterogeneous because of the surface and the openings to the park and the memory. This is a linear contemporary city which includes in itself collective core and history of Gwangju. A vast urban wood dedicated to meditation and relax. Expositions, public halls, metro and parking lot are placed underneath. Among the historical buildings the permeable organic nature of the wood turns into stone, square and consolidated geography. A metaphysical site full of nature and art, not simply a central park but a wood for meditation in Gwangju. All the streets that surround the project area cross and rejoin like the geometrical pedestrian tracks inside the natural amphitheater.

Park and contemporary inhabited wall are the symbols of the collective sense of the project and the centre of the democratic cultural values of a community. The fountain is the geometrical centre and the symbol of the over mentioned balance.

PUBLIC

ENCLOSURE

BOUNDARY

PROXIMITY

MIXTURE

GEOGRAPHICAL CULTURE

INTERCONNECT

SITE ANTHROPOLOGY

VOID-SPACE

COLLECTIVITY

NATURE

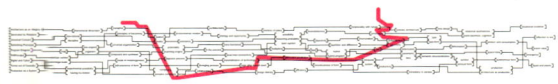

아시아문화전당
현상설계 2등작

ALBERTO FRANCINI + ANDREA BOSCHETTI
Metrogramma

1. Sculpture Garden
2. Car Park
3. "Glass" Buildings
4. "Terracotta" Buildings
5. Gateway/ Plaza
6. Pedestrian Access
7. Road Circulation
8. Forest
9. Infrastructure
10. Road Axes

SITE PLAN

MODEL VIEW

Alberto Francini + Andrea Boschetti
Metrogramma, Italy
New Headquarters of Lombardia District
Milano, Italy
2004

신 롬바르디아 정부 청사 프로젝트는 밀라노와 롬바르디아의 현대적 기념비와 편집되지 않은 상징, 도시가 생성한 공간을 정의하고자 한다. 공간적 측면에서 이는 시민광장이자 업무공간이며, 지방의 상징이자 도시 인프라여야 한다. 건물 자체는 롬바르디아 풍경의 복잡성과 웅장함, 산세, 평야의 고요를 담아낼 수 있는 혁신적인 지역 모델을 제시한다. 또한 개념적으로는 현대적 메트로폴리스와 그 불확실한 영역에서 뚜렷한 제도적 표식 역할과 정체성에 대한 고찰을 유발한다. 세 개 타워는 지상층에서 공공공간을 정의하고, 하늘 바로 밑에 위치한 넓은 파노라마 테라스를 지지한다. 사회적 · 명상적 공간이 100m 높이에 떠 있고 주위 파노라마 경관이 펼쳐진다. 지상층에서는 기하학적 공공공간이 가리발디 캠퍼스와 대중교통 네트워크와 연결된다. 땅에 박힌 보석처럼 강당과 건물 홀이 공공 공원에 위치한다.

외부 컨설턴트: Milanoprogetti, Ing. 기술 설치: Giovanni Cernuschi 파사드 시스템: Permasteelisa 조경: Arch. Maurizio Ori 비디오: e Arch. Andrea Antonelli

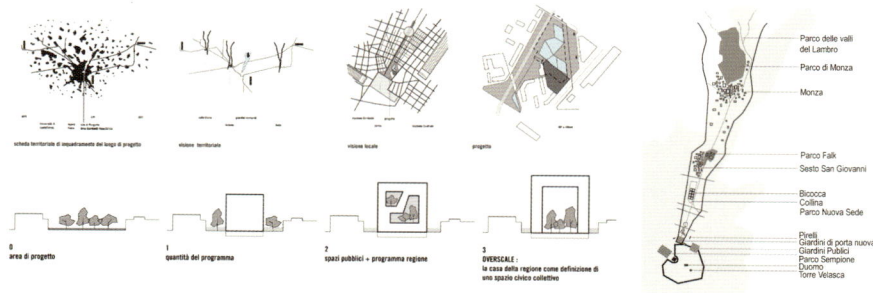

신 롬바르디아 정부 청사 주변 항공사진 AERIAL VIEW OF SURROUNDING AREA

Infrastructure

Duel C

Network

Complexity

Symbolism

Programmatic Density

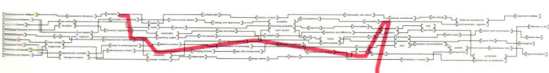

Alberto Francini + Andrea Boschetti
Metrogramma

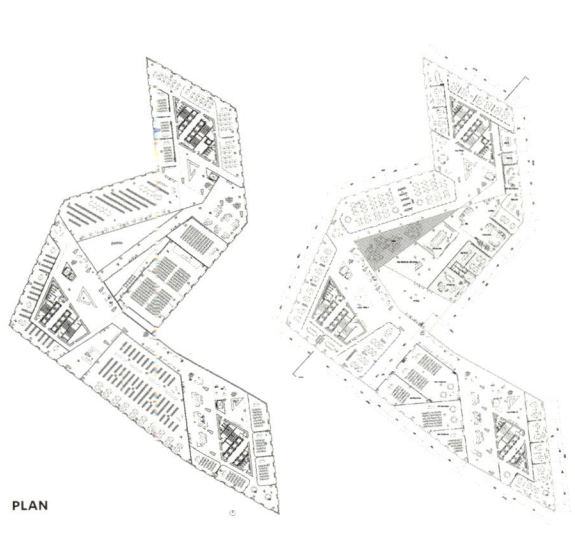

PLAN

저층부가 개방된 타워 LOWER OPENING IN THE TOWER

The project for the Lombardia Government new headquarters aims at the definition of a contemporary monument for Milan and Lombardia, of an inedited sign, and of a space generated by the city. Of a space, which should be both a civic square and a work place, a territorial sign and an urban infrastructure. The building presents itself as an innovative territorial model, as a body, which is capable of absorbing the complexity and the majesty of the Lombardy landscape, as well as the strength of its mountains and the horizontal calm of its plains. The building conceptually develops a consideration on the role and the identity of an institutional distinguishing mark in the contemporary metropolis and in its uncertain territories. The three towers frame the public spaces on the ground level and support the wide panoramic terrace, placed right under the sky. Social, contemplative spaces are suspended 100 meters above the territory and benefit of a panoramic view of the surroundings. On the earth level the geometry of public space is connected to the Garibaldi campus and to the network of public means of transport. Like precious stones in the ground, the auditorium and the building hall are situated in the public park.

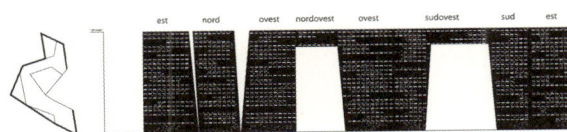

입면의 전개 UNFOLDING OF ELEVATION

Space Group + Beom Architects, Korea
Sanglim Lee, Haecheon Seo + Youngkern Park
Speed Dome
Gwangmyeong, Gyeonggi-do, Korea
2006

경륜은 아마추어 선수들의 기록경기가 아닌 프로 선수들에 의한 순위 경쟁으로, 사회적·시대적 흐름에 따라 그 성격과 이미지가 변하고 있다. 경륜문화가 정착되고 성숙해짐에 따라 경륜 자체가 기존의 폐쇄적이고 격리된 특정 대상을 위한 경기가 아니라, 건전하게 즐길 수 있는 엔터테인먼트의 한 부분으로 자리잡아가고 있다. 이러한 인식 변화와 사회적 맥락을 전제로 광명돔경륜장은 기존 경기장이 갖는 기념비적이고 폐쇄적인 이미지를 벗어나 도심 속 복합문화시설로 이용될 수 있도록 계획하였다.

현대사회는 다양한 형태의 매스 엔터테인먼트를 요구하고 있다. 광명돔경륜장의 경우 경주가 없는 날에는 인필드에서 다양한 공연이 열리고, 기타 다목적 공간에는 지역 주민들의 커뮤니티 행위가 가능할 것이다. 일상에서 일탈을 꿈꾸는 현대인의 욕구 충족만이 아닌, 대형 무주공간의 특성을 활용하여 지역사회의 다양한 프로그램을 수용하는 커뮤니티의 장이 될 것이다.

도심 속 복합문화시설로서의 스피드 돔 전경 PERSPECTIVE VIEW OF SPEED DOME AS CULTURAL COMPLEX

©KIHWAN LEE

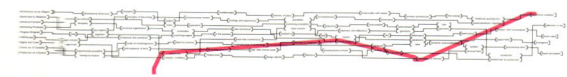

스피드 돔   공간그룹 + 범건축

이상림, 서해천 + 박영건

지원시설 부분 전경 PERSPECTIVE VIEW OF SERVICE BUILDING

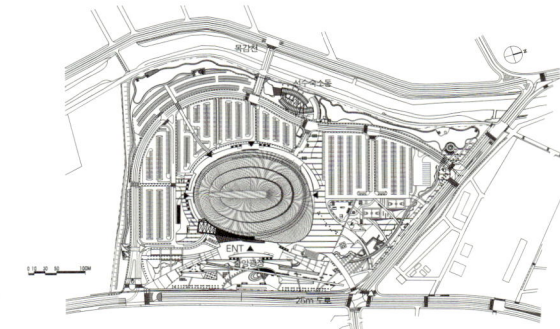

SITE PLAN

The Gwangmyeong Velodrome Dome Stadium was planned as a multipurpose building located in the downtown area, with a friendly and bright image, a departure from the imposing and closed image of existing stadiums. The dynamism and symbolism of the pista or race track of the velodrome are expressed in the continuity of volume of support facilities in the periphery of the stands, while the domed stadium's mass was developed into a form by which one can feel a sense of rhythm and speed.

Inside the stadium, support facilities that include various cores, balloting booths, stands, TV viewing rooms, a hall and atrium in the perimeter were allocated to improve the efficiency of functional routes, such as entries and exits and a balloting route. In particular, the stand was divided into upper and lower parts, with the main entrance level for general spectators as the center, allowing people to see all the facilities when entering the stadium, and openness in the internal and external facilities was emphasized.

Modern society requires mass entertainment in various forms. It is hoped that the cycle race dome, a large space without columns, could be used not only for cycle races but also for mass entertainment by accommodating diverse social and cultural programs.

내부 전경 INTERIOR VIEW

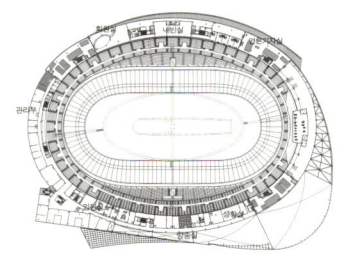

5F PLAN

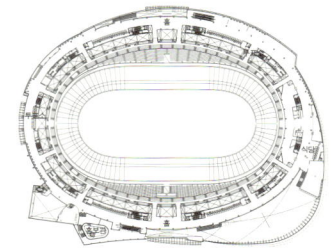

3F PLAN

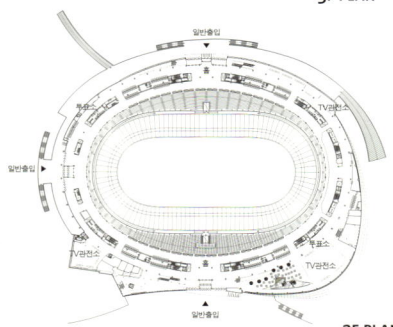

2F PLAN

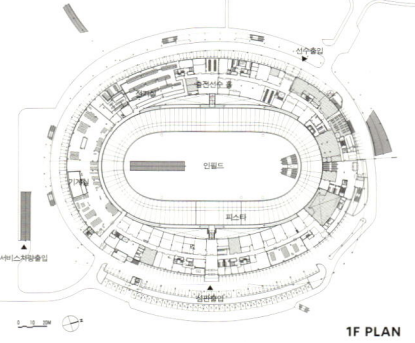

1F PLAN

스피드 돔   공간그룹 + 범건축
이상림, 서해천 + 박영건

주진입 부분 MAIN APPROACH | 스피드 돔의 야경 NIGHT SCENE

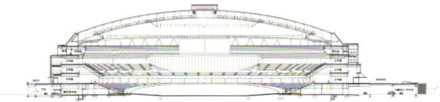
HORIZONTAL SECTION

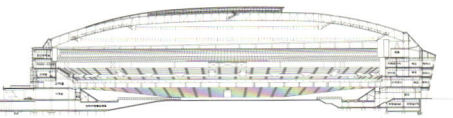
LONGITUDINAL SECTION

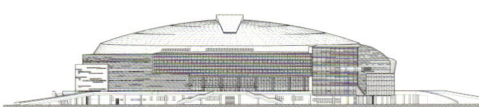
SOUTH ELEVATION

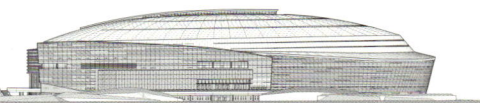
EAST ELEVATION

가벼움과 속도의 은유 | 스피드 돔의 동쪽 전경 METAPHOR OF LIGHTNESS AND SPEED | EAST FACADE OF SPEED DOME

시스템으로서의 건축 생산

최왕돈
국민대학교 건축대학

'시스템으로서의 건축 생산' 이라는 주제를 한국 현대건축의 지형을 그리고자 하는 시대적 해석에 포함할 것인가는 논란이 있었다. 대형 설계회사가 9개의 탐침이라는 유형 구분에서 주로 주목한 아틀리에 중심의 작가주의 성향 건축가와는 격이 맞지 않으며, 이 회사들이 양산하는 프로젝트들이 상업성의 한계를 가진다는 이상주의적인 의견 때문이었다. 하지만 만약 작가주의 성향 건축가만으로 제한한다면 한국 건축의 현실을 상당 부분 놓칠 수 있으며 대형 설계회사에서 생산하는 프로젝트들의 가치를 수용해야 한다는 현실주의적인 의견이 맞섰다. 이 과정에서 현재 우리나라에서 발주되고 있는 대부분의 대형 공공 프로젝트를 이 회사들이 독점하고 있는 상황을 고려할 때 대형 설계회사의 시스템으로서의 건축생산이야말로 한국 건축계의 현실이라는 의견이 더 설득력을 얻었다.

우리나라에서 공간건축이나 정림건축 등 40년 이상의 역사를 가진 설계회사들은 한국의 모더니즘을 이끈 세력이다. 1970년대 중반 이후 서울건축, 삼우설계 등 대기업 계열사 개념의 설계사무소가 등장하면서 대형 설계회사가 형성되기 시작했다. 그후 1980년대 경제호황과 서울올림픽, 신도시 건설에 힘입어 몇몇 설계사무소가 규모의 경제를 추구하면서 1990년대 중반 이후 독자적인 대형 설계회사들이 다수 등장하게 되었다. 이들 대형 설계회사는 수적으로는 일부에 불과하지만 시장점유율은 극히 높게 나타나고 있다. 최근 들어서는 대형 건축회사와 중소 규모 건축사무소의 양극화 현상이 심화되고 있다. 사무소의 생애주기로 볼 때 대형 설계회사 창업세대들의 은퇴 시점이어서 세대교체 상황으로 전이되고 있다.

MEGA-SCALE
COMPLEX CONDITIONS
DESIGN IDENTITY
ARCHITECTURE AS PRODUCTION
SYSTEM
COLLABORATION
PRODUCTIVITY
EFFICIENCY

대형 설계회사의 특성

한국의 경제 규모 확대에 상응하여 건축시장 규모도 커졌다. 이에 건축설계시장 분야가 확장되어 부문별로 더 많은 전문성을 요구하게 되었다. 현대 건축시장의 양태가 그러하듯, 설계업무 범위의 확대와 전문화 요구는 상당한 규모의 인적 자원이 확보되어야 가능해지므로 필연적으로 설계조직의 대형화가 이루어진다. 조

Architectural Production as a System

Wangdon Choi
School of Architecture, Kookmin University

At the earlier stage of publishing, it was much debated whether the theme of "Architectural Production as a System" should be included in the interpretation to delineate the topography of contemporary Korean architecture. It resulted from the idealists' view that large architectural firms stand on a different level with the other categories of "atelier" architects who are the main target of interest in the 9 probes, and the projects produced by the firms have the limit of being too commercialized. On the other hand, there was another realists' view that the contemporary architectural map of Korea cannot be grasped if the firms are excluded and also the value of the projects produced by the firms should be accepted. However, considering most of big-scale public projects are virtually monopolized by the large architectural firms, it has had more persuasive power that architectural production as a system by the large firms is an unneglectable reality of Korea, and future-oriented diagnoses and proposals should be made based on such recognition of the reality.

Some architectural firms such as Space Group and Junglim Architecture established almost 40 years ago were the main power to lead Korean Modernism. It seems that large architectural firms began to set up in Korea after mid-seventies when architectural offices including SAC and Samoo affiliated with conglomerates appeared. Thereafter some offices expanded their size to pursue scale economics with the help of economic boom in eighties, Seoul Olympic Game and construction of new towns; then several independent large firms made their appearance after mid-nineties. These large architectural offices are small in number but their market shares are extremely high, and recently the bipolarization between large firms and small and middle-sized firms are getting deeper and deeper. From a viewpoint of office life-cycle, since the generation of firm founders has already been retired or is at the stage of retirement, generation shifts are being progressed.

Characteristics of Large Architectural Firms

The size of architecture market has increased in proportion to the expansion of domestic economy. The realm of architectural practice has therefore expanded and been diversified to require more specialization. On the other hand, as overall

포스코역사관 | 간삼파트너스
THE POSCO MUSEUM |
GANSAM PARTNERS

아시아문화전당 | 희림건축
ASIAN CULTURE COMPLEX |
HERIM ARCHITECTS & PARTNERS

직이 대형화되면 다양한 프로젝트와 다수의 프로젝트 수행 경험을 통해 축적된 설계정보의 장점을 이용할 수 있다. 설계의 전반적인 영역에 대한 사업화가 가능하므로 설계업무 범위가 확대되며 이는 다시 매출 규모 확대로 이어진다. 대형 설계회사는 법인 형태로 운영되므로 다수의 주주로부터 자금지원을 받을 수 있으며 금융권의 자금조달도 상대적으로 용이하다.[1] 다른 한편으로 전문화된 시장 및 다양한 분야의 시장을 대상으로 하므로 포트폴리오를 다양하게 구성할 수 있어 특정 시장의 침체에도 다른 시장의 활성화로 사무소를 원활하게 운영할 수 있다. 특히 일부 대형 설계회사가 추진하고 있는 해외진출은 국내 경기 변동이라는 불확실성에 효율적으로 대처할 수 있으며 조직의 능률을 배가시킬 수 있는 장점이 있다.

한 연구에 의하면[2] 우리나라 대형 설계사무소의 조직 특성으로 종합서비스 제공, 팀 중심의 직능별 부서조직, 운영관리 비체계화, 정체성 확립의 필요성과 같은 네 가지가 언급되었다. 대형 설계회사 중에는 공동주택부문에 설계 영역이 편중된 경우가 있으나, 대부분 광범위한 시장에 대응하는 다양한 프로젝트 수행 능력을 보이며 설계 외에도 기획, 감리, 시설관리 등에 이르는 업무범위를 포괄한다. 그리고 프로젝트 단위에 따른 팀 형식을 프로젝트 매니저가 운영하는 매트릭스 체제이며 직능에 따른 부서에 의해 구성된다. 또 관리인원이 과다하거나 운영 관리 조직이 체계적으로 확립되지 않은 회사도 적지 않다. 마지막으로 대부분의 대형 설계회사가 30년 내외의 기간 동안 소규모 회사에서 점진적으로 성장하였기 때문에 회사의 정체성을 점검해볼 필요가 있다.

외국 건축설계회사와의 협업

최근 들어 타사와의 수주경쟁에서 경쟁력 우위를 확보하기 위하여 능력과 지명도를 갖춘 외국 건축가 혹은 설계회사와 협업을 하는 경우를 대부분의 대형 프로젝트에서 발견할 수 있다. 그간 아예 특정 외국 건축가를 건축주가 지명하거나 외국 건축가 및 건축회사와의 협업을 응모조건으로 내건 국제현상이 있기도 했다. 이제는 공공기관이 발주하는 턴키 프로젝트에서도 계획설계는 외국 회사에서, 기타 기본 및 실시설계는 국내 대형 회사에서 담당하는 이원 체제가 요구되기도 한다.

외국 설계회사와의 협업 역사는 한국 모더니즘 초기부터 건축문화의 의타성을 키웠다. 미국의 킹아키텍츠어소시에이션이 설계한 USOM 청사(1961)가 완공된 이래, 1970년대는 호텔건축과 상업건축을 중심으로 주로 일본 설계회사와 협업

[1] 정태용 '국내 대형 설계사무소의 조직 특성에 관한 연구' 대한건축학회논문집 계획계, 21권 1호 2005년 1월 73쪽.
[2] 정태용, 같은 책 76, 77쪽.
[3] 김영환, 권종욱 '국내 고층건물에 있어 외국건축가와의 설계협력작업에 관한 연구' 대한건축학회논문집 계획계, 21권 10호 2005년 10월 16쪽.

phenomena in the pattern of architectural market, projects have become so gigantic and complex that the size of architectural firms has become bigger correspondingly. The expansion and specialization of architectural practice also requires the large-sized organization of architectural firms by necessity since the firms should increase the personnel over a certain size in order to do business without any trouble. Once the organization of firms is enlarged, the realm of architectural practice is expanded and then results in the growth of total sales because they can utilize the past design information, use available hands and make business easily in the wider realm of architectural practice. In addition, as large architectural firms are operated as a form of corporation, they can be easily financed by financial institutions as well as shareholders. They also make so diverse portfolio since they do business at specialized and diverse markets that they can manage themselves through the activation of another market in spite of the depression of a particular market. Especially overseas expansion some firms are actively promoting has a merit of coping with the uncertainty of domestic economic fluctuation and doubling the efficiency of the organization.

According to a study, the characteristics of large architectural firms in Korea are mentioned as follows: provision of total service, team-centered departmental organization, unsystematic management and administration, need for identity making. Most of firms, except a few centering on a housing sector, show the comprehensive ability of diverse project management dealing with wide range of market and give service in the realm of planning, supervision, and facility management besides design. They have a matrix system in which a project manager operates his own team according to each project, and are comprised of departmental organization. Also, several firms have too large management section, or do not have systematic management system. Finally it has to be pointed out that they should check their own identity since most of firms have gradually grown from small firms to large firms for almost 30 years

Cooperation with Overseas Architectural Design Firms

Recently it has been frequently found that large firms try to cooperate with overseas architects or architectural firms with ability and fame in order to be more competitive in getting order. Of course, there were some cases in which the client nominated overseas architects from the beginning, or international architectural competition in which cooperation with overseas architects or architectural firms are the prerequisite for an entry. Now an dualistic system has been often required in that, even in turn-

은평뉴타운 2지구 | 토문건축
EUNPYUNG NEWTOWN 2ND DISTRICT
APARTMENT | TOMOON ENGINEERING
& ARCHITECTS

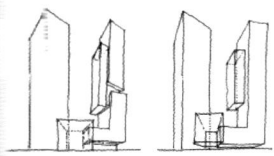

G 프로젝트 | 진아건축
G PROJECT | JINA ARCHITECTS

이 이루어졌다. 1980년대부터는 한국 근린의 문화였던 일본과의 협업에서 미국 설계회사와 협업으로 전환되는 양상을 보였다. 1990년대에 접어들면서 특히 도심의 스카이라인을 결정짓는 사무소 건축에 미국 건축설계회사의 진출이 급증했다. 건물 유형 역시 수도권 신공항, 경부고속철도역사 현상설계 등의 공공건물, 주상복합건물, 연구시설 등 다양한 프로젝트로 변화하였다. 엘러비 베켓, 퍼킨스앤윌, KPF, RTKL, HOK, SOM, 라파엘 비뇰리, ADP, 번스앤도넬, DMJM, KMD, 칼리슨 등의 회사가 주로 참여했으며 리차드 로저스, 니콜라스 그림쇼, 테리 파렐, 마리오 보타 등 유럽 건축가의 참여도 증가하였다. 이중에서도 특히 SOM, HOK, KPF 등 미국의 대형 설계회사에 편중되는 현상이 나타났다.[3]

2000년 이후 설계협업의 범위를 볼 때 과거에 비해 다른 양상이 나타난다. 외국 설계회사가 설계 주도권을 장악하고 계획설계 및 기본설계를 담당하고 한국의 설계회사는 프로젝트 뒷부분과 실시설계를 담당하는 형태가 과거 모습이라면 이제는 로컬 건축가로서 종속관계에서 탈피하여 점차 국내 설계회사가 계획 초기 단계부터 적극적으로 참여하는 양태이다. 그러나 저작권 범위가 모호해지며, 원작자의 정체가 흐려지는 경우가 많다.

최근 동남권이주상가 광풍과 그와 비슷한 시기에 발주되었던 굵직굵직한 턴키 프로젝트가 그러했다. 상당수의 설계안이 외국의 계획설계와 MD에 의존했다. 물론 과거에 비해 외국 회사에 대한 디자인 의존도가 약화되었다고는 하지만 기술 이전이나 새로운 노하우 습득보다는 그때그때 임시방편으로 아이디어를 외국 설계회사에서 구입하여 수주를 위한 방편으로 삼았다는 혐의가 짙다. 그러다보니 지난번 프로젝트를 위해 구입해왔던 설계안을 다른 프로젝트에서 다시 구입하는 소극도 벌어진다.

디자인의 익명성

비슷비슷한 주제를 수백 번 고쳐 써먹었다고 비난받는 비발디도 문제지만 비발디와 바그너를 오락가락 하는 예측불허의 창작도 문제다. 디자인에 관한 한 대형 설계회사들의 가장 심각한 문제는 각 설계회사의 디자인 정체성이 모호하다는 점이다. 프로젝트가 바뀔 때마다 전혀 새로운 디자인 어휘들이 등장하는가 하면 사무소는 다른데 설계안의 스타일은 거의 유사하기도 하다. 전자는 다양한 창의성의 표현이라기보다는 정체성의 혼란과 상실로 비추어지며 후자의 경우는 몰개성으로

key based projects public sectors order, overseas firms take charge of schematic design and local firms do the design development and construction documents.

The history of the cooperation with overseas firms has fostered foreign dependency of architectural culture. Since USOM building(1961) was constructed by the design of US-based King Architects Association, cooperation was mainly made with Japanese architectural firms centering around hotel and commercial buildings in 1970s. In 1980s, likewise, hotel and commercial buildings were designed by the cooperation with overseas firms, and the cooperation pattern gradually shifted from the cooperation with Japanese firms, which showed similar pattern in architectural culture, to the cooperation with American firms. Entering 1990s, American firms rushed into Korea, especially in the area of office buildings which determine a skyline of the urban center. Building types also shifted to various projects including a new metropolitan airport, KTX stations, mixed-use buildings of residential and commercial purposes, research facilities etc. The firms included Ellerbe Becket, Perkins & Will, KPF, RTKL, HOK, SOM, Rafael Vignoly, ADP, Burns & Donnel, DMJM, KMD, Callison, and also increased the participation of European architects such as Richard Rogers, Nicholas Grimshaw, Terry Farrel, Mario Botta. Among these, American architectural firms such as SOM, HOK, KPF were too much concentrated.

Since 2000, different aspects from the past have been observed from the viewpoint of the range of design cooperation. It seems that, in the past, the overseas firms held the initiative of the design and took change of schematic design and the early stage of design development, and then local firms gradually participated in the later phase of design and took charge of the later stage of design development and construction documents; now, the work range of local firms expanded and actively participated at the early stage of design. However, there are many cases in which the range of copyright is getting obscure and the existence of original designer is not clear.

It was applied to a raging wind of the recent development project of Southeast Shopping Center in Seoul and other big turn-key based projects ordered in the similar period. The majority depended upon the schematic design and MD planning of overseas firms. Admittedly, dependency on overseas firms is getting weaker and weaker, but they are suspected to purchase an idea from overseas firms and use it as a temporary makeshift to get order rather than as technology transfer and know-how acquiescence. A comedy happened in which a design which have been used in a project was later bought for another project.

읽힌다. 아웃소싱한 계획설계 때문인가? 최근의 건축 경향에 지나치게 민감해서인가? 설계본부별로 돌아가면서 디자인하다보니 그런 결과가 빚어지는 것일까? 물론 비슷비슷한 템플레이트를 두고두고 써먹는 동어반복보다 낫기는 하다. 그러나 설계안으로 봐서는 도무지 설계회사를 짐작할 수 없는 상황이 바람직한지 자신있게 말할 수 없다. 각 설계회사가 나름대로 디자인 아이덴티티를 구축하지 못하고 있다는 사실은 창작성보다는 단지 생산성으로서 건축을 지향하는 게 아닐까 하는 염려를 만든다.

또 다른 문제는 저작권이다. 현재 대부분의 대형 설계사무소에서 생산한 작품의 저작권은 대표이사, 혹은 대주주가 차지하는 게 사실이다. 파트너들의 저작권을 인정해주는 풍토가 확립되어야 대형 설계사무소의 인적 자원에 안정성과 발전이 있을 것이다. 자신의 작품을 인정받지 못하는 대형 설계사무소의 파트너들은 중소기업의 회사원일 뿐이다. 건축가로서 자긍심도 기대하기 힘들고 조직을 향한 충성심을 계속 기대하기도 힘들다.

설계사무소의 경쟁력은 전적으로 인적 자원에 달려 있다. 그러나 설계조직이 대형화될수록 소수의 인재에게 의존하는 것은 위험할 수 있다. 역량 있는 소수에 의지하기보다는 경쟁력 있는 시스템을 갖추고 그 시스템에 의해 디자인을 진행하는 것이 분명 옳다. 그러나 대형 설계회사들이 종합선물세트가 아니라 각각의 분야에 전문성을 구축한다면, 그리고 회사 내 디자인 리뷰 시스템이 제대로 돌아가고 충분한 여유와 고민을 가지면서 디자인이 진행된다면 디자인의 정체성에 관한 한 긍정적인 방향을 모색할 수 있을 것이다. 그렇게 된다면 대형 설계회사들은 각각의 개성을 살리면서 전체 설계시장에서 다양성의 스펙트럼을 구성할 수 있다. 그것이야말로 대형 설계회사가 시스템으로서의 건축생산을 지향하면서 놓치지 않아야 할 부분이 아닐까. 그것이 존재의 이유가 되어야 하지 않을까. 최근 정림건축, 공간그룹, 희림건축, 삼우설계, 건원건축 등 한국의 대표적인 대형 설계조직이 개념적 창발과 대형 설계기술을 종합하려는 지향이 돋보인다. 진아건축 등이 차세대 CEO로 조직을 개편하며, 대형설계의 생산성만이 아니라 적극적인 창작 욕구를 보이는 것이 그러하다.

경쟁력 향상을 위한 노력

한 건의 설계를 수주하기 위해서 적어도 3, 4개 이상의 설계 컨소시엄이 중복 설계

Anonymity and Copyright of Design

Neither Vivaldi who re-composed hundreds of musical works with similar themes is desirable, nor unpredictable production ranging from Vivaldi to Wagner is desirable. The most serious problem of large architectural firms in terms of design may be ambiguous design identity of each firm. Sometimes completely new design vocabulary appears whenever a new project appears. Or, very similar designs are produced in spite of difference of each firm. The one seems to be the confusion and loss of identity; the other seems to be personality-absence. Is it because of the schematic design from outsourcing? Is it because they are too sensitive to the contemporary architectural trends? Is it because of the departmental organization in which design is alternately produced? Of course, it is better than tautology which exhaustedly repeats similar templates. But I am not confident to say that the situation in which one cannot say "which firms are which" with their designs is desirable. The fact that the firms cannot make their own identity makes a concern that they may head for production as a system at the expense of creativeness.

Another problem is the issue of copyright. It is true that the copyright of design works produced by large architectural firms is exclusively possessed by a CEO, or one or two heavy stockholders. The climate in which the copyright of partner architects are acknowledged will promise the stability and development of human resources in the firm. The partner whose work are not acknowledged as his own work are merely salary workers in medium and small enterprises. It is hard to expect the pride as an architect and the continuous loyalty to the organization. The competitiveness of architectural firms entirely depends on human resources. However the bigger the organization becomes, the more dangerous the dependency upon a few talented persons is. Rather than depending on them, it is obviously better to equipped with a competitive system and to carry out design projects with such a system. If large architectural firms construct, not a comprehensive gift set, but specialization in each field, if they actively acknowledge the copyright of their partners, if an internal review system is properly operated and design is processed with sufficient time and efforts, I dare to say that a positive way for design identity will be discovered.

Then, large architectural firms can develop their individuality and make a diverse spectrum in the market of architectural design. It is the very point the firms cannot miss. And it is their raison d'etre. Recently top-ranked architectural organizations in Korea such as Junglim, Space, Heelim, Samoo and Kunwon are showing a desirable tendency of synthesizing conceptual emergence and design

공간 웹진 VMSPACE.COM
WEBZINE VMSPACE OF SPACE GROUP

원도시세미나
POSTER OF WONDOSHI SEMINAR

정림 뉴스레터
NEWS LETTER OF JUNGLIM ARCHITECTURE

를 진행하며, 심지어 사인 시스템까지 중복 디자인하는 현재의 고비용 저효율 체제에서는 결코 경쟁력이 있을 수 없다. 이는 설계회사만의 문제는 아니며 턴키 기반 프로젝트 발주방식이라는 기형화된 시스템을 고쳐야 하는 문제이기는 하다. 그러나 턴키방식을 '그들만의 리그'로 만들면서 기득권을 누려온 대형 설계회사들도 전혀 책임이 없다고 하기는 힘들다.

경기 악화로 인하여 민간투자가 현저하게 위축되는 바람에 대부분의 설계 용역이 공공기관이 발주하는 턴키사업 아니면 아파트먼트 설계를 중심으로 이루어지고 있다. 그리고 민간 기업의 반복 수주에 많은 부분을 의존하던 일부 설계회사들은 상당한 어려움에 봉착해 있기도 하다. 이러한 시장 구조에 원활하게 대처하기 위해서는 설계회사의 업무를 기존 범위에서 확장하여 각종 기획, 개발 등 시장의 외연을 넓히는 한편 더 적극적으로 해외 설계시장을 개척해야 할 것이다. 물론 인적 자원 양성에 중점을 두어 생산성을 향상시킴으로써 경쟁력을 키우는 것이 전제되어야 한다.

일반적으로 모든 기업마다 자신의 문화가 있듯, 대형 설계회사에도 문화 차이가 드러난다. 정림건축은 정림문화재단을 따로 가지고 포럼, 건축아카데미, 학생건축상, 청년세대 지원 프로그램 등을 운영한다. 원도시건축이 만드는 정기 세미나는 그 질량에서 학술적 가치가 있다. 희림건축이 표방하는 희림문화 등은 자본의 사회 환원만이 아니라 자신의 문화 역량을 키우는 일일 것이다. 원래 문화적인 생태로 탄생한 공간그룹은 잡지발간과 출판을 포함하여 판화비엔날레, 국제학생건축상, 국제학생실내건축상 등으로 사회 지원을 계속하고 있다. 건원건축도 건원문화를 말하며 주로 세미나를 통해 학자들과 건축을 소통하고 있다.

한국 건축계에서 대형 건축회사가 차지하는 비중은 막중하다. 처음에 언급한 나머지 부류를 다 합쳐도 그 영향력과 파급효과는 시스템으로서의 건축생산을 지향하는 대형 설계회사에 결코 미치지 못할 것이다. 그런 만큼 한국 건축계에서 그 역할과 사명에 대한 진지한 자기반성과 성찰의 자세가 필요할 것이다. 궁극적으로 대형 설계회사들의 선전에 힘입어 한국 건축계가 국제적으로 비약할 수 있기를 기대한다.

technology. So is the Jina in that the next generation CEO reorganized the firm and shows an active appetite for creation as well as the productivity of large firms.

Endeavor for Competitiveness

The present system of high cost and low efficiency in which several consortiums overlap designs, even at signage system, in order to receive an order a project, can never have competitiveness. Architectural firms are not solely to blame for this but a deformed method of turn-key based project order should be to blame. However, it is true that large architectural firms which have made the above-mentioned method "a league of their own" and have enjoyed vested rights are not entirely innocent.

Due to economic recession and thereby generated decrease of private sector's investment, most design projects are being made centering around public sector's turn-key based projects and housing projects. Some firms depending on repeated order of private enterprises are confronting with considerable difficulties. In order to cope with this market situation smoothly, the firms should expand their range of work including planning and development and actively explore overseas market. It is presupposed that they should increase their productivity focusing on the training of human resources and therefore be more competitive.

Generally large firms shows their cultural identity as every enterprise has its own culture. The Junglim Architecture has Junglim Foundation and Forum, Architectural Academy, Students Architectural Awards, and New Generation Sponsoring Program. The seminar organize by Wondoshi Group has an academic significance. "Heelim Culture" by Heelim is useful for Heelim itself as well as the social return of the capital. The Space Group which originally aimed for cultural enlightenment is sponsoring International Students Architectural Design, International Print Biennale, International Students Interior Design including magazine and publication. The Kunwon Architects is mentioning "Kunwon Culture" and tries architectural communication with academics.

The weight large architectural firms have are enormous in Korea. The influence and ripple effect above-mentioned "atelier" architects of all the other categories have cannot match those of large architectural firms pursuing architectural production as a system. That is why they need reflection and self-examination on their role and mission in Korean architectural world. Ultimately, I hope Korean architecture will make rapid progress in international architecture world owing to good performance of large architectural firms in Korea.

JNA Architects, Korea
Jaejin Bu + Moohyun Kim + Sanghoon Bu
Underground Campus and International Dormitory, Sogang University
Seoul, Korea
2006

프로젝트의 구상은 전통적인 개념의 캠퍼스 중앙공원에서 시작되었다. 지하캠퍼스의 지붕과 국제기숙사의 서쪽 입면은 서강대학교의 새로운 중앙공원 구획을 만든다. 도시 속 캠퍼스이기 때문에 도시적 경계를 만들고, 대학 주변의 시민들과 대학 관계자들의 정체성을 구분할 필요가 있었다. 여러 단으로 구성된 거대한 계단들과 캔틸레버 형태의 기숙사 매스는 캠퍼스 입구의 기념비적인 모습을 만들어낸다.

서강대학교는 예수회재단에서 설립한 학교로 가톨릭교회의 오랜 역사를 지니고 있다. 오랜 역사를 통해, 스테인드글라스는 교회건축 안에서 강력한 언어의 한 요소로 존재했다. 유리라는 요소를 이용하여 스테인드글라스와 같은 빛과 컬러 스펙트럼의 특성을 강조하였고 댄 플래빈과 도날드 저드 (미니멀아트 작가)의 작품에서 영향을 받았다.

새로운 중앙공원을 구성하는 지하캠퍼스의 지붕과 국제기숙사의 서쪽 입면
ROOF OF UNDERGROUND CAMPUS CONSISTING OF NEW CENTRAL PARK AND WEST ELEVATION OF INTERNATIONAL DORMITORY

서강대학교 지하캠퍼스 및 국제학사

진아건축
부대진 + 김무현 + 부상훈

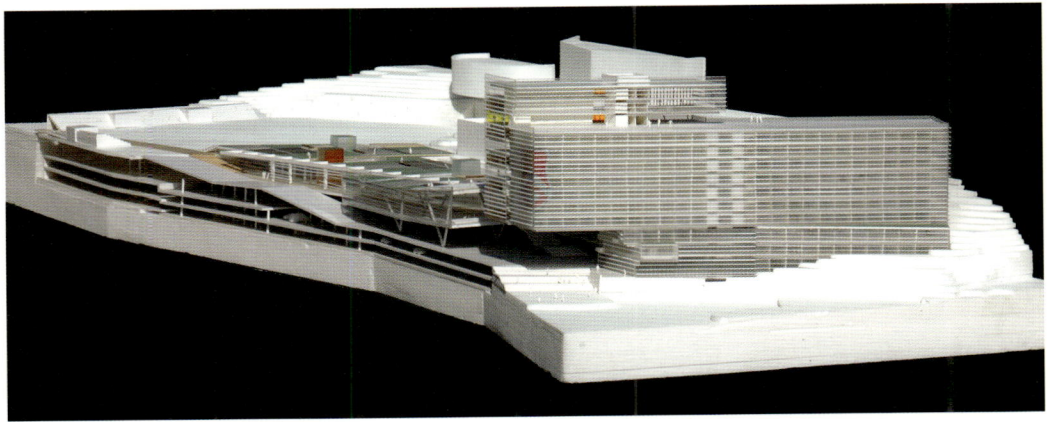

국제기숙사의 동쪽 전경 EAST VIEW OF INTERNATIONAL DORMITORY

캠퍼스의 배치 컨셉 LAYOUT CONCEPT OF CAMPUS

The traditional notion of the Campus Quadrangle or Quad generated the project's conceptual start. The roof of the underground campus and the dormitory's west elevation create Sogang University's newest Quad. As an urban campus the need to create an urban boundary and to reinforce the "town and gown" identities created the dormitory's edge. A large cascading stair and a cantilevered dormitory mass create a monumental gateway into the campus.

As a Jesuit institution, the school has a long history with the Catholic Church and the Jesuits. The precedence of the stained glass is a powerful vocabulary in ecclesiastical structures. We were inspired by Dan Flavin and Donald Judd, among others, in seeking to create an identity consistent with the elements of glass, light and the color spectrum.

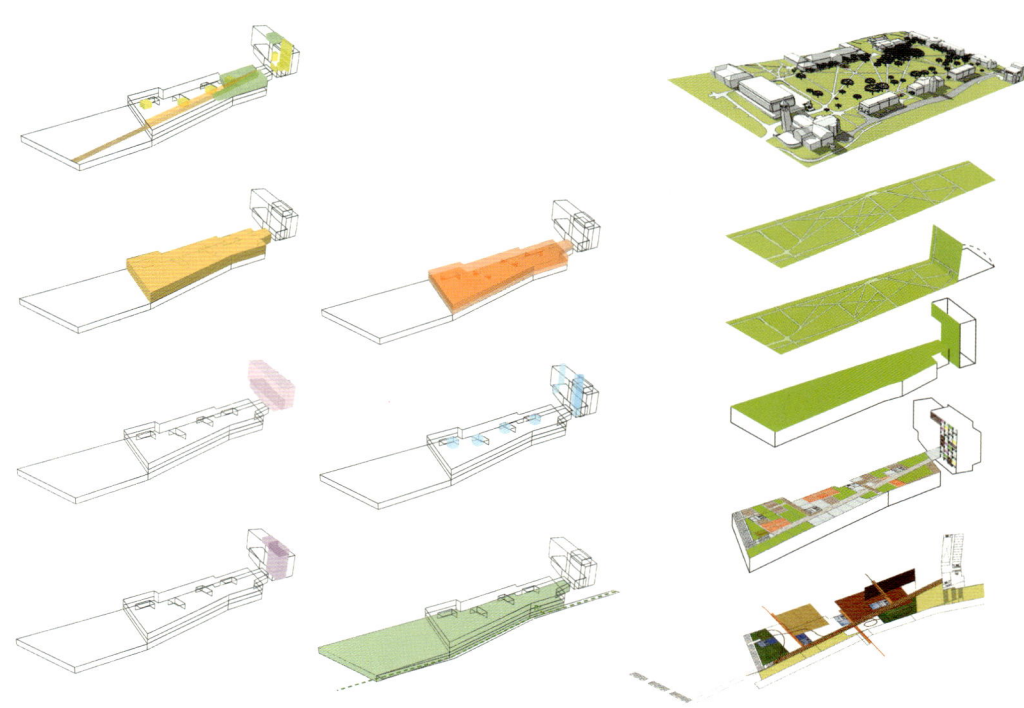

지하주차장과 기숙사의 켜 쌓기
STACKING THE LAYERS OF PARKING UNDERGROUND AND DORMITORY ABOVE

매스 다이어그램 MASSING DIAGRAM

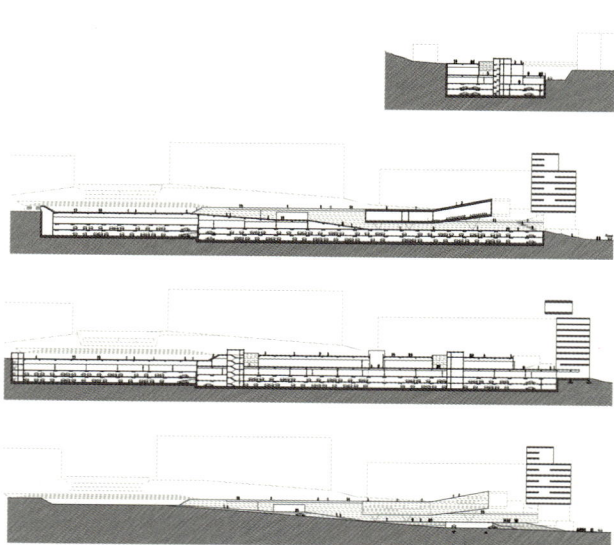

SITE PLAN

SECTION

서강대학교 지하캠퍼스 및 국제학사

진아건축
부대진 + 김무현 + 부상훈

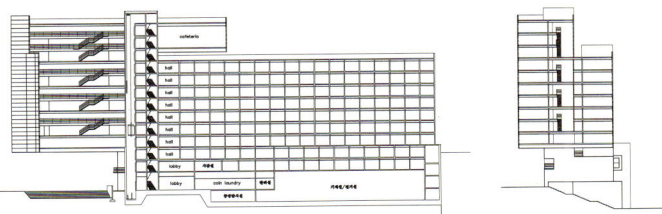

기숙사 입면 DORMITORY ELEVATION

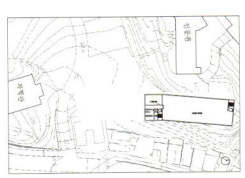   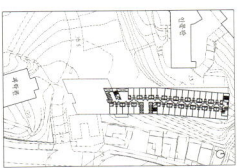

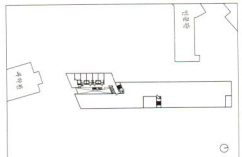 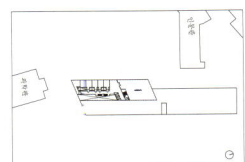 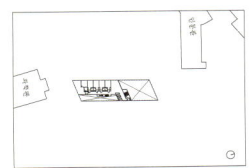 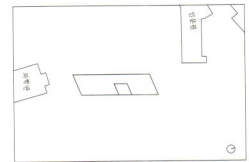

PLAN

지하캠퍼스 주진입 부분 MAIN ENTRANCE OF UNDERGROUND CAMPUS

Junwon Architects Planner Engineers + Da Group, Korea
Jongkook Kim, Insun Hahm + Hyunho Kim + Yongkeun Yoon
The First Town
Multi-Functional Administrative City, Korea
Competition Project_Winner
2006

한국 사람들은 이부자리를 펴고 잔다. 낮 동안 장롱에 들어가 있는 이부자리에는 잠자리가 접혀 있는 것이다. 전통 한옥의 방은 침실과 식당, 거실이 접혀져 있으며, 대청마루에는 실내와 실외가 접혀져 있다. 한옥은 마루와 온돌이 접혀 있는 세계 유일의 주거 형태이다. 우리는 한옥이나 전통 마을처럼 삶의 다양한 행위가 접혀 있는 공간, 자연과 인공물이 접혀 있는 마을을 제안한다. 접혀 있음은 단순한 병치가 아니다. 그것은 서로가 서로를 머금고 있는 방식이다. 수평선이 바다 밑 지형을 머금고 있듯이, 우리가 제안하는 판platform은 원래 지형과 생태를 둘러싸고 있다.

판 외곽선은 전체 도시의 얼개에 조응하는 선과 원지형을 충돌시켜 얻어지며, 이들 사이로 커뮤니티시설들이 스며든다. 자연과 인공이 공존, 사적 공간과 커뮤니티의 공존을 통하여 공동체성의 회복이 이루어질 것이다.

Korean people sleep in "Ibujari". What they sleep in is folded and stacked inside closets in the daytime. A bedroom, a dining, and a living room are folded in "Bang" of "Hanok". And on "Maru", outdoor space and indoor space are folded. In Hanok, floor and heating system are folded, thus making it unique housing style in the world. We propose a village with folded form of the natural and the artificial structures alike, which accommodates spaces with diverse activities, just like Korean traditional villages. Folded things are not just simply juxtaposed, but embrace each other. As underwater topography lies beneath the horizon, the plates we propose embrace the original topography and ecology.

Outline of the plates are drawn out by colliding the original topography with the corresponding lines within the City. Community facilities are infiltrated between these plates. The true recovery of the community shall be realized upon symbiosis between the natural and the artificial.

Ecology

Coordination on Program

Korean Traditional Villages

Embracing

Platform

Pinwheel

Topography

Zipper

Community

Symbiosis

Weaving the Program

New Lifestyle

마스터플랜 모형 MASTER PLAN MODEL

첫마을

다기능 주거도시, 행정중심복합도시 현상설계 당선작

건원건축 + DA그룹

김종국, 함인선 + 김현호 + 윤용근

MASTER PLAN

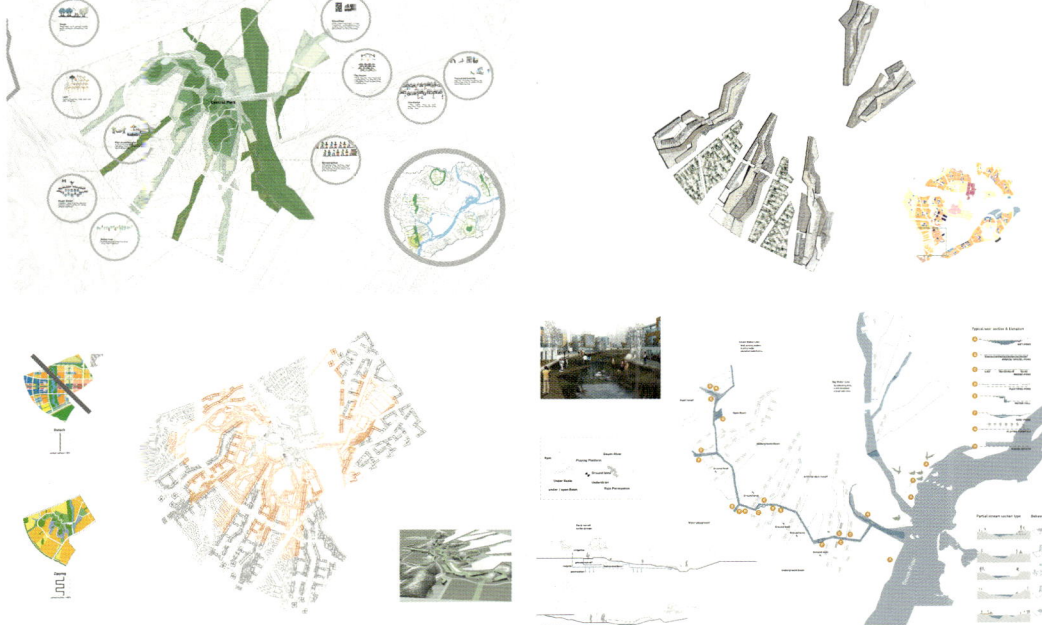

지형으로부터 흐르는 공간 구조 | 첫마을의 배치 컨셉 LAYOUT CONCEPT OF THE FIRST TOWN

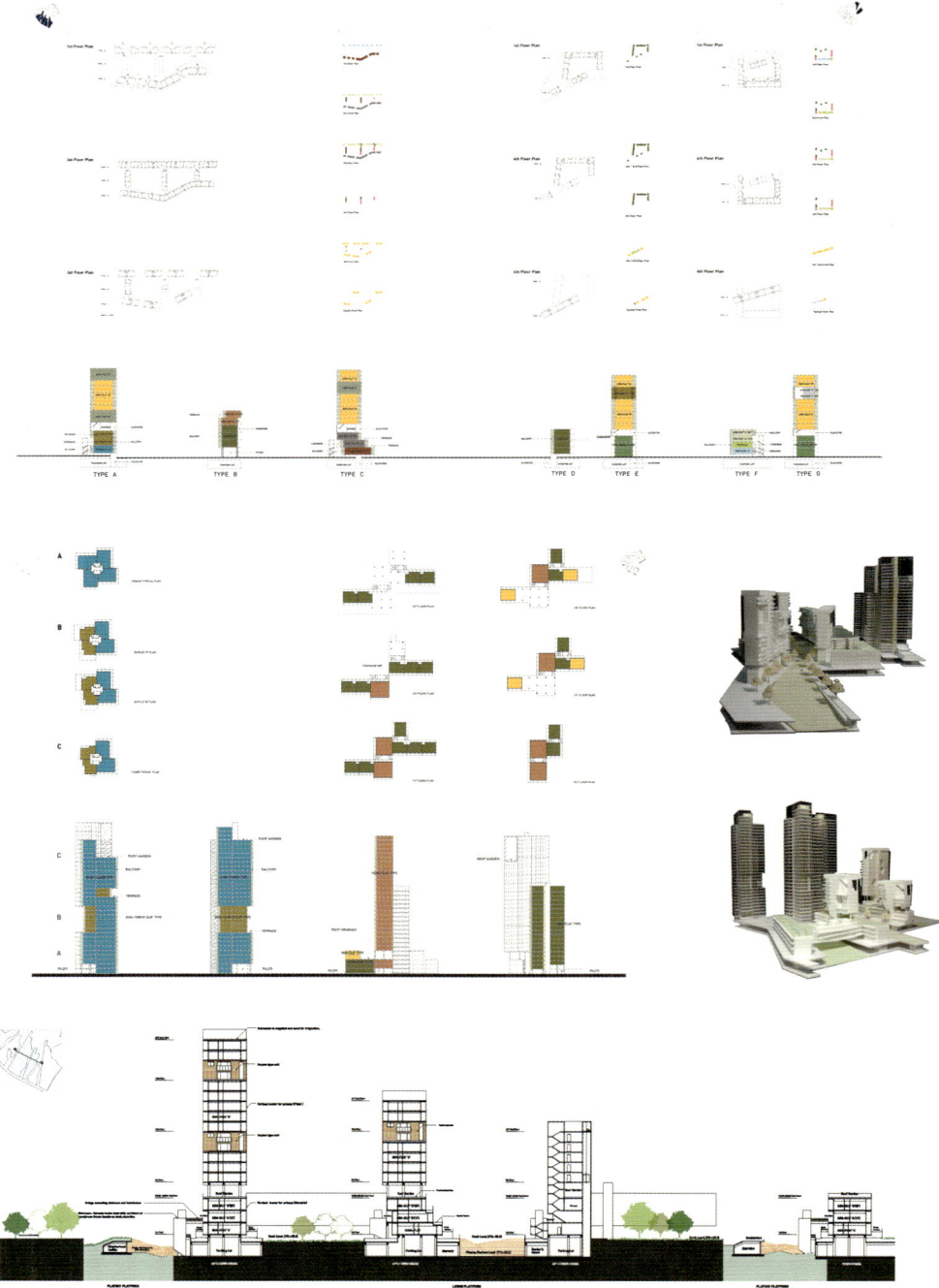

여러 가지 볼륨 타입의 혼합 MASSING, ELEVATION SECTION

첫마을

다기능 주거도시, 행정중심복합도시 현상설계 당선작

건원건축 + DA그룹
김종국, 함인선 + 김현호 + 윤용근

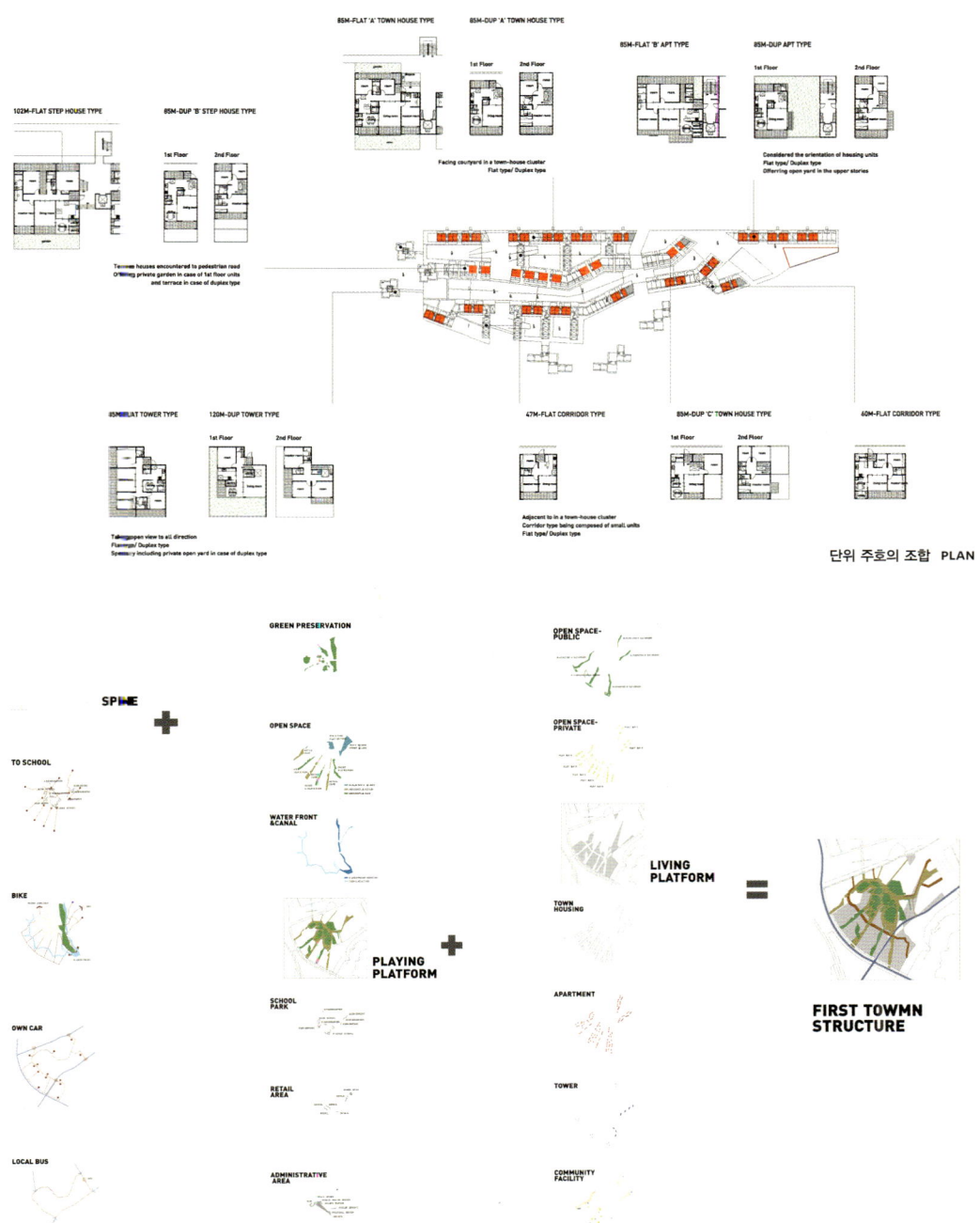

Samoo Architects & Engineers, Korea + SOM, U.S.A.
Tower Palace III

Seoul, Korea
2002-2004

타워팰리스 III의 평면과 매스 계획은 세 개의 날개를 지닌 나뭇잎을 모티프로 삼았다. 자연에서 영감을 얻은 유기적 형태로 무미건조함을 덜고자 한 것이다. 단조로운 빌딩 숲 가운데 치솟은 한 그루 나무와 같은 랜드마크가 될 것이다. 하늘을 향해 뻗고 지상을 향해 리드미컬하게 떨어지는 나뭇잎 이미지는 초고층 건물의 인상을 강화한다. 녹색빛 유리와 메탈릭 실버의 알루미늄 표면이 자아내는 투명성은 첨단 테크놀로지를 표현하는 동시에 입면의 입체감을 강조한다.

타워팰리스 III의 Y자 평면은 낮에는 남쪽의 태양을 향해 열리고 밤에는 도시의 불빛을 향해 열린다. 윤곽의 요철 면을 따라 광합성 면적을 극대화하는 나뭇잎처럼, 세 갈래로 뻗은 Y자 평면의 더욱 넓어진 외벽 면적은 최고의 전망과 채광을 생활 속으로 끌어들인다. 코어의 공용 공간은 나뭇잎 중심에서 순환을 담당하는 잎맥이 된다. 1, 2층 저층에 연회장, 놀이방, 독서실 등 주민복지공간을 배치하고, 코어공간에는 이동동선을 배치했다. 이같은 평면은 리드미컬한 변화감으로 거대한 매스의 위압감과 지루함을 덜어줄 뿐만 아니라 각 실의 독립성과 각 공간의 기능적 독립성을 극대화한다.

최근 주거 수요는, 중간층을 선호하고 방위를 중시하던 과거와 달리 훌륭한 전망과 정서적 만족감을 더욱 중시한다. 공원과 사람들이 어우러지는 근경, 도시의 파노라마가 펼쳐지는 원경을 즐길 수 있는 집을 원하는 것이다. 변화하는 라이프스타일과 다양한 욕구를 충족시킬 수 있는 새로운 평면을 제안했다.

High-Rise Living

Sky Garden

State-of-the-Art

Landmark

Natural Space

Organic

Dynamic

Diversity

Fashionable

Skyline

Emotional Satisfaction

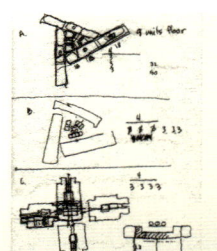
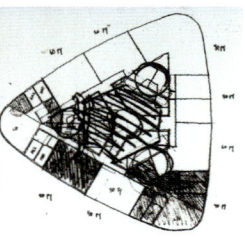
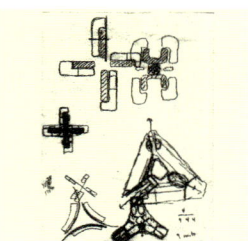
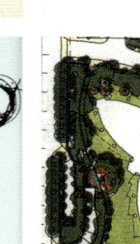

나뭇잎 형태의 평면 개념 LEAF SHAPED PLAN CONCEPT

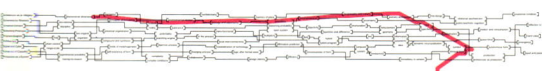

타워팰리스 III   삼우설계 + SOM

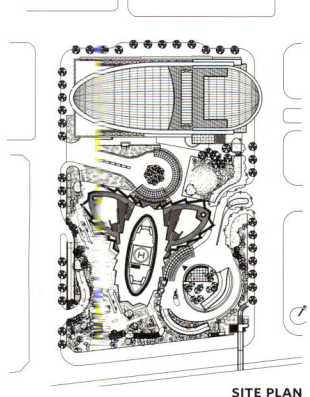

SITE PLAN

녹색 유리와 메탈릭 실버 알루미늄의 입면
ELEVATION WITH GREEN GLASS AND METALLIC-SILVER ALUMINUM

Tower Palace III was adopted from the leaves so it's shape is natural & organic. This remarkable building breaks the monotony in and around the site. The tower will be clad in a light blue-green tinted high performance glass to allow for maximized light transmission and bright views with metallic panels to provide sparkle to the solid surfaces.

    The tower's form is intended to add a truly unique shape to the skyline. This form is characterized by a series of radiating planes, which will provide a dynamic appearance as the sun moves across the faces. These planes provide a strong sense of verticality to the tower and allow for the discreet movement of air through the units. Each unit is arranged centering around the centralized core space. Differently from the existing parallel plate type layout, this layout type maximizes the area of outer wall for each household. Differently from the past trend preferring the middle floor and putting the importance on the standing direction of household, the recent needs for residential space are diversified, such as emotional satisfaction from good view, convenient plane layout, etc.

유리와 알루미늄의 입면 디테일 ELEVATION DETAIL OF GLASS AND ALUMINUM

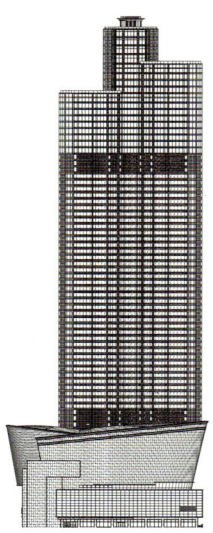
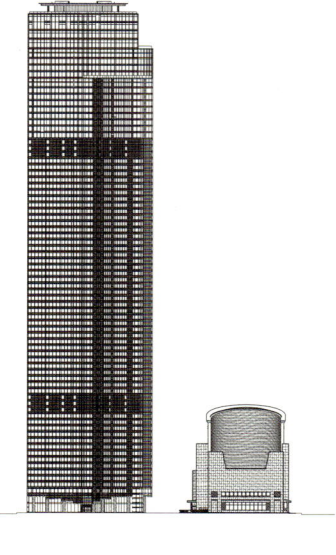
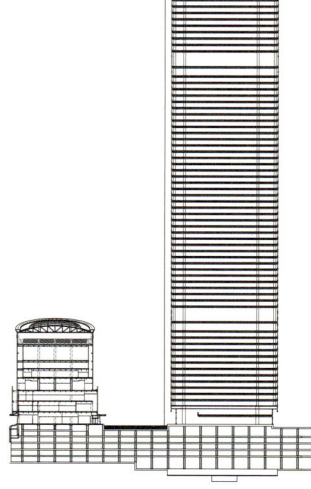

ELEVATION   SECTION

타워팰리스 III   삼우설계 + SOM

TYPICAL PLAN

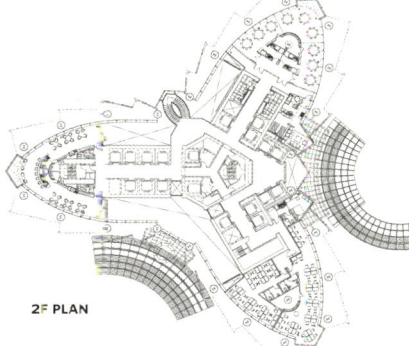

2F PLAN

1F PLAN

앞마당 및 로비 전경 FRONT YARD AND LOBBY

MOOYOUNG ARCHITECTS & ENGINEERS, KOREA + DESTEFANO & PARTNERS, U.S.A.
GILWON AHN + CHANGSEOP CHOI
DANANG ADMINISTRATION CENTER
DANANG, VIETNAM
2006-

이 프로젝트는 다낭 시 한강과 인접하여 매개공간으로 워터프론트 활성화에 기여하도록 하였다. 진입 광장은 한강과 성곽 터 및 박물관으로 진입이 유연하도록 하였으며 시청사, 박물관, 공원, 소프트테크 센터 건물의 조화로운 배치를 추구하였다. 저층부는 시민들에게 열려 있는 공용공간으로, 광장을 이용하는 시민들을 위해 편의시설도 마련하였다. 타워부는 일반사무실로 구성되었고, 강한 바람에 구조적으로 유리하도록 원형 평면을 가지고 있고 360도 조망이 가능하여 다낭시 중심에서 전체를 바라볼 수 있도록 하였다.

외관은 높이 153.3m로 다낭 시의 최고 높이 건물로서 상징성을 가지며 투명성, 공정성, 역동성을 나타내고 있다. 안정적인 현재의 다낭 시를 반영한 저층 기단부와 역동적인 모습의 미래를 상징하는 타워 부분은 현재와 미래가 조화를 이루는 모습을 상징화하여 다낭 시의 미래인 21세기 새로운 도약의 시작이 될 것이며 첨단의 설비들은 업무환경 개선과 효율적인 업무 행정을 가져올 것이다.

Since the site is located near Han River in Danang, the location itself would take a role to introduce the water front activities as an intermediate space. The approach square is also introduced for comfortable access to Han River, Ancient Castle Mall and Museum. This placement is originally intended to harmonize the composition of City hall, Park, Softech Center. The low level is designed for a public space which is open to pedestrian using the citizen square along with convenient facilities. The upper level, the tower is consisted of a general office. The circle shape plan is proposed to deal with a lateral force (wind) structurally. In addition, a sense of 360 degree outlook makes the opportunity to view out overall Danang city.

The building height is 153.3m, the tallest building for city of Danang, which is symbolized as landmark for city of Danang. The building design is based on the concept of See-Through, Fairness and Dynamic. The podium is designed to reflect a stable context of Danang city and the tower shape is represented the dynamic activities for the future. The building will be a symbol of the take-off stage of 21 century in terms of harmony between now and future. In addition, the high tech, MEP equipments will help to increase the quality of work environments and efficient administration service.

WATER FRONT

LANDMARK

SEE-THROUGH

FAIRNESS

DYNAMIC ACTIVITIES

SYMBOL

HARMONY BETWEEN NOW AND FUTURE

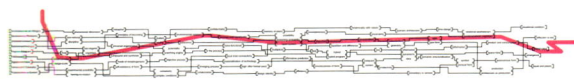

베트남 다낭시청사

무영건축 + 데스테파노
안길원 + 최창섭

원형 매스의 시청사
TOWER OF ROUND SHAPE

수변에 위치한 시청사 타워 전경
VIEW OF THE TOWER AT WATERFRONT

SITE PLAN

시청사 전경 BIRD'S-EYE VIEW

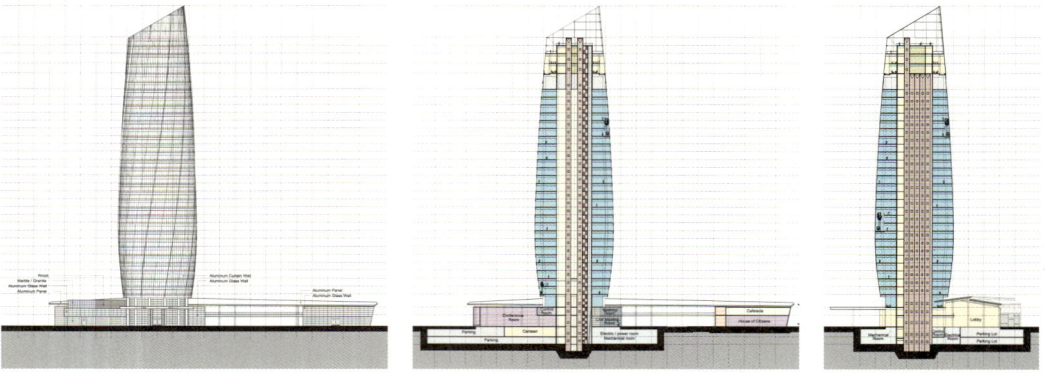

ELEVATION　　　　　　　　　　　　　　　　　　　SECTION

베트남 다낭시청사 　 무영건축 + 데스테파노
안길원 + 최창섭

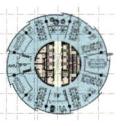  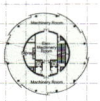 

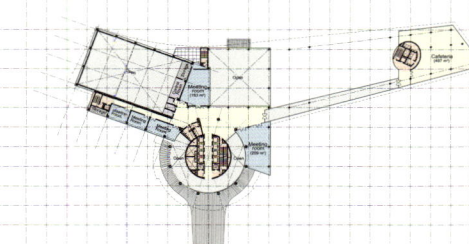

2F PLAN

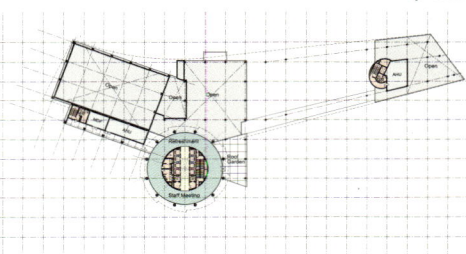

4F PLAN

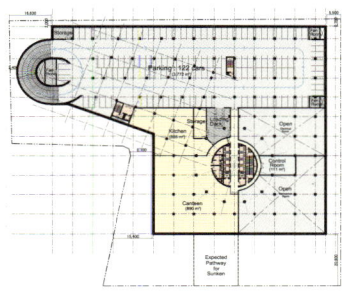 

B1 PLAN

3F PLAN

내부 부분 투시도 INTERIOR PERSPECTIVES

Gansam Partners, Korea
Boheon Building
Seoul, Korea
2005

북촌마을은 서울 유일의 전통 한옥 밀집지역으로 경복궁, 창덕궁, 금원(비원)과 인접한 북악산 기슭에 자리잡고 있다. 전통 한옥들 사이에 현대적인 업무시설을 건축하는 것은 기존 마을 질서 전체를 훼손할 수도 있어 보헌빌딩은 주변 경관을 해치지 않으면서 기존 질서와 자연스럽게 어우러지도록 설계하는 데 주안점을 두었다. 한옥과 콘크리트 건축의 이질적인 충돌을 전통적인 건축 재료와 휴먼스케일의 내부 공간 구성을 통해 해결하고자 했다. 형태적으로만 전통 한옥을 흉내내기보다는 업무시설 기능을 충족하면서도 전통 한옥의 공간 구성 원칙에 따르는 평면 계획을 통해 사용자에게 친숙하게 다가서고자 했다. 곳곳에서 한옥의 절제된 자연 질감의 전통 재료를 사용하여 주변 한옥 건물군과 조화를 시도하였으며, 나지막하게 둘러친 건물 주변의 기와 담장은 보헌빌딩이 북촌마을에 적극적으로 적응하려는 자세를 보여준다.

중정에서 올려다본 하늘의 모습은 한옥마당에서 바라본 하늘과 매우 흡사하다. 건물 1층 전체를 감싸는 복도는 지하 1층의 중정을 향해 열려 있고 필로티에 의해 시선의 연속성을 가져 마치 정원을 걸어다니는 기분을 느낄 것이다. 도로에 면한 외벽은 절제되면서도 자유로운 입면이 되도록 하여 대지에 꽉 찬 매스의 비대감을 약화시키고 있다. 중정을 향해 열려 있는 내부 외벽은 커튼월로 처리하였다.

절제되면서도 자유로운 입면 **MODERATE BUT LIBERATED ELEVATION**

KOREAN TRADITICNAL HOUSINGS

HISTORICAL CONTEXT

EXISTING CONTEXT

TRADITIONAL MATERIALS

HUMAN SCALE

ADAPTATION

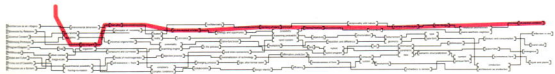

보헌빌딩　간삼파트너스

전경 PERSPECTIVE VIEW

주변 한옥과의 관계성
RECIPROCITY WITH NEAR TRADITIONAL HOUSE

SITE PLAN

Bukchon Village is the only currently existing traditional Korean housing cluster in Seoul. It is located on the edge of Mt. Bukak and the center of the historical context among Gyungbok, Changduk Palace, and Secret palace. Since the design of modern official building in the context of traditional Korean housing could be regarded as the destruction of the existing order, our main design topic is to preserve the existing context and fit into it very naturally without any distortions.

In other words, the conflict between traditional Korean housing and the tectonics of concrete building has to be resolved. The use of traditional materials and human scale interior space composition are the design alternatives for this problem. Knowing that the imitation of the self-restrained Korean order is more dangerous to the existing context, Boheon building has to fulfill the requirements for the office and fundamentals of space composition of the traditional building for more human-oriented space planning simultaneously. As described above, the design intention and attitude for an active adaptation to the traditional context is witnessed on the application of the moderate use of natural materials from the existing context and surrounded wall of traditional tiles.

한옥의 마당을 연상시키는 중정
**COURTYARD REMINDING OF MADANG,
KOREAN TRADITIONAL OUTDOOR SPACE IN DWELLING**

중정과 복도 **COURTYARD AND CORRIDOR**

ELEVATION

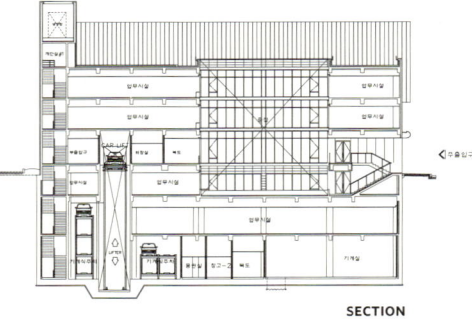

SECTION

중정 전경 **COURTYARD VIEW**

DETAIL

355 보헌빌딩 간삼파트너스

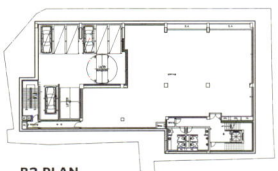
B2 PLAN

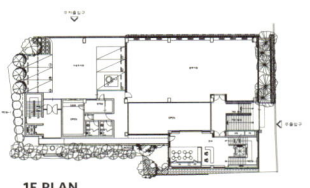
1F PLAN

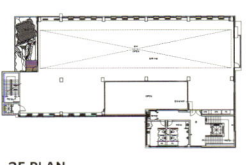
3F PLAN

B3 PLAN

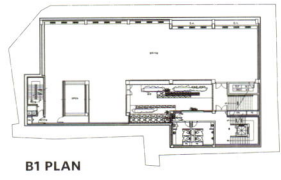
B1 PLAN

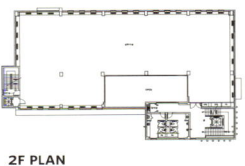
2F PLAN

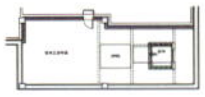

Wondoshi Architects Group, Korea
Yong Byun
Gangdong Culture Art Center
Seoul, Korea

이 프로젝트는 문화활동의 주체와 객체가 서로 쉽게 교류하는 공간이자, 그들 각자의 문화예술 활동에 이바지하는 합리적인 인프라 공간이다. 이는 사회 변화에 적극적으로 대응하면서 문화예술 실현의 중추 역할과 새로운 문화현상 창조를 위한 실험 무대라고도 할 수 있다. 문화 본질에 관한 물음과 함께 대지가 품고 있는 잠재력에 귀기울이며 다음과 같은 접근방법을 제시한다.

Community-NEST 문화란 기본적으로 만남을 전제로 한다. 일종의 주고받는 과정에서 생성되는 합의된 방향성의 부산물인 것이다. 이를 위해 누구나 언제든지 편안하게 참여할 수 있는 기회의 보금자리를 만든다.

Culture-NEST 교류의 산물인 문화가 잉태되고 자라고 확대되기 위한 열린 공간이다. 예술 창작의 물리적 시스템에 주목하며 기능성을 극대화한 편리한 공간이다.

Creative-NEST 창조의 원천은 상상일 것이라는 가정에서 창조적 상상을 적극적으로 유도하는 우연성과 일상성의 공간이다.

- 도시와 자연, 문화가 어우러지는 이벤트 마당으로서 어반 로비
- 도시와 자연이 상호보완 공존하며 하나되는 랜드스케이프 스테이지
- 도시와 자연의 경계, 문화로의 초대를 상징하는 어반 게이트
- 창조적 상상을 일상에서 체험할 수 있는 플로잉 테라스

INFRASTRUCTURE
CULTURE
COMMUNICATION
ART
ECOLOGY
TOPOGRAPHY
FLOW
URBAN SPACE
LANDSCAPE
OUTLINE
CONTINGENCY

강동문화예술센터 전경 PERSPECTIVE VIEW

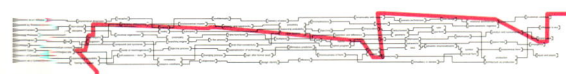

URBAN LOBBY AS EVENT SPACE

C+Nest Our proposal is to create a space as a useful infrastructure, where the actors and audience of cultural activities interact with ease, and which contributes to the cultural and artistic activities of each. It is a space that actively responds to social changes while serving as the central forum for culture/art activities and enabling creation and experiments in new cultural phenomena. To create such a space, we provide the following three main approaches while pondering the essence of culture and remaining attentive to the potential of nature.

Community-Nest Culture is basically predicated upon encounters. It is the by-product that is agreed upon and created through the process of give and take. To allow such exchange, we need to create a nest of opportunities where anyone can participate anytime with ease.

Culture-Nest It is an open space where culture, the product of exchange, will be born, developed and expanded. It is designed as a highly functional and convenient space that pays keen attention to the physical systems needed for artistic creation.

Creative-Nest It is a space of the accidental and the quotidian, designed to stimulate creative imagination under the assumption that "imagination is the mother of creation".

- **Urban Lobby** a backyard for events, where the urban, the natural and the cultural converge.
- **Landscape Stage** where the city and nature become mutually complementary and inclusive.
- **Urban Gate** which signifies the boundary between the city and nature, or invitation to culture.
- **Flowing Terrace** where the creative imagination can be experienced in one's everyday life.

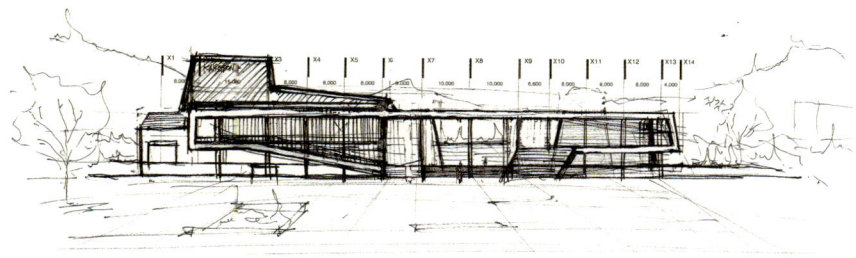

ELEVATION SKETCH, EMBRACING OUTDOOR SPACE

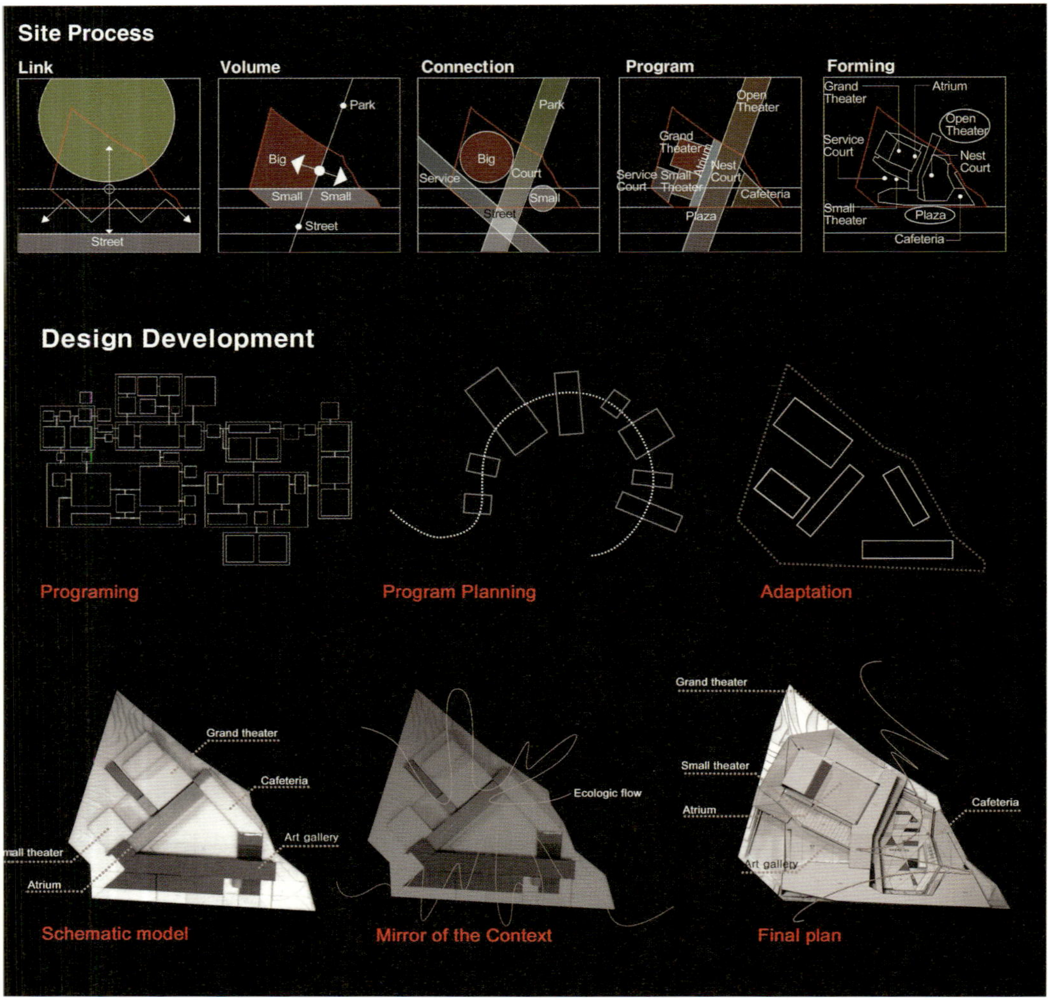

Site Process
Link | Volume | Connection | Program | Forming

Design Development
Programing | Program Planning | Adaptation

Schematic model | Mirror of the Context | Final plan

CONCEPT

강동문화예술센터　　원도시건축
변용

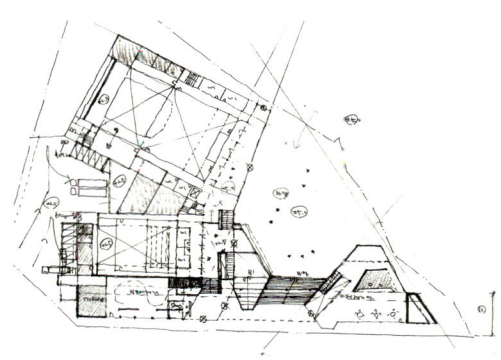
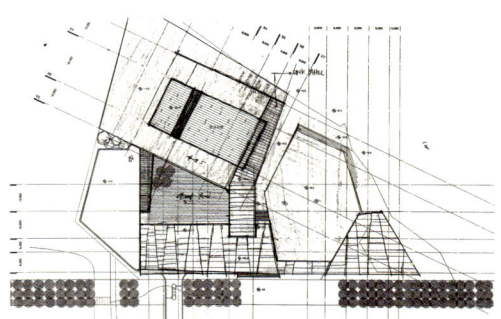

PLAN

 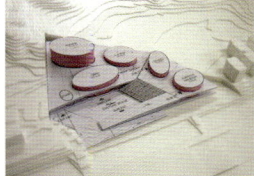 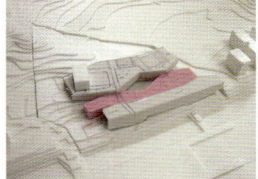 
 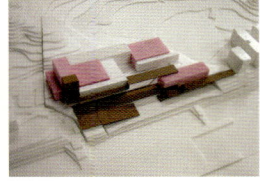 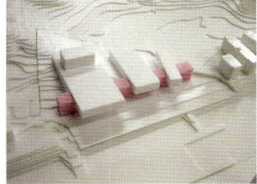 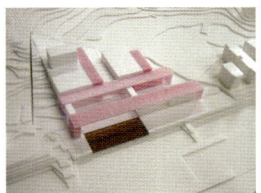
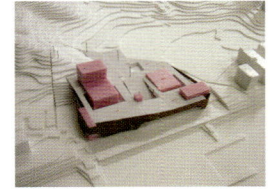 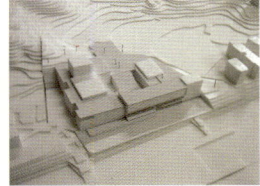  

매스 스터디 과정　MASSING PROCESS

MODEL

NONYANG ARCHITECTURAL DESIGN GROUP + HAEAHN ARCHITECTURE, KOREA
DONGCHAN LEE + SEHAN YOON
OSONG BIO-HEALTH SCIENCE TECHNOPOLIS
CHUNGWON, CHUNGCHONGBUK-DO, KOREA
2004-2008

오송생명과학단지 안에 세워질 바이오산업 관련 4대 국책기관은 자연의 흐름에 순응하여 여유로움과 쾌적함을 체험할 수 있다. 연구원들은 데크를 통해 커뮤니티를 활성화하게 된다. 4개 기관은 중앙광장을 중심으로 유기적으로 연결하여 개방감과 확장성을 고려하였다.

경사진 대지 지형에 따라 연속된 선 흐름으로 건축물을 배치하고, 단지는 전체적으로 산 능선보다는 낮게 건축물을 계획하여 자연 속에 녹아들도록 하였다. 프로그램 특성상 다소 독립적인 성격의 단지를 외부인과 공유할 수 있도록 대지 중심부는 열린 공간으로 구성하였다. 독립된 각 기관들의 동선을 연결시키기 위해 새로운 판을 설치하여, 연구원 상호간 커뮤니티 활성화를 위한 기관별 소광장 및 임대형의 폴리 요소를 적절히 배치함으로써 연구원들의 몰을 구성하였다. 이는 사색과 여유의 공간을 제공한다. 데크 하부는 서비스주차, 코어, 민원 관련 사무행정, 설비배관 등의 시설이 마치 혈관과 같이 각 기관을 유기적으로 연계한다.

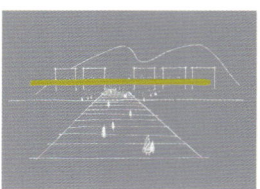 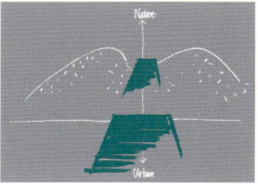 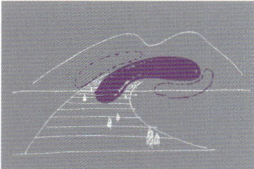

Four national institutions related to the bio industry to be built in the Osong Bio-Health Science Technopolis conform to the flow of nature, allowing people to experience tranquility and pleasant feelings. Researchers can invigorate a community through decks. The four institutions have been designed to be organically connected with the central plaza as a center, taking consideration of openness and scale.

Buildings have been deployed in accordance with the continuous flow of lines following the sloped topography of the site and overall, to make them blend with nature, buildings in the complex were planned to be situated lower than ridge lines of mountains. In order to make the complex, which is of a rather independent nature due to its programs' characteristics, be shared with outsiders, the center of the site was planned as an open space. Meanwhile, in an effort to connect movement lines of independently located institutions new plate decks have been installed to appropriately deploy small plazas and rental poly elements by institutions, thereby creating malls for researchers. This provides a space for reflection and relaxation, while the lower part of the decks will house facilities, including a serviced parking area, cores, administrative office, and plumbing, which connect the institutions organically like blood vessels.

CITY AND HUMAN

THE LINK OF NATURE

MERGE

LAYERS

DOUBLE GROUND

OPEN

CONNECT

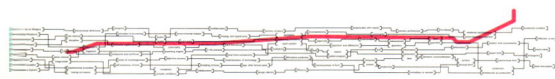

오송생명과학단지 　원양건축 + 해안건축
이동찬 + 윤새한

경사 대지의 해석과 공간의 위계 설정 INTERPRETATION OF SLOPED SITE AND SPATIAL HIERARCHY

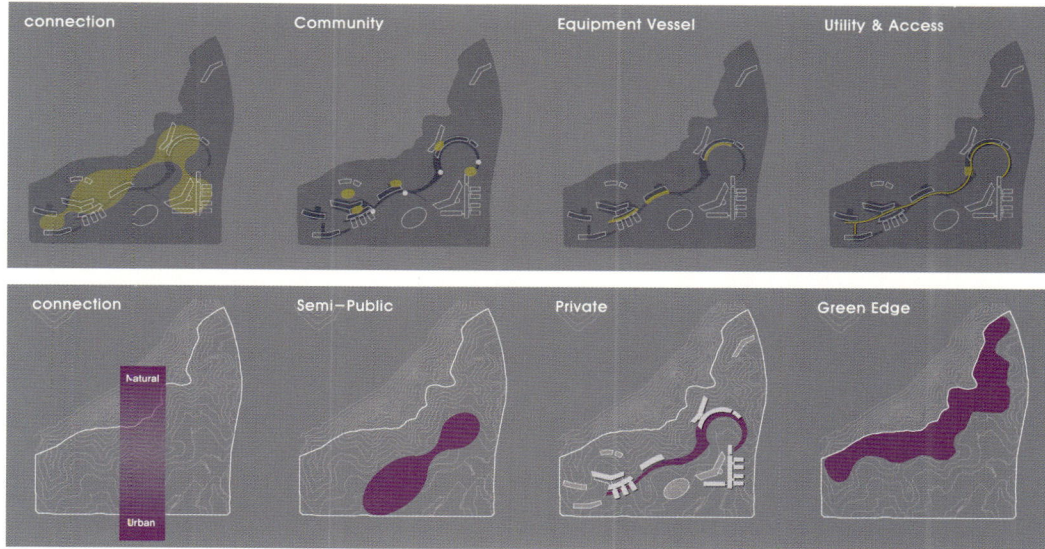

배치 개념도 LAYOUT CONCEPT

커뮤니티 활성화를 위한 새로운 데크 설정
NEWLY DESIGNED DECK FOR COMMUNITY

질병관리본부 전경
CENTER FOR DISEASE CONTROL AND PREVENTION

원형 가든을 둘러싼 국립독성연구원 전경 NATIONAL INSTITUTE OF TOXICOLOGICAL RESEARCH BUILDING SURROUNDING THE ROUND GARDEN

오송생명과학단지

원양건축 + 해안건축
이동찬 + 윤새한

식품의약품안정청의 매스 MASSING OF KOREAN FOOD & DRUG ADMINISTRATION

질병관리본부의 매스
MASSING OF CENTER FOR DISEASE CONTROL AND PREVENTION

식품의약품안정청의 브리지
BRIDGE OF KOREAN FOOD & DRUG ADMINISTRATION

숙소동 전경 VIEW OF DORMITORY

Homoon Engineering & Architects + UnSangDong Architects, Korea
Hamsoo Han, Dooho Choi, Kichul Choi + Yoongyoo Jang
Design Center of Gwangju
Gwangju, Korea
2003-2005

광주디자인센터는 디자인 교육 및 지원기능 관련 업무시설과 전시·홍보관까지 겸비한, 디자인 인프라를 A부터 Z까지 총망라한 건물이다. 때문에 각 프로그램이 상충되지 않고 맞물려가는 시스템이 필요하다. 전체를 한꺼번에 다루기보다는 프로그램에 따라 공간을 나누고 각각을 디자인하여 이것을 조립하는 방식을 통해 형태의 단조로움을 피하고 공간은 풍성해진다. 이때 중요한 것은 각 요소를 접합하는 공간적·형태적 메커니즘이 무엇인가이다.

이 프로젝트는 인터랙티브 맵을 공간화하는 작업을 제안한다. 기본적으로 제시된 디자인센터의 프로그램을 선형으로 적층하고 이것을 서로 연결하는 시스템을 구성한다. 문화, 자연, 정보, 디자인이 광주라는 도시와 연결될 수 있도록 디자인센터는 프로그램을 재구성하는 디자인 네트워크 중심에 놓인다. 이곳은 문화적 네트워크와 자연적 네트워크를 만드는 교류의 장으로서 역할을 수행한다. 총체적인 랜드스케이프 개념을 도입, 평평한 대지에 입체적인 인공대지를 조성하여 건물 내외부 공간이 하나로 연결되는 랜드스케이프 포디움을 제안한다. 대지의 입체적 연장인 랜드스케이프 포디움은 디자인센터의 주기능인 전시공간과 로비공간을 여러 다른 프로그램과 레벨로 재구성하는 새로운 대지가 된다. 이는 평지인 계획대지의 풍경을 입체적으로 구성하기 위해 대지적 장치로서 의미가 있다. 외부 공간인 광장과 내부 공간인 전시를 포함한 공공 성격의 공간이 분리되지 않고 연속적이고 입체적인 데크로서 랜드스케이프 포디움을 구성하게 된다.

내부 프로그램이 즉각적으로 입면화하는 개념으로 프로그램 스킨을 제안한다. 내부 프로그램을 외부 스킨에 반영하여 공간 기능에 따라 재료를 분리하여 적용하는 개념을 도입한다. 프로그램에 따라 서로 다른 재료와 물성을 가지고 대지 전체의 조경 개념에도 적용하여 프로그램적 조경이 되도록 구성한다. 선형의 조경 요소에 각기 다른 프로그램을 제공하는 프로그램 필드의 개념을 적용한다.

SYSTEM
STRUCTURAL, TECHNICAL
DESIGN INFRASTRUCTURE
PROGRAM
INTERACTIVE BOX
DESIGN NETWORK
LANDSCAPE PODIUM
PROGRAM SKIN

프로그램의 적층 ELEVATION SHOWING THE STACKED PROGRAMS

광주디자인센터

토문건축 + 운생동
한남수, 최두호, 최기철 + 장윤규

랜드스케이프 포디움 **LANDSCAPE PODIUM**

The Design Center of Gwangju is a building that encompasses everything about design infrastructure ranging from design education, support facilities to exhibition and promotion facilities. Therefore, there is a need for a system that facilitates and harmonizes programs with one another, rather than being in conflict. Rather than dealing with the whole at once, spaces are divided and designed according to different programs and then assembled, thereby avoiding monotonous form and enriching the spaces. Here, the important element is what spatial and formative mechanism is in place that connects the different elements.

This project proposes a work that transforms an interactive map into space. A system will be in place in which basically proposed programs of the Design Center are laminated in a linear manner and are interconnected. The Design Center will be placed in the center of a design network that restructures programs that connect culture, nature, information and design with the city of Gwangju. "Landscape Podium" is proposed in which, by introducing a total landscape concept, three-dimensional artificial land will be created on top of a flat piece of land, thereby interconnecting internal and external spaces of the building. The landscape podium, which is a 3D extension of the site, constitutes a new piece of land that recomposes exhibition space and lobby space, which are the main functions of the Design Center, with different programs and levels. This bears significance as a land device to make the landscape of the planned site more 3D-like. The landscape podium is established as a continuous 3D deck where the plaza, which is an external space, and the space with public nature, which is an internal space including exhibition area, are not separated.

As a concept to make interior programs immediately be elevated to a 3D-level, "program skin" is proposed. The interior program is reflected in the exterior program, thus depending on spatial functions, materials are divided and applied. The use of different materials and physical properties is also applied to the overall landscaping concept for the entire land, making it program-wide landscaping. A concept of "program-field" is applied where different programs are offered to linear landscaping elements.

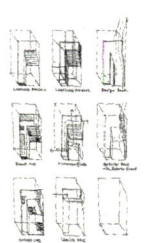
LANDSCAPE PODIUM

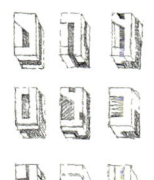
DESIGN COMPOSITION

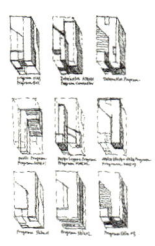
DESIGN PROGRAM

DESIGN ATRIUM

개념 다이어그램 **CONCEPT DIAGRAM**

로비 전경 VIEW OF LOBBY SPACE

광주디자인센터

토문건축 + 운생동
한남수, 최두호, 최기철 + 장윤규

연속적 로비공간 CONTINUOUS LOBBY SPACE

전시공간과 로비 EXHIBITION AND LOBBY SPACE

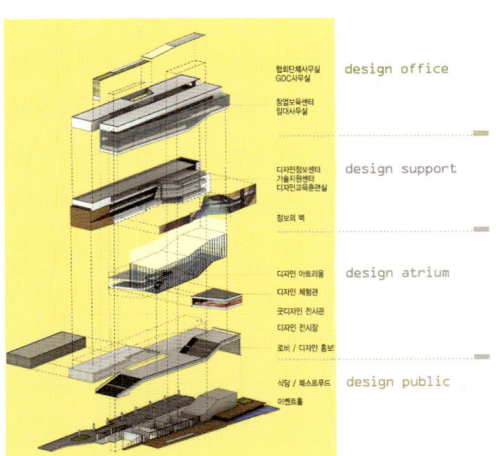

프로그램 스킨 PROGRAM-SKIN

Changjo Architects, Korea + Hillier Architecture, U.S.A.
LG Electronics Seocho R&D Campus
Seoul, Korea

자연환경에 접해 있는 서울 서초구 남단 외곽에 시각적으로 서초의 관문이 될 디지털 연구소이다. 아날로그 감수성을 그대로 가지면서도 디지털 시대에 걸맞은 첨단 IT연구소는 어떤 모습이어야 하는가. 24,201㎡의 대지 위에 연면적 124,579㎡의 첨단 전자연구소 프로그램은 규칙적이면서 융통성 있는 공간의 고층 연구소 건물과 저층 포디움에 수용된 방문객센터, 전시공간 및 후생복리공간으로 구성되어 있다.

초기 계획단계에서 기하학의 단순한 형태의 건물군과 경사면을 주제로 한 역동적인 형태로, 변화하는 주변환경을 다각도로 투영하는 타워 건물을 검토하였다. 이것은 연구소가 가져야 할 절제된 공간구조를 중요시하여 판상형 타워가 주제가 되는 건물군으로 발전한다. 판상형의 지나치게 육중한 외형을 지양하기 위해 타워 양단에 틈새공간을 삽입한 두 개의 날렵한 판을 만들고, 다시 이를 미끄러뜨려 두 개의 접합된 판을 더 적극적으로 표현하게 된다.

유리를 주제로 한 투명면과 금속판으로 된 불투명한 면으로 구성된 외벽은 이질적인 두 재료의 물성이 그 볼륨감을 상쇄시키고 날렵하면서도 강렬한 느낌의 디지털 게이트를 형상화하고 도약하는 기업의 이미지를 부각시키고 있다.

역동적 형태의 타워 DYNAMIC FORM OF TOWER

- DIGITAL GATE
- VISUAL EXPOSURE
- DIGITAL AGE
- ANALOG SENSIBILITIES
- DYNAMIC FORM
- CONTEXT
- ADVANCED TECHNOLOGY
- HIGH STANDARDS

LG전자 서초 R&D 캠퍼스

이질적 재료의 접합 GRAFT OF HYBRID MATERIALS

The New Electronic Seocho Research & Development Campus located in the southern outskirt of Seoul will serve as a gate to the Metropolitan Seoul, contributed by its strong visual exposure to the Seoul-Busan Expressway. The program for the new campus to be developed by LG Electronics calls for a high-rise tower of research laboratory spaces and podium buildings to accommodate visitor and exhibition spaces, on a 24,201m² site with total gross floor area of 124,579m².

The design of the campus was begun with the question in mind that what would be the right image of a state of the art research campus in the digital age that still manifests the inherent analog sensibilities. In the earlier design stage, a group of simple form buildings including a tower building of more dynamic form employing inclining surfaces responding to the rapidly changing context of the neighborhood area were explored. However, recognizing the importance of keeping laboratories in a disciplined spatial structure, the tower is developed into a simple slab form. The bulky slab tower is articulated by splitting and sliding it to two thin plates.

The enclosure system developed for the tower consists of two distinctive systems and materials, the band of vision glass panels on each floor with the recessed glass panels underside, framed by solid metal panel walls at the both side and the top of the tower, is intended to create a powerful image of digital gate. This vocabulary is consistently applied to the podium structures to give homogeneous quality throughout this campus to demonstrate LG's commitment to advanced technology and high standards of work environment.

주변 모습
VIEW OF SURROUNDING AREA

절제된 외부 공간 MODERATE OUTDOOR SPACE

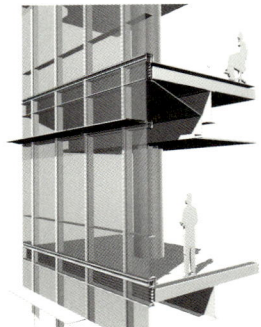

유리벽 디테일 CURTAIN WALL DETAIL

디지털 이미지의 외벽 ELEVATION REMINDING OF DIGITAL IMAGE

CONSILIENT
NINE PROBES FOR A*

Director	장윤규	Yoongyoo Jang
Curator	한상일	Sangil Han
Exhibition Coordinator	한상길	Sanggil Han
	엄해용	Haeyong Um
	허종철	Jongchul Heo
	노종헌	Jonghun No
	김송태	Songtae Kim
Assistant Coordinator	성치훈	Chihun Sung
	김대욱	Daeuk Kim
	박윤선	Yunsun Park
	박주현	Joohyun Park
	설종한	Jonghan Seol

MAPPING
HITECTURE IN KOREA

EXHIBITION

일시 2006.11.28~12.23 **장소** 토탈미술관
28.11.2006~23.12.2006 @ Total Museum of Contemporary Art, Seoul

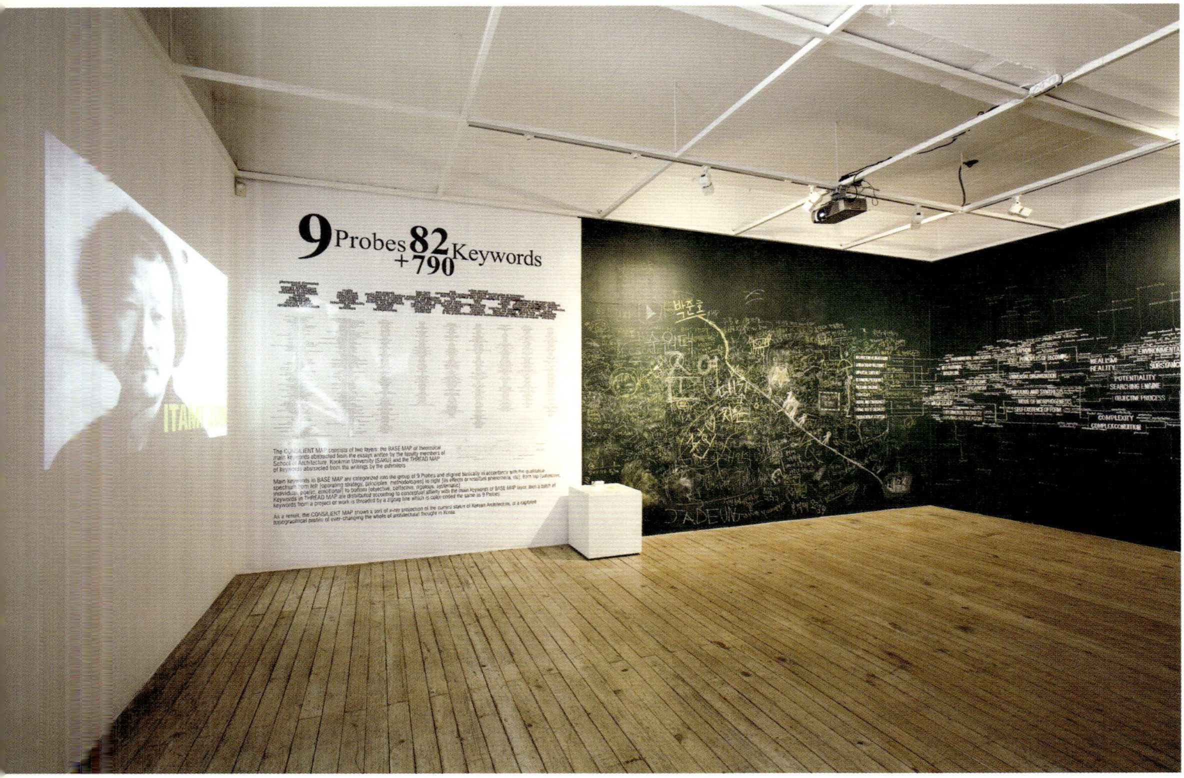
9 Probes 82 +790 Keywords

CONSILIENT

NINE PROBES FOR AR

#1 한국 현대건축의 프로세스

DESIGN PROCESSES OF KOREAN CONTEMPORARY ARCHITECTURE

김우일 국민대학교 건축대학
배형민 서울시립대학교 건축학부
김영준 김영준도시건축
김준성 건축사사무소 hANd
박승홍 정림건축
최문규 연세대학교 건축학부
박길룡 국민대학교 건축대학

Wooill Kim SAKU
Hyungmin Pai Dept. of Architecture, University of Seoul
Youngjoon Kim yo2 Architects
Junsung Kim hANd Architecture
Seunghong Park Junglim Architecture
Moongyu Choi Dept. of Architecture, Yonsei University
Kilyong Park SAKU

#2 신종을 위한 패러다임

PARADIGMS FOR NEW SPECIES

이경훈 국민대학교 건축대학
권 영 국민대학교 건축대학
김일현 경희대학교 건축전문대학원
닐 리치 건축가, 이론가
김찬중 시스템 랩
김주령 도시건축연구소 SaiA+

Kyunghoon Lee SAKU
Young Kweon SAKU
Ilhyun Kim Graduate School of Architecture Kyunghee University
Neil Leach Architect & Theorist
Chanjoong Kim _System Lab
Jooryung Kim Urbano Architectural Research Group, SaiA+

SYMPOSIUM

일시 2006.12.1
장소 토탈미술관
주관 국민대학교 건축대학

1.12.2006 @ Total Museum of Contemporary Art, Seoul
Organized by School of Architecture, Kookmin University (SAKU)

한국 현대건축의 프로세스

박길룡 '통섭지도: 한국 건축을 위한 아홉 개의 탐침' 전시에 대한 결론을 내기 위해 키워드 매핑 작업과 심포지엄을 준비했다. 이번 전시는 개개인의 작품보다는 한국 현대건축의 전체적인 이야기에 의의를 두었다. 현대건축을 바라보면서 발견되는 특징을 아홉 가지 선정하고 그 특징에 얽혀 있는 작가들을 선별하였다. 분명 9개 키워드로 현대건축을 모두 이야기할 수는 없다. 그 한계성은 인정하지만, 그러한 이해 속에서 풍부한 이야기를 끌어낼 수 있으리라 생각한다.

김우일 먼저 주제로 삼고 싶은 것은 건축의 사고 프로세스에 대해서다. 이와 관련하여 배형민 선생께 모두冒頭발언을 부탁한다.

배형민 프로세스, 좀 더 정확히는 디시플린(discipline, 기율)에 관심이 많다. 이것이 바로 '건축'의 역사를 연구하는 기본 통로라고 생각하기 때문이다. 이런 맥락에서 서양의 20세기 현대건축은 디시플린의 측면에서 상당한 이해가 있다고 생각한다. 문제는 한국 현대건축이다. 한국 현대건축은 적어도 서양에서 받아들였다거나 서구의 (예를 들어 변증적인) 역사 인식의 틀에서 보기에는 힘든 부분이 많다. 어떤 면에서는 아직 역사가 부재하다고도 볼 수 있고 그렇기 때문에 나 역시 '역사를 만들고' 있다.

먼저 우규승 선생의 스케치를 보자. 우규승 선생은 컴포지션composition을 한다. 컴포지션이라는 것은 서구의 고전적 디시플린을 기반으로 하는 설계방법으로, 한국 건축가로서 이것을 방법으로 삼는 경우는 우 선생이 거의 유일하다고 본다. 이것을 특징적으로 보는 이유는 디시플린이라는 것은 스스로 터득하는 것이 아니라 배우지 않으면 익힐 수 없는 것이고 1960, 70년대 한국의 건축학교와 건축사무실에서는 건축을 (하나의 방법에서부터 세계관까지) 체계적으로 배운다는 것 자체가 힘든 일이었기 때문이다. 누군가에게 무엇을 배운다는 점에서 승효상 선생이 김수근 선생과의 관계를 명백하게 드러내어 스스로를 노출시켰다는 것 자체가 그가 한국 현대건축에 기여한 긍정적인 면이라 하겠다. 이런 점에서 오늘 논의는 추상적이기보다는 구체적이었으면 한다.

김우일 디시플린이라는 것이 조형언어의 일관성을 이야기하는 것은 아니지 않는가?

배형민 조형언어의 일관성은 아니다. 승 선생의 경우만 보더라도 조형적 언어 또는 그 결과물의 언어가 다르다 할지라도 디시플린 자체는 일관성을 가질 수 있고 또 상당 부분 그러하다고 생각한다.

김우일 그 이야기는 결국 태도라는 입장을 말하는 것인가?

배형민 태도를 달하는 것이기도 하지만 가주 기술적인 이야기가 될 수 있기도 하다. 선을 어떻게 긋는가, 창을 어떻게 내는가, 아주 세밀한 디자인의 움직임어 서조차도 디시플린이 발견될 수 있다. 하지만 오늘의 큰 주제에 비추어 생각해본다면 그렇게 세밀한 과정보다는 보다 넓은 의미에서 프로세스와의 곤계를 생각해봐야 할 것 같다.

김우일 다른 분들의 의견은 어떠한가?

박승홍 나는 바이틀 포뮬러vital formula라는 말을 떠올렸다. 살아 있는, 생동하는 방정식이라 할가…. 예전에 한 강연에서 라파엘 모네오와 리차드 마이어를 비교하면서 리차드 마이어는 바이틀 포뮬러를 갖고 있지 않다고 비판한 적이 있다. 아시다시피 마이어의 건물은 어떤 프로그램의 건물이든 어떤 입지에 놓이든 모두 똑같은데, 모네오의 경우는 어느 땅에 위치하느냐에 따라 모두 다르고 거기에는 부정할 수 없는 일종의 원론과 같은 비전이 깃들어 있다. 그러면서 서로 다른 상황을 만나 각각의 경우가 생동하는 결과를 만든다. 이와 같이 내부 논리와 상황들이 만나 작용하는 것을 바이틀 포뮬러라고 할 수 있다.

김영준 실제로 디시플린이라는 것을 생각해본 적이 없다. 어떤 면에서는 프로젝트마다 생각을 달리하자는 것이 내 입장이었다. 그런데 여기에는 배형민 선생이 말하는 것과는 다른 하나의 요소가 있는데, 자신만의 언어를 찾는 과정에서 '일관성'이라는 것에 너무 집착했던 것은 아닌가 하는 생각도 든다. 지금 일관성을 유지하기가 힘든 것은 결국 정보량이 너무 많고, 다른 이들이 하고 있는 작업을 서로가 모두 알고 있기 때문이다. 그러므로 '다르게' 하는 것이 관건이 되었다. 이것이 바로 예전에 디시플린이라는 것을 가지고 작업하던 시대와는 달라진 점이 아니겠는가. 그런 점에서 대응 방식 자체를 다르게 만들어야 한다고 생각한다. 디시플린이 있어야 한다는 생각보다는 오히려 빌딩 프로토타입에 관심있다.

최문규 덧붙이자면, 차이를 만들기 위해서일 수도 있지만 단순한 호기심 때문이기도 하다. 예를 들면 역사적 맥락에서 지금 어떤 프로그램이 만들어졌다면 앞으로 그것이 어떻게 될 것인가 하는 문제에 대한 단순한 호기심이라든지, 주어지는 여러 현실적인 조건들을 가지고 앞으로 내가 무엇을 할 수 있을 것인가에 대한 호기심과 같은 것들이다. 프로세스라는 것은 결국 지금까지 우리에게 알려져 왔던 것들에 대해 그것과는 다른 어떤 가능성을 탐색하는 방법이 아니겠는가.

김우일 박승홍 선생이 라파엘 모네오 얘기를 했듯이 김준성 선생은 알바로 시자의 영향을 많이 받았다. 특히 건축은 '현상에 대한 반응체'라는 점에서 그러한데, 그것을 어떻게 적용해야 하는가?

김준성 건축은 본질적인 현실에 대한 번역이다. 현실을 건축을 구성하는 현상이라고 볼 수 있다. 하지만 건축의 물리적인 측면뿐만 아니라 비건축적인 측면도 건축을 결정하는 많은 부요소가 된다. 그런 무수한 요소들이 서로 부딪히고 결정하는 과정이 건축이라고 생각한다.

김우일 건축이 본질에 대한 번역이라면, 굳이 '다름'이라는 가치를 추구할 당위성은 없지 않은가?

김준성 우리는 시각이라는 감각기관에 의지한 결과, 현상에 대한 반응을 외부 형체로 규정하곤 한다. 하지만 다른 감각기관(후각, 촉각, 청각 등)을 통하여 다양한 느낌으로 건축을 대할 수 있다. 즉 형체는 중요한 가치가 아니다. 어떤 절차를 거쳐 결과물이 생성되었는지, 어떤 의미를 갖고 있는지, 나아가서는 그 형체라는 것이 갖는 냄새가 어떤지, 촉감이 어떤지 등이 중요한 가치인 것이다.

배형민 디시플린이라는 말을 꺼내서 건축가의 사고과정이나 작업과정을 황폐화시키려는 의도는 전혀 없다. 김준성 선생의 말씀처럼 시각을 넘어 말로 설명하기 힘든, 그래도 표현을 해야 하는 그런 영역들이 있다는 것을 인정한다. 디시플린에 대한 탐구가 탐구 대상을 축소시키거나 건축이 갖는 풍부한 차원들이 간과되어서는 안 된다. 다만 왜 지금 이 시점에서 프로세스에 대한 관심이 부상하는 것일까 하는 의문은 계속 갖게 된다. 일전에 프로세스에 관한 토론을 진행한 적이 있는데, 그 당시에도 토론자로 최문규, 김영준 선생을 떠올렸다. 두 건축가가 왜 스스로를 드러내려고 하는지 궁금했기 때문이다. 이들은 의식적으로 자기 프로세스를 드러내려 한다. 궁극적으로는 형태가 중요하지 않다는 내용과도 결부되어 있고, 새롭고자 하는 생각과도 관련있는 듯하다. 그런 점에서 왜 드러내고 싶어하는가, 왜 드러내야만 할까를 묻고 싶은 것이다.

김영준 어떤 결과물은 반드시 어떤 과정에 의해서 생겨난다. 그 작업을 정리하는 것은 중요한 가치가 있다. 실제 작업 진행에서 프로세스는 많은 가능성이 있고 자신 고유의 작업임을 증명하는 또 다른 증거일 수 있다. 도시의 문제에서는 특히 프로세스가 중요하게 작용한다. 건축과 도시의 중간 입장에서 작업을 지향하면서 프로세스가 프로젝트의 해석과 목적을 변화시켜준다는 사실을 경험하였다. 결과물의 활용을 건축주가 결정하는 현실에 비추어볼 때 최종 결과물에만 국한된 건축적 정의에 집착하는 것은 무의미하다. 이 문제 역시 프로세스에 가치를 두면서 극복할 수 있었다. 기존의 많은 건축가들이 결론 위주의 건축을 이야기했다. 그것을 각성하는 의미에서 프로세스를 공개하기 시작한 것이다. 나의 건축과정을 남에게 알리고 커뮤니케이션을 통해 더 나은 발전 가능성을 얻고자 한다.

김준성 건축에 프로세스를 도입하는 것은 자신의 기존 관념을 타파하고 새로운 가능성을 찾으려는 노력의 일환이다. 하지만 여기에 큰 함정이 있다. 기존의 틀과 조건하에서 진행되는 프로세스는 자신을 속이는 부분이 존재하기 때문이다. 그래서 프로세스의 도입 여부보다는 도입된 프로세스를 어떻게 진행하느냐가 더 중요한 문제라고 생각한다.

김우일 대지에 반응하는 건축에서 프로세스를 진행하는 것은 대단히 어려운 것으로 알고 있다. 대지의 반응을 어떠한 방식으로 건물 내부까지 끌어들일 수 있는가?

박승홍 대지라는 것이 건축물 외부의 땅만 포함하는 것은 아니다. 내부와 외부를 똑같은 관점에서 보고 양쪽 모두 건축에 반영해야 한다. 프로그램과 건축물의 성격에 따라 한 가지 원칙만으로 설명하기는 불가능하다. 나는 프로세스를 자신과의 대화라고 생각하여 중요한 가치로 둔다.

배형민 프로세스를 왜 드러내는가라는 질문에 대한 답변을 들으면서, 흥미롭기도 하고 일견 질문 자체가 무리한 것이 아니었는가 싶다. 승효상 선생의 건축에 대한 책을 쓰게 된 발단은 〈빈자의 미학〉을 번역하는 작업에서부터 시작되었는데, 이 책을 자세히 들여다보면서 대단히 불가사의한 텍스트라는 것을 발견하게 되었다. 승 선생 세대에는 자신의 건축 바탕이 될 경험을 스스로 만들어내야만 했다. 디시플린을 스스로 쟁취해야만 하는 처절한 상황, 즉 제대로 교육을

받지 못한 세대이자 정보도 적었던 시대에 건축을 배운 세대로서 말이다. 《빈자의 미학》을 보면 많은 경우 출처불명의 파편화된 이미지와 정보들을 차용하여 스스로의 것으로 흡수하고 동화시키는 독특한 방식들을 볼 수 있다. 이것이 바로 역사 부재의 상황, 디시플린 부재의 상황에서 자신의 위치를 만든 한 방식이라고 생각한다. 어떤 면에서 지금과는 많이 다르다 해도 그런 방식들 역시 건축가 자신을, 그리고 자신의 건축 프로세스를 드러내는 한 방식과 하나의 논리가 아니었나 생각한다.

김우일 조금 다른 맥락에서 박길룡 선생께 질문하겠다. 감성과 시적 감흥 등이 담겨 있는 대지와 반응하는 무엇인가를 언급했는데, 건물을 건물 자체로 보지 않고 내부 공간의 연출이 어떻게 만들어지는지에 중점을 두고 있는 것 같다. 어떤 이야기를 하고 싶은 것인가?

박길룡 지금 우리가 이야기하고 있는 주제는 '디자인 프로세스'인데, 9개 탐침 중 전반부에 언급한 탐침에 해당된다고 할 수 있다. 뒤이어 열릴 '신종을 위한 패러다임'이라는 심포지엄이 그 후반부 탐침에 해당된다. 그런데 내가 제시한 '알레고리로서 건축'은 이 두 심포지엄 어디에도 해당되지 않는 주제라고 생각한다. 이 주제가 프로세스가 아니라는 단서에서 시작했기 때문이다.
어떤 건축가들은 프로세스가 무의미하거나 그 과정이 무시된 채 또는 초월된 채 건축이 이루어지는 경향이 따로 있다는 전제에서 알레고리라는 말을 쓰기 시작했다. 거기에 해당하는 건축가가 이타미 준, 민현식, 조성룡, 승효상이라 생각한다. 무엇의 프로세스인가 그 정의가 필요하겠지만 여기서 프로세스라는 것은 최소한 앞과 뒤가 있는 자기 추론에 계층적 사고구조를 끼운다고 생각하는 듯하다. 나는 반드시 그렇지만 않다는 데 주목하고 싶었다. 그러다보면 프로세스라는 것보다 알레고리적인 태도가 먼저 작용하게 되고 결국 마지막 결과물까지 그것이 이어지는 태도를 보아왔기 때문이다. 물론 거기에는 좀 더 감성적이거나 시적 메타포가 능해지거나 또는 대지나 환경적인 건축을 물상보다는 현상으로 보거나 하는 특징들이 있다.
하지만 사실 좀 더 먼저 세대이지만 이타미 준이나 승효상 등의 프로세스라는 것이 자신의 건축을 좀 더 극적으로 만들어야 한다는 조바심, 초조함도 작용했던 것 같다. 그것이 좀 더 초월되고 한꺼번에 결과를 이룰 때까지 손에 쥐지 못하느냐는 의문은 아직도 가지고 있다.

김우일 그렇다면, 박승홍 선생은 조형을 다룰 때 어떠한 관점을 취하는가?

박승홍 내용이 형태에 반영되는 디자인을 지향한다. 물론 구조나 주변 여건 등이 그 내용에 포함되고 많은 부분 결과물을 결정한다.
조형을 위한 조형을 생각한다든가 요즘처럼 형태론적인 방법에는 익숙지 않다. 조형에 대한 내 생각은 건물의 속내가 표출되는 것이었으면 좋겠다는 것이다. 특히 요즘 표피에 대한 것이나 구조적인 감성이 뛰어나기 때문에 그 것들을 어떻게 함께 할 수 있을지 생각한다. 내 작업 중 이런 방향에 가장 근접한 프로젝트가 '청계천문화관'이다. 대지가 좁고 길어서 전시장을 만들면 복도가 나오기 힘들 정도의 조건이었는데, 경사면을 전시장으로 만들어 동선을 따라 이동하면서 관람할 수 있게 했다. 그것은 그 전시장의 성격, 즉 동적인 성격에 적합한 것이기도 했다. 따라서 청계천문화관의 조형은 그렇게 동선에 의해 형성된 경사면과 공간을 표피와 함께 표출한 것의 결과라고 할 수 있다.

김영준 프로세스는 조형을 만드는 근원이 된다. 프로세스는 감성적인 접근과는 다른 것으로, 조형적 작업(스케치)과 좀 더 분리해서 생각할 수 있는 힘을 준다. 자신이 가지고 있는 것(경험 등)에서 벗어나기 위해 프로세스를 따르는 것이다. 어떤 결과물이 도출될지 모르기에 프로세스에 집착한다. 예전에 도시를 계획할 때는 모델 방식으로 형태를 만들고 결론을 만들었다. 하지만 지금은 어떤 기준을 갖고 끌고 나가는 크라이테리언criterion적인 방법과, 형태적인formal 솔루션이 아니라 연계적인relations 솔루션을 만드는 것이다. 그것이 프로세스에 집착하는 이유이다. 결국 프로세스에서 얻어야 하는 것은 '기준'이다. 그 기준이 다시 프로세스를 만들어가면서 이루어진다.

결국 건축적 논의를 이끌어내기 위해서는 토론이나 혹은 이런 전시를 통해 서로 깨달음을 나누어 가질 수 있어야 하고 그리고 그것이 나아가 건축적 이론, 역사가 될 거라는 생각에서 일단은 프로세스에 매달리는 것이다.

최문규 내가 프로세스를 드러내는 이유는 소통-대화하기 위해서다. 대부분의 건축가가 나름의 프로세스를 가지고 있지만 그것을 드러내는 사람은 많지 않다. 프로세스를, 조형을 만들기 위한 것이라 말하는데 나는 건축을 구성하는 내용을 만드는 것과 연관지어 보고 싶다. 이를 위해 단순하고 명료한 생각의 틀을 만들고 그 속에서 새 질문을 만들어 구성을 하는 것이 옳다는 생각이다. 다만 다이어그램은 건축 과정을 단순화하는 면에서 좋은 것이지만 그것이 수단이 아니라 목적이 되는 것은 우려한다. 다이어그램이 바로 건축화되는 것보다는 다이어그램이 가지고 있는 가능성들이 나중에 다양한 공간과 형태를 만들어낼 수 있다고 생각한다. 건축을 하는 사람뿐 아니라 다른 이들과의 합리적이고 소통 가능한 커뮤니케이션과 차후 결과를 예측하기 위해 프로세스가 필요하다는 입장이다. 그 과정이 서로 소통 가능하도록 발표되어야 한다.

김우일 내부 공간의 원리와 스킨 공간의 이원화가 어떤 면에서 독립적 연구에 의해 오버랩되어 나타난다는 의견에 대해서는 어떻게 생각하는가?

최문규 내부와 외부가 연관성이 있다고 말하지만 사실 대부분의 건물은 내부는 내부대로 외부는 외부대로 이루어진 경우가 많다. 그러다보니 표피라는 것이 내외부를 나누는 수단으로 제대로 이해하고 있는 것인가, 또 외부의 표피가 내부로 연속되거나 내부를 보여줄 필요가 있는가 의문이다. 조금 더 생각이 진행되어야 하겠지만 나는 건물의 내부와 외부가 한 건축가에 의해서도 완전히 분리될 수 있다고 생각하고 이러한 건물을 계획하고 있다.

김영준 프로세스가 필요한 이유는 최종 표피의 결단보다는 내부와 외부의 해석으로 결론이 만들어질 수 있다는 것이다. 그 해석이 어디에서 오는 것인가를 알기 위해 프로세스가 필요한 경우도 있다.

김우일 그런 논리라면 빌바오 구겐하임 같은 작품은 어떻게 나오는 것인가?

최문규 내가 논리적인 프로세스를 가지고 건축을 하고 싶다는 말이다. 건축가들의 다름을 인정해야 한다. 게리처럼 개인의 감성과 공간과 외부에 대한 이해를 정확하게 보여줄 수 있는 사람이라면 프로세스를 꼭 주장할 필요는 없다. 결과는 비슷하지만 그 시작이 어디에서 왔는지 모르는 경우 자신의 내부를 드러내는 것은 필요하며 그것이 우리가 '프로세스'라고 부르는 것을 만들어 드러내고 보여주는 것이다.

김준성 프로세스가 중요한 것이라는 의견에는 동감한다. 그렇지만 건축을 프로세스에 국한시켜 생각하는 것은 위험하다. 한 형태에 대한 본질을 경시하는 것 같아 염려스럽다.

박승홍 학생들과 함께 구겐하임 빌바오를 보면서 그 작품이 '상식의 건축Common Sense Architecture'이라고 얘기 나눈 적이 있다. 커먼센스라는 것은 논리적 경험에 의한 것이 아니라 일상의 경험에서 얻은 좋은 감각이다. 구겐하임 빌바오를 '상식의 건축'이라고 표현한 이유는 건축교육을 받고, 설계 프로세스를 이해하지 않고도 이 건물을 보면서 누구에게나 좋다는 감흥을 줄 수 있다는 이유에서였다. 프로세스에는 감성적인 부분이 함께한다고 얘기하고 싶다.

배형민 학생들 작품을 보면 프로세스에 대한 집착이 위험 수준에 이르렀다는 생각을 하게 된다. 학생들이 제출한 패널들을 보면 온갖 다이어그램이 난무하고 있다. 하지만 이 문제보다는 다시 알레고리의 주제로 돌아가고 싶다. 일례로 민현식 선생의 작품을 알레고리로 바라보는 것은 매우 타당하다. 실제로 가서 보면 많은 것을 느끼게 된다. 다만 민현식 선생의 건축이 왜 알레고리로서 힘을 갖는가에 대한 이야기가 더 필요하다고 본다. 승 선생이나 민 선생에 관해 연구를 하면서 이분들이 특히 젊은 날부터 죽음에 관한 생각을 하고 있었다는 것을 느꼈다. 그러니 알레고리적일 수밖에 없는 것이다. 알레고리라는 것이 본래 죽음을 지연시키는 신화적 서술이기 때문이다. 죽음과의 관계에서 보면 다르다 할지라도, 프로세스를 드러내는 것 역시 서술(스토리텔링)의 한 방식이라고 생각한다. 양자가 상호배제될 이유는 전혀 없다.

김영준 도시 차원에서 작업을 하다보면 외국과 한국의 현실이 다름을 알 수 있다. 따라서 현재 우리 상황의 건축가로서 작업해야 하는 것을 고민해야 한다. 그렇기 때문에 프로젝트를 우선으로 놓아야 한다고 생각한다. 현실과의 접목을 충실히 다루어야 하는 게 한국 현대건축의 과제이다.

SYMPOSIUM #1

DESIGN PROCESSES OF KOREAN CONTEMPORARY ARCHITECTURE

KILYONG PARK The School of Architecture at Kookmin University is planning to compile work texts sent by several architects and publish them in this book. Although we have set nine probing issues, I hope that in today's symposium participants are not restrained by them and rather have free discussions.
Then, what is the significance of today's symposium? Rather than placing the emphasis of its significance on works by individuals, we are focusing on the overall story of Korea's contemporary architecture. We selected nine characteristics among those discovered when professors view contemporary architecture and chose architects associated with each characteristic. Based on their works and literature, we extracted keywords and defined their relationships and impacts. We are certainly aware of the fact that it is impossible to uncover all aspects of contemporary architecture through mere nine keywords, thus we concede the limitation. However, based on this understanding, I think we can sufficiently discuss the theme.

WOOIL KIM I would like to suggest process of architectural design as the first issue to talk about. Professor Pai, would you please begin with this issue?

HYUNGMIN PAI Precisely speaking, I am interested in discipline rather than process. I think it is one of the basic categories of tracing a history of "architecture". In this sense, I do have a general comprehension of the discipline of modern architecture in the west since the 20th Century. The problem is modern architecture in Korea. It is not easy to understand Korean architecture through the historical framework (for example, dialectically) of the West. In a certain sense, it could be said that there is no history, I thus perceive my work as a "building up of history". Here are sketches of the architect Kyusung Woo. As you see, the sketches show the traces of composition. He is almost the only architect who practices composition, which is originally one of the western traditional disciplines. His case is very unique because during the 1960s and 70s there was no well organized educational program or architectural office that really taught of architecture in Korea. In case of H-Sang Seung, by elucidating his relationship with his mentor Swoogeun Kim, he has contributed by exposing his own disciplinary formation. I will have more to say about this later. Instead, I would like to hear somebody else's opinion about topic of today. Let's talk in as concrete sense as possible about the topic.

WOOIL KIM Aren't you saying that discipline is the consistency of design language?

HYUNGMIN PAI No, discipline does not necessarily entail consistency of design language. In the case of Seung, his discipine can be consistent in spite of differences in the design language of his projects.

WOOIL KIM Then, is discipline the overall attitude in designing?

HYUNGMIN PAI Discipline can be a matter of overall attitude but it can also be about very detailed technical aspects of design. We can find discipline in individual line drawing, a fenestration, in very minute details. But considering today's topic, it would be better to talk about its more comprehensive meaning.

WOOIL KIM I would like to hear others' opinion, too. How do you think, Mr. Park?

SEUNGHONG PARK Now I am thinking about the term "vital formula", that is to say, ever-changing equation or so. When I did a lecture comparing Rafael Moneo with Richard Meier before, I criticized Richard Meier that he does not have vital formula. In comparison with his buildings which all have similar aesthetic characteristics, Rafael Moneo has designed every buildings different from each other in terms of its location, its program, and so on;

nevertheless, there is a certain formula. This common but vital formula or inner logic of design generates different results in accordance with different circumstances.

YOUNGJOON KIM I have not thought about discipline. In a certain sense, my basic attitude is rather to make differences in every projects. There is another point which is opposite to what professor Pai talked about. In terms of consistency, I think I sticked too much to consistency when I was seeking for my own language. But now, it is not easy to be consistent in design process because we have too large amount of data. We know almost in real time what other architects do today, so the point is how to make "different". This is what differs from the era in which architects designed their works with discipline. In this reason, I think we should make the other mode of operation. Rather than discipline, I am interested in what the building prototype is.

MOONGYU CHOI In addition, it is not only because of efforts to make differences but because of simple curiosity that I am interested in process. For example, I have curiosity about how a certain program based on historical context will be in future, or what I can make with given conditions of the real. In my opinion, process is a method for searching possibilities that is not known yet.

WOOIL KIM I would ask to Mr. Junsung Kim. I think Alvaro Siza had influences on you like Rafael Moneo on Mr. Park. What is your opinion about the issue that architecture is a reactant of a phenomenon and how should it be applied?

JUNSUNG KIM Architecture is a translation of fundamental reality. It can be said that reality is a phenomenon comprising architecture. However, not only physical aspects of architecture, but also non-architectural aspects also serve as numerous secondary factors determining architecture. I think architecture is a process in which the numerous deciding factors are confronted with one another and decisions are made among them.

Wooil Kim You mentioned that architecture is a translation of essence. Then, is there no necessity to pursue the value of "being different"?

Junsung Kim We rely on the sensorial organ of vision. As a result, we often define reaction to a phenomenon in the form of external forms. However, architecture can be accessed with various feelings through other sensorial organs (sense of smell, touch, and sound). In other words, forms are not an important value. What is more important is through what procedures outcomes have been produced. In other words, an important value is what meanings they have.

Wooil Kim At the beginning, professor Pai suggested to us discussion in concrete sense. Do you have anything to say more about that?

Hyungmin Pai By posing the question of discipline, I have absolutely no intention of devaluating the poetic aspects of the architectural process. I agree with Mr. Junsung Kim on his remark that there is something that exists beyond the simple visual dimension, that nevertheless must be expressed. Discipline does not confine the range of our discussion or neglect the rich dimensions of architecture. The question that I would ask at this moment is why and how we have come to be interested in process. A few months ago, I had the opportunity of coordinating a symposium on process in architecture. At the time, I had also considered inviting Mr. Moongyu Choi and Mr. Youngjoon Kim as panelists because I wanted to ask them why they wanted to expose themselves. I would surmise that their intentions are to demonstrate that form is not important, or they want to be different or new. In this sense, I want ask them why they want to expose themselves.

Youngjoon Kim An outcome is certainly generated by a certain process. Organizing the work is a task with an important value. Also, while proceeding with actual work, the process has many possibilities in it and can be further evidence that proves the work is one's own unique work. Especially in urban issues, process is an important factor. While pursuing work from an intermediate position between architecture and cities, I experienced that process alters the interpretation and purpose of project. Being obsessed with the architectural judgement only by final outcome, given that in many cases clients decide on the utilization of the resulting buildings, is meaningless. This was an issue that had been resolved while I was placing value on the process.
Many existing architects talk about architecture only with its final results. As a starting point of awakening to such trend, I disclosed my process. I disclosed the building process to others and communicate about it with others, thereby gaining better possibility for development.

Junsung Kim The biggest reason to introduce process in architecture is a part of efforts to break away from traditional ideas and explore new possibilities. However, for architecture, the introduction of process has a big pitfall. The process which proceeds under existing frameworks and conditions has an aspect that is self-deceiving. Therefore, I think it is more important to decide on how to proceed with the introduced process rather than whether to introduce process in architecture.

Wooil Kim Regarding architecture reacting to land, it is extremely difficult for the process to proceed. How do you proceed with it? How can one draw reaction of land into interiors of buildings?

Seunghong Park Land does not merely include land that is extraneous to buildings. Land both internal and external to buildings should be viewed from the same perspectives and both of them should be considered in the context of architecture. It is impossible to explain in terms of a single principle because of the different programs and the nature of sites. I consider process as dialogue with myself, and thus regard it as an important value.

Hyungmin Pai To hear the answers to the question about exposure is quite interesting. But I think the question itself may have been a bit absurd. The reason I began to write a book about the architecture of Mr. Seung is because I had decided to retranslate his Beauty of Poverty. Looking at this book carefully I discovered that it was an almost indecipherable text. It can be said that his generation had to create their own experience by and for themselves. In other words he was part of a generation that really were not educated and had little access to information. I found in Beauty of Poverty a particular way of absorbing and constructing his own concepts from the fragmented images and dispersed information. I think it was a way of positioning himself in the absence of history and discipline. This was a way he exposed himself, his process, even though it was very different from what is happening these days.

Wooil Kim In a little different sense I would ask to professor Park. You mentioned something relating to land encompassing emotions and poetic metaphors, and it appears that you don't see buildings as buildings in themselves but rather focus on how presentation of internal space is made. What can you tell us about this?

Kilyong Park What you are asking now is part of the story of the evolution of the design process. In fact, the first and second sessions today are divided according to the nine themes we constructed. Among these, the former session is about design process and the latter session is about "Paradigms for New Species" in which recent youthful thinking and concept-oriented discussion will be conducted. Among the nine themes what I suggested is the story of allegory.

The reason why people believe that architecture as allegory is not categorized into design process or paradigms for new species is based on their belief that allegory is not a process. Architects started to use the word "allegory" with preconception that process is meaningless or there exists separate trends of architecture where the process is ignored or transcended. I think that architects who agree with this viewpoint include Itami Jun, Hyunsik Min, Sungyong Joh, H-Sang Seung. Although there is a need to define a process of "what" when we talk about a process, at least here the process is believed to insert hierarchical thinking structure into self-reasoning with the order of front and back.

However, I want to point out that this is not necessarily so. If we stick to such perceptions, allegorical attitude acts first rather than process. and I often witnessed that such attitude persisted until the final outcome is produced. In this there are characteristics like becoming more emotional or good at poetic metaphors, or like viewing land or environmental architecture as a phenomenon rather than an object. Although this example is that of the previous generation and the process purported by Itami Jun and H-Sang Seung was dubbed as "Exhibitionism" by architect Moongyu Choi. However, I suspect that it was partly attributable to the fact that they were anxious and nervous to make their architecture more dramatic, and that they were doubtful whether it could be transcended more and could be grasped until the final outcome was achieved.

Wooil Kim What perspectives do you adopt when you deal with formative works?

SEUNGHONG PARK I pursue a direction of design where content is reflected in the form. Of course, structures and surrounding conditions are parts of the contents and in many ways determine the final outcome.

I am not familiar with formative ideas for the sake of forms or morphological methods, which are popular in recent days. My idea about forms is that it would be good to express the content. These days, new ways of using materials and structural sensibilities are outstanding and I try to find ways to express them best. Among my works, one of the closest to this is the "Cheonggyecheon Museum." As the site was narrow and long, initially it was difficult to create even a corridor if a museum would be constructed on the site. Thus, slopes were made into an exhibition hall where visitors can view displays while moving along the sloped movement line. Needless to say, such design suited the nature of the exhibition and the site. My approach to the formative aspect of the building was to harmonize the movement line of the ramp with the surface to express the contents.

YOUNGJOON KIM Process serves as the foundation for creating forms. This is different from emotional approach and process offers one the power to think, separated from formative work (sketch). Process is created departing from Exhibitionism or something like that and people follow process because they want to depart from what they have (experiences, etc.). Although one is not sure what will come out finally by following this, yet when we look at the ways of building cities in the past, they make forms first based on models and then reach a conclusion. Today, relations-oriented solutions are sought, rather than formal solutions by criterion methods in which the current standards are used to reach the conclusion. They are the very reasons why one sticks to process. After all, what should be learned from process is the standard, which is being obtained while creating process.

In sum, in order to inspire architectural discourse, architects have to share their own knowledge or comprehension in the chance of discussion or exhibition like this case. This kind of sharing will becomes the very theory and history of architecture; it is the reason why I, first of all, adhere to process.

MOONGYU CHOI The reason why one reveals process is for the communication between us. Everyone has his or her own process, yet there are few people who disclose their process. It is said that process is designed to create forms, but I think it would be proper to create diagrams and mix programs or make new questions and create structures while viewing programs. Although I believe that diagrams are an excellent means, I am concerned that they become the end, rather than the means. If diagrams are not architecturalized, but rather possibilities possessed by diagrams exist in space, I believe it can create diverse spaces and forms later. I think process is necessary to facilitate communication in architecture and reflect the future results.

WOOIL KIM What do you think about the argument that the principles of internal spaces and dualization of skin areas are overlapped in independent studies in some ways?

MOONGYU CHOI Although somebody says that there are linkages between interiors and exteriors, in fact most buildings have separate interiors and exteriors in terms of their design. Under the circumstances, I have questions over whether the skin is properly understood as a means to divide interior and exterior and whether the skin needs to intrude into the interior or to reveal what happens in interior. Still having to think over, I believe that in reality the interior and exterior of buildings can be separated even in a building designed by an architect and I am planning to design such a building.

YOUNGJOON KIM The reason why I need process is that consequent results can be reached by interpretation between interior and exterior rather than the decisive and final skin design. Sometimes, I need process in order to elucidate where the interpretation stems.

WOOIL KIM Then, how do Gugenheim, Bilbao fit into this picture?

MOONGYU CHOI I am just saying my case; I want to keep a logical process. We should accept the individual differences of architects. If one is a person who can accurately portray emotions of individuals with perfection like Frank Gehry, one does not need to argue for process. In cases where the actual reality is similar yet one is not sure from where it comes, process needs to be created and disclosed to reveal what is inside of himself/herself.

JUNSUNG KIM I agree that process is important. However, thinking that architecture is only viewed from the viewpoint of process is dangerous. I am concerned about negligence of essence of a form.

SEUNGHONG PARK I had an informal discussion with students once, while watching Guggeheim Bilbao. There, I described the building as a "Common-Sense Architecture". Given that common sense comes from good sense out of daily experiences, not from educated understanding, I believe that the building belongs to that category in which anyone viewing the building has good emotional interaction with it. I believe that in a process, emotive aspect exists together with others.

HYUNGMIN PAI In the case of student work, the logic of process appears to have reached dangerous levels. When I observe works by students, all I see are diagrams. But for now, I would rather talk about the issue of allegory. I believe it is proper to see the works of Mr. Hyunsik Min, for example, in terms of allegory. But we need to be more specific in talking about why we have certain allegorical feelings about his architecture. What I have noted from my studies about Mr. Seung and Mr. Min is that they are architects who in their youths had already begun to think about death. That is why we feel there is allegory in their work, for is not allegory itself the myth of storytelling, that is the delay of death. Although it differs in terms of how we relate to death, exposure of process is also one way of storytelling. Thus I believe that there is no inherent contradiction between allegory and process.

YOUNGJOON KIM While working on urban levels, it is easy to recognize that realities of foreign countries and Korea are different. So we have to think over the fact that we should work as architects here in Korea and work in the very present moment. I believe that is why projects should be given priority. Faithfully dealing with integrating architecture with reality is a task of Korea's contemporary architecture.

Symposium #2

신종을 위한 패러다임

김일현 현재 '뉴패러다임'이라는 용어는 어떤 의미로 사용되고 있는가? 그리고 어떠한 차원에서 현재의 현상이 새롭다고 말할 수 있는가? 일단 이 용어를 중심으로 이야기를 시작하기로 하자.

권영 새로운 종의 탄생이라는 것은 의문을 남긴다. 기본적으로 종을 규정할 수 있는 상황인지조차 의심스럽다. 파라메트릭, 테크놀로지의 새로운 해석, 물질주의, 자연친화적 건축 등은 서로 개별적으로 존재하는 것이 아니라 중첩되어 있고 역사적인 경계를 통해 세상에 등장했다. 건축에서의 새로운 종이라는 것은 실체를 가진 본질화된 영역으로 구분 가능하게 존재하기보다는 그런 것들이 그물망처럼 이루어져 있다고 보는 것이 타당할 것이다. 건축가들이 주장하는 하나의 단어가 종을 실체화시키는 오류를 범하는 것은 위험한 생각일 수도 있다. 이런 의미에서 이 자리에 모인 건축가들의 의견을 들어보는 것은 중요하다. 무엇이 새로운 것이고, 그렇다면 왜 지금인가?

김주령 실체적인 건축을 이야기하는 것이 아님을 전제로 하고 싶다. 테크놀로지에 대한 논의에 국한되는 것이 아니라 다양한 맥락에서 이해해야 할 것이다.

김찬중 툴의 변화가 있었다고 할 수 있다. 역사적으로 툴은 많이 변해왔고, 다른 산업에서는 이미 범용화되어 따로 논의되지도 않는다. 툴의 변화로 인해 사람들이 추구하는 욕망 자체가 커졌다.

김일현 그렇다면, 범용화와 표준화를 적용 단계에서 어떻게 다르게 대해야 하는가?

김찬중 표준화는 기본적으로 사람들이 원하는 것을 갖기 위한 요구라는 것이 단편적인 생각이다. 일반적으로 요구사항이 기능이 있느냐 없느냐의 문제였다. 여기서 점점 다원화되면서 사람들은 스스로가 달라지고자 하는 욕구가 생겼다. 초기 산업화에서는 기존 인프라를 건드리지 않으면서 다원화할 수 있는 방식이 추진되었다. 건축의 기본 속성이 표준화가 아닌가라는 문제가 제기될 수 있고 공공장소와 같은 생산적인 기능의 건축에서 그러한 새로운 관점이 노출되고 있다고 생각한다.

권영 산업생산의 일환으로 건축이 대량생산되기 시작한 것은 1920년대 이후부터다. 표준화와 규격화는 이미 경험하였고 진행되고 있는 현실이다. 문제는 지금의 시기가 기술발전으로 인한 생산양식에서 이미 그 단계를 뛰어넘고 있다는 것이다. 결국 건축이라는 것은 규격화되어 '지어지는' 것이라기보다는 공장 제품과 같이 '생산되는' 것이라고 보는 것이 더욱 타당할 것이다. 이 모든 것을 포스트포디즘적 건축 생산방식의 징후로 볼 수 있는데, 결과적으로 생산양식의 발전에 따라 매스 프로덕션으로부터 매스 커스터마이제이션이 강제되는 방식으로 나타나게 된다. 그러나 이러한 생산방식의 변화에 전제되어야 하는 것은 건축물이 더 이상 미적 감수성에 호소하는 예술적 오브제가 아니라, 끊임없이 기호로서 소비되는 일상적 소비재의 하나로 변모하고 있다는 점이다. 자본주의체제 하에서 건축을 대하는 개인의 취향은 기본적으로 미적이나 도덕적이라는 영역을 떠나 물질화되고 가치화될 수 있다는 관점에서만 이해 가능하다고 생각한다.

김일현 생산양식의 변화에 대해서 김주령 선생은 문화의 변화와 건축의 관계에 대해 어떤 견해를 갖고 있는가?

김주령 테크놀로지는 계속 변형, 진화되어왔다. 테크놀로지를 긍정적으로 디자인에 활용하는 입장과 테크놀로지 사용을 비판적으로 대하는 입장 이 두 가지는 공존해왔다. 1990년대 초부터 지금까지 근 15년간 테크놀로지의 사용은 상당히 긍정적이었다. 1960년대 경제공황 등과 같은 침체기를 겪으면서 사람들은 테크놀로지를 그러한 불황기에서 구해줄 수 있는 어떤 미래지향적인 것으로 보았다. 이러한 점은 지금 상황에서도 읽을 수 있다. 현저 국제적으로 뒤틀어지고 기발하고 예상을 뒤엎는 디자인들은 바로크적인 매너리즘에 빠진 것으로 간주된다. 이제 이러한 상황을 되짚어보는 것이 중요하다.

닐 리치 '신종을 위한 패러다임' 이라는 이 심포지엄의 제목이 상당히 좋다. 종이라는 것이 동적이며 변종적인 것을 포함할 수 있기 때문이다. 나의 오해가 때로는 생산적일 수도 있을 것이다. 새로운 툴은 사고방법에도 영향을 미친다. 예를 들면 발터 벤야민의 경우 모더니즘에서 사람들이 어떻게 변해가는지 아주 잘 이해하고 있고 기술지배적 사회에서 이러한 영향에 잘 접근하고 있다. 근대주의에 기계적 이성으로부터 컴퓨터에 의한 물 흐르듯 유연한 사고방식이 어떻게 영향을 미쳤는지 알아보고자 한다. 여기에 있는 학생들을 변종이라고 부르고 싶은데, 우리 세대와는 달리 디지털에 의해 생활방식이 바뀌어 있기 때문이다.
우리가 경험한 마지막 변화는 1968년의 혁명이다. 이것은 서구의 관점이기는 하지만 사회의 각종 영역을 근본적으로 바꾸어놓았다. 하지만 건축에서 이러한 변화는 그렇게까지 근본적이지 않다. 내가 쓴 책 중에 20세기의 중요한 관점을 추적한 〈Rethinking Architecture〉라는 것이 있다. 이 책을 처음 출판했을 때 뉴욕에서는, 흥미로운 생각들을 모으긴 했지만 한물간 이론이라는 비판을 받았다. 맞는 얘기다. 사람들에게는 의미가 별로 중요하지 않고 새로운 세계관을 갖는 것이 더 중요하다.

김일현 이러한 변화와 맥락에서, 실무를 하고 학생들을 가르치는 건축가 입장에서 가장 중요하다고 생각하는 개념과 의미는 무엇인가? 그리고 생산의 차원에서 한국의 여건이 외국과는 다르고 테크놀로지를 적용하는 데에도 제한이 있다. 이에 대해서 어떤 대처방안이 있는지를 개인적 경험이나 신념을 통해서 말해보자.

김찬중 디지털건축에 대한 답변은 상당히 난해하다. 작업을 하면서 가장 먼저 드는 생각은 새로운 것을 해야 한다는 것이다. 그 다음 생각이 스스로의 정체성에 대한 것이다. 혼성의 영역에 머물러 있던 시기가 길었기 때문에, 우리는 다른 좋은 결과물을 모방할 수도 있다. 그러나 어느 시기부터는 테크놀로지에 대한 것이 디자인 언어와 묶여져서 결과물이 되기 때문에 쉽게 모방할 수 없게 되었다. 내가 테크놀로지에 관심을 가진 이유는, 테크놀로지를 어떻게 실현할지 더 고민하지 않는다면 우리는 단지 표현에 국한된 공중에 뜬 디자인을 하게 될 것이기 때문이다. '내가 디자인한 것은 나밖에 할 수 없는 것'이라는 개념이 성립되어야 한다고 생각한다. 그래서 테크놀로지에 집중하게 된 것이다.

김주령 지금 이 논의는 우리나라 교육 실정을 감안할 때 중요한 문제이다. 디자이너가 테크놀로지를 이용하는 것과 엔지니어가 테크놀로지를 이용하는 것은 다르다. 디자인에 테크놀로지를 적용한다는 것은 디자이너가 더 많은 생각을 하게 된다는 것이다. 파라메트릭 디자인은 디자인 프로세스는 느려지지만 생산체계를 빠르게 해주는 것이다. 디자인의 실현 가능성에 대해 좀 더 자신감을 가져야 한다. 파라메트릭 디자인이란 변수의 관계성을 디자인하는 것이다. 이것은 굉장히 동시성을 가진 다이내믹한 상황이다.

김일현 어원적으로 프로젝트란 앞으로 일어날 일의 예시를 의미한다. 이러한 변화의 상황에서 우리가 프로젝트라고 부르는 것 자체의 의미를 재정의할 필요가 있지 않을까. 계획을 예시할 뿐만 아니라 미래의 가능성에 대해서도 개방적이어야 하지 않을까. 건축가 상에 대해서도 마찬가지다. 그 역시 수많은 가능성을 선택하는 입장에 있다. 건축가는 데이터, 프로세스를 제어하며 구조를 만드는 사람일 수도 있다.

닐 리치 한국은 중국이나 일본과는 좀 다른 아주 특이한 상황이라고 할 수 있다. 건축의 경우에는 외국의 유명 건축사무소에서 일했던 경험을 가진 건축가들이 많다. 파주출판도시를 가보았는데 거기에 있는 건물들은 유럽의 건축가가 설계했는지 한국 건축가가 했는지 구분이 안 될 정도였다. 이 전시 또한 그러하다. 이는 매우 흥미로운 사실인데, 아마 비슷한 컴퓨터 프로그램을 사용하고 있기 때문일 것이다. 비슷하게 보이는 것은 문제가 되지 않을 수 있지만 이것은 한국의 정체성을 찾느냐 아니면 렘 콜하스와 같이 보편적인 것에 가치를 두느냐에 관한 문제이다. 건축 형태에 집중할 것이 아니라 그 건축물이 놓여 있는 상황에 어떻게 적응하고 있는지가 더 중요한 질문일지 모른다. 케네스 프램튼이 말했듯이 "서로 다른 건물은 서로 다른 장소가 있어야 한다." 건축가의 역할에 대한 생각은 매우 낙관적인데, 그 역할은 감소되고 있지 않다. 건축가들이 컴퓨터를 이용하는 것을 보고 상상력과 나머지 조건에 관한 것과 타협하는 것이 아니냐는 질문이 있을 수 있다. 하지만 컴퓨터를 건축 제어에서 처음 사용한 칼 추의 경우만 해도 상상력이 부족한 것이 아니고 그 상상력을 표현하는 방법으로 컴퓨터 프로그램을 이용했을 뿐이다.

김주령 프로세스는 학생들을 위한 언어다. 건축주를 만나야 하는 건축가 입장에서는 프로세스보다는 결과물에 집중할 수밖에 없다. 굉장히 급진적인 형태의 디자인에 관해서 우리나라 사람들은 본 적이 없기 때문에 불안해하는 반면 중국이나 유럽에선 오히려 본 적이 없기 때문에 가지고 싶어 한다. 하나의 스타일을 깨는 비평적인 입장이 있어야 자신의 디자인에 대해 긍정적인 생각이 생기는 것이다. 스케치를 하면서 생각하는 내용과 컴퓨터로 작업하면서 생각하는 내용은 질적으로 차이가 있다. 지금은 1980년대 이전과는 달리 종합적이고 일시적인 방식들이 주를 이루기 때문에 사고방식 차이에서 미디어와 맞물리는 경향이 있다.

권영 디자인에서보다는 생산과정에서의 테크놀로지 사용이 오히려 건축가에게 더 중요하다고 생각한다. 이런 면에서 건축가는 이전과는 달리 더 이상 계획을 하는 사람이 아니라 실험과정을 제어하는 사람이다. 결정론적인 예측이라는 것이 모더니즘 안에서 실패를 맛본 후 건축가들이 스스로 그 해결책을 내놓는 것이 아니라 결론에 도달하기 위한 과정을 조정하는 역할을 부여받는다. 그 역할 때문에라도 건축가들은 테크놀로지 자체를 수용할 수밖에 없는, 그래서 건축 생산 자체가 건축설계 과정에 내재되어야 한다고 생각한다. 그런 면에서 보다 본질적인 면에서 건축가의 사회적 책임을 요구하는 시대가 도래했다. 페라리보다는 GM과 같은 생산구조가 좀 더 건축가의 역할에 가까워지고 있는 사례라고 본다.

김찬중 페라리와 GM 모두 하이엔드 상품이다. 중요한 것은 시행착오이다. 이 시행착오 속에서 남은 것들이 다음 세대에선 일반적으로 수용될 수 있는 테크놀로지가 된다. 지금 이야기들은 어떠한 방식이 맞다 틀리다의 문제라기보다는 계속 실험하고 있다는 입장인 것 같다. 이러한 실험을 계속하는 가운데 어느 정도 보편화라는 것이 형성되는데 이 보편화가 일반화될 수 있는 확률이 큰 테크놀로지들이다. 건축가는 옛날과는 달리 '마스터 빌더Master Builder'보다는 '마스터 컨트롤러Master Controller'의 개념이 강하다. 건축의 영역을 제한적으로 생각하기보다는 좀 더 넓게 볼 필요가 있다.

이경훈 우리가 논의하고 있는 테크놀로지는 구체적으로는 디지털 테크놀로지이고 이것이 문제가 되는 것은 디자인 프로세스에 도입되었기 때문이다. 이전에는 상상도 못했던 형태들이 디지털 테크놀로지를 통해 시각적으로 실현될 수 있었다. 그런데 지금 기술적으로 이것을 실현시킬 수 있는 능력은 없는데 교육하는 입장에서 어떻게 하야 하겠는가?

김주령 베이징건축비엔날레에서 느낀 것인데, 가장 기괴하고 실현 가능성이 없어 보이는 형태들이 지어지는 곳이 바로 아시아다. 특히 일본 같은 나라가 그러하다. 엔지니어링의 테크놀로지와 디자인의 테크놀로지는 분명 틀린 것이다. 훌륭한 구조전문가들이 있기 때문에 건축가들이 발현을 하는 것이다. 독일 건축이 발전했던 것도 독일 엔지니어들의 뒷받침이 있었기 때문이다. 한국에서 특이한 것은 건축가가 자신의 역할뿐만 아니라 엔지니어 역할도 하고 코디네이터 역할도 한다는 것이다. 문제는 엔지니어가 부족한 것이 아니라 여러 건축 제반요소들을 통합할 수 있는 마인드가 부족한 것이다.

닐 리치 디지털 교육에서 말하고자 하는 것을 보면, 교육이 실무보다도 더 보수적이다. 많은 프로그램들, 즉 마야나 맥스 같은 경우 교육에서보다는 실무에서 효과적으로 쓰이는 경우가 더 많다.

김일현 지금부터는 몇 가지 논제와 관련하여 개별적으로 질문하겠다. 닐 리치 선생은 '새로운 자연'이라는 말을 한 적이 있는데, 예측불허의 돌변하는 자생적 구조체를 논하면서 왜 굳이 자연에 준거를 두어야 하는가?

닐 리치 자연 특히 생물에 기대는 것은 이유가 있다. 1920년대와 1930년대 스타일로 이해되었던 것과 요즘 말하는 생물학적 준거는 다르다. 그때가 '스타일'이었다면 요즘에 말하는 생물학적 준거는 '프로세스'에 더 집중되어 있다. 자연에서 얻는 교훈은 형태들이 물질적인 계산으로부터 나왔다고 생각하기 때문이다. 건축 또한 디지털의 계산이 그 물질적인 계산을 따르고자 하기 때문이다.

김일현 디지털 문화와 관련해서, 예전의 사이버스페이스와 지금의 유비쿼터스는 많은 차이를 보여준다. 이러한 상황에서 김주령 선생은 건축이 장식된 헛간이 되어야 한다고 보는가 아니면 오리가 되어야 한다고 생각하는가?

김주령 오리가 되어야 한다고 생각한다. 나는 건축의 외적인 부분까지 모든 것을 조정하면서 자신을 피곤하게 하는 스타일이다.

김일현 김찬중 선생은 어떤 쪽에 가까운가? 이토 도요인가? 프랭크 게리인가?

김찬중 이토 도요다. 이것은 아이덴티티의 문제라고 생각한다. 프랭크 게리처럼 자신이 디자인하고 엔지니어에게 넘기는 방식이 아닌 디자인 자체에 테크놀로지가 묻어 나와서 결코 모방할 수 없는 것을 지향한다. 그럼으로써 정체성을 높이는 것이다.

김일현 양식론에 의한 근대건축운동까지의 분류는 명확한 반면 포스트모더니즘, 해체주의 이후의 현상에 대해서는 이렇다 할 해답을 제시하지 못한다. 닐 리치 선생은 현재 경향을 '디지털 텍토닉'이라고 칭했는데 그 이유는 무엇인가? 이것을 새로운 전위라고 할 수 있을까?

닐 리치 포스트모더니즘 이후 건축이 정의할 수 없는 어려움에 빠져 있는 것은 사실이고 이 책임의 대부분은 찰스 젱크스에게 있다. 모더니즘 이후를 전형적인 것과 퇴행적인 것으로 볼 수 있는데 전형적인 것은 해체주의이고 그렇지 않은 것은 역사적인 이론을 장식으로 차용하려 했던 포스트모더니즘일 것이다. 해체주의를 찰스 젱크스는 모더니즘으로 본다. 처음에는 형식적으로 해체주의라는 것은 러시아 구성주의에 기대는 바가 크고 이것은 신구성주의라는 제목으로 시행되려 했었다. 단지 De-constructivism이란 말이 더 선정적이기 때문에 채택되었다고 생각한다.
'디지털 텍토닉'은 내가 처음으로 사용한 말인데 케네스 프램튼의 책 〈건축의 텍토닉〉에 대한 비판의 뜻을 담고 있다. 즉 디지털건축이 건축의 외적인 부분에 침범을 가하고자 한다는 그의 견해를 비판하고자 이 말을 사용했다. 나는 그것이 디지털이든 아날로그이든 기본적으로 재료에 관심을 갖고 있다. 프라이 오토 같은 건축가들은 재료에 관심이 많았고 디지털건축의 즉물적인 성격에 대해서도 관심을 가지고 있다.

권영 디지털 환경의 도래라는 것이 1968년의 변화 시기만큼 격렬한가라고 했을 때 닐 리치 선생은 그렇지 않다고 말했다. 나도 이 점에 동의한다. 디지털 툴을 이용해 화려한 형태를 만들어내는 것이 새로운 종이라고 할 수는 없을 것이다. 이것보다는 건축가들의 새로운 감수성이 급격하게 침투할 수 있는 것이 만들어졌을 때 이 새로운 종으로 건축이 만들어질 수 있다고 생각한다.

박길룡 이번 전시회를 준비하면서 느낀 점이 있다. 심포지엄을 통해 좀 더 분명해졌는데, 확실히 한국 건축은 종의 다양성에서 성공했다. 30년 사이의 일이니, 경이로운 사실이다. 또 한 가지 한국 건축이 세계와 동시적으로 소통되고 있다는 점이다. 동시성으로 이야기할 수 있는 가장 큰 인자는 컴퓨터다. 이 사실은 또 다른 국제주의로 이해된다. 이전에 에스페란토로 국제주의가 이루어졌던 것과 현재 마야 언어로 이루어지는 국제주의는 차이가 있는가? 이 국제주의라는 것을 거침없이 받아들여도 좋은가?

닐 리치 1968년 혁명의 교훈은 모든 것에 대해 비판적인 것이다. 디지털하다는 것에 대해 비판적일 필요가 있다. 디지털 테크놀로지에 너무 치중하거나 비판적으로 거부할 것이 아니라 이것을 이해하고 인간을 위해 사용할 수 있도록 유도하는 것이 더 중요하다. 이번 전시회는 지난 30년간의 한국 건축의 변화를 보여주고 있어 가치가 있다. 그러나 더 중요한 것은 이 전시 자체가 비판적으로 질문을 던지고 있다는 점이다. 일반화, 국제화, 세계화의 문제는 매우 보편적인 것이 되고 있다. 그렇지만 이 일반화의 과정 중에서도 분명히 변별적인 지역적 차이는 존재한다.

Symposium #2

Paradigms for New Species

Ilhyun Kim What I would like to ask first individually is why the term "new paradigm" is used and in what meaning, and why this phenomenon can be accepted "new"? Mr. Kweon, could you start discussion with this issue?

Young Kweon The birth of new species leaves some doubt. It is even doubtful whether now it is a situation where basically species can be defined. New interpretation of parametric technology, Materialism and eco-friendly architecture do not exist individually, but are overlapped and have emerged into the world through historical boundaries. Although new species in architecture have a tendency of actual entities, yet they are comprised like networks. What could be dangerous is that a single word uttered by architects may make an error of transforming a species into an actual entity. In this sense, I think it is important to listen to opinions of architects who are present here. What is new and why now?

Jooryung Kim I would like to talk based on the premise that we are not confined to talking about architecture in actual forms. This is not confined to technical discussions, but should be understood in various contexts.

Chanjoong Kim It can be said that there has been a change in means used. Historically, means have undergone tremendous change. In other industries, means have been already been made universal so that there is even no need to discuss them. With the change of means people's desire itself has become augmented.

Ilhyun Kim If then, please explain how differently we should deal with the application of universalization and standardization.

Chanjoong Kim Basically, standardization was

encapsulated by demand by people wishing to possess what they desire. In general, the issue was whether something had desired functions or not. As things have become more diversified, people started to have desires to become different from others. Ways for diversification without altering existing infrastructure were adopted in the early industrialization. A question may be raised regarding the issue of whether basic attributes of architecture is standardization, and new perspectives are being exposed for architecture with productive functions like public places.

YOUNG KWEON Mass production of architecture started since the 1920s as part of industrial production. Standardization has already been experienced and is also valid at the present. But the problem is that our time begins to exceed the stage of standardization in terms of the mode of production by the development of technology. We should consider architecture not as what is to be standardized and built but as what is to be produced like a factory product. This can be defined as post-modern mode of architectural production; as a result, the mode of production will be shifted from mass production to mass customization. But also there should be a premise considered that architecture is no longer than artistic object appealed to aesthetic eyes but one of the everyday consumed goods or volatile signs. In the capitalist economic system, the personal taste of architecture can be accepted only when it is materialized and evaluated beyond the traditional aesthetic or moral dimension.

ILHYUN KIM Regarding the change in production modes. Now, Ms. Jooryung Kim, would you talk about change of culture and its relevance to architecture?

JOORYUNG KIM Technology has continuously evolved. In terms of viewpoints on technology, the two positions, namely how to utilize technology positively in design and being critical about the use of technology have coexisted. For past 15 years, since the early 1990s, the use of technology has been quite positive. In the 1960s while people experienced economic depression, people considered technology as something futuristic that could save them from the depression. I think such tendency can be detected even in the current situation. Currently, designs which are internationally distorted, novel, and unexpected are considered as having been immersed in Baroque-style mannerism. I believe it is important to rethink objectively on this current situation.

NEIL LEACH I think that the title of this session "Paradigms for New Species" is a considerably good thing, because species could include something dynamic and changeable. My misconceptions can be sometimes productive. New tools may affect ways of thinking. For instance, Walter Benjamin had a very proper understanding of how people change in terms of Modernism and approached such impact in technology-controlled society very well. I intend to explore how from mechanical reasoning to a flexible way of thinking, like water flows by computer affected Modernism. I wish to compare students here with us mutants as their lifestyles have been changed by digital technology unlike our generation.
The last change we experienced was the revolution in 1968. Although it is a Western viewpoint, basically it fundamentally altered various realms of society. However, in terms of architecture such change was not as fundamental as in other areas. One of my books is titled "Rethinking Architecture" which traced impor-

tant perspectives of the 20th century When this book was first published, I was in New York and I was criticized that although the book had collected interesting ideas, it was already old-fashioned. This is true. For people, meanings are no longer very important and what is more important is to have new viewpoints for seeing the world.

ILHYUN KIM I would like to ask architects more concrete questions. First, in doing education and practice in this changing context, what are the ideas you think most importantly and what are their meanings? Second, conditions in Korea are different from those of foreign countries and in Korea there are limitations in applying technological aspects. Please tell us about your personal experiences and faith on how you respond to such situations.

CHANJOONG KIM Answering questions on digital architecture is quite difficult. The first thoughts I have while doing work are that I have to try something new and then share my ideas about my identity. We can copy other good buildings. This was possible as we stayed in the realm of composition for a long time. However, from a certain point, it became impossible to do this. We cannot copy easily as something about technology is combined together with design language, which leads to final outcome. The reason why I take interest in technology is that if we do not ponder upon the issue of how to realize any longer, we could make design, which is not based on reality, but only focuses on the issue of expression. I came to concentrate on technology as I thought that the concept that "what I design can be only designed by myself" should be established.

JOORYUNG KIM I think the discussion we are having now is also important in terms of the educational reality of Korea. Designers using technology is different from engineers using technology. Utilizing technology in design is to make designers think further through technology. Parametric design may slow down the design process, yet it makes production systems go faster. We need to have more confidence on realization possibilities of design. Parametric design is to design relationships of variables. This is a dynamic situation with tremendous simultaneity.

ILHYUN KIM A project -in etymological sense- is to predict what happens in the future. I think we should redefine the meaning of a project itself in consideration of changing atmosphere. My position is that projects are not merely to predict what happens in the future but open possibilities and make a choice among numerous possibilities. Here the position of architects may be the same. Architects may be those who control data and process and create structures.

NEIL LEACH I would like to talk about what is different in Korea from other countries. Korea is different from China or Japan, having a very unique situation. What is very peculiar in architecture is that in Korea there are many architects who worked for famous architectural offices abroad. Yesterday, I visited Paju Book City and it was difficult to distinguish who designed buildings there whether by European architects or Korean architects and also with their exhibits. This is very interesting for me. I think the reason might be that they use similar computer programs. Although what appears to look similar may not be a problem, but it is an issue of whether to pursue Korean identity or placing value on something universal like the works of Rem

Koolhaas. It might be a more important question how buildings are getting adjusted to situations they are in, rather than focusing on the forms of buildings. As Kenneth Frampton stated, different buildings must have different places.

I am very optimistic about the roles of architects. Architects' roles are not on the decline. Concerning architects using computers, there might be a question whether architects are compromising over their imagination and remaining conditions. Carl Choo used computers in the control of architecture for the first time and it does not necessarily mean that he lacked imagination. Rather, his way of expressing imagination is presented through computer programs.

JOORYUNG KIM Process is a language for students. From the position of clients and people who experience buildings, importance is given to final outcome rather than process. As Korean people have rarely seen very radical designs, they are having a sense of insecurity over this. On the other hand, in China or Europe people want to have such designs because they have not seen them before. Positive thoughts on their own design come only when there is a critical perspective that breaks a single style.

The quality of content is different when the media itself make a sketch and when it works on computer. Unlike the 1980s and before, nowadays synthetic and simultaneous ways are dominant thus in terms of ways of thinking there is a tendency for the media to interfere with the process.

YOUNG KWEON Ms. Jooryung Kim made a good point here. In using technology, for architects, the use of technology in the course of production rather than in the design process is more important. In this light, architects are those who are no longer planning like in the past, but who control the process of experimentation. After experiencing failure in their deterministic prediction in Modernism, architects did not present solutions, but rather were endowed with authority to coordinate the process to reach a conclusion. Partly due to such role, I think architects have no choice but to accommodate technology itself and that architectural production itself should be embedded in architectural design process. In this sense, an era fundamentally requiring social responsibility of architects has dawned. GM-style production structure, rather than that of Ferrari, nowadays appears to be closer to the roles of architects.

CHANJOONG KIM Ferarri and GM are extremely high-end products. What is important is trial and error. What we learn from such trial and error becomes technology that can be generally accommodated during next generations. What we are discussing now is the ongoing continuum of experimentation, rather than saying that certain ways are right. While undergoing such experimentation continually, universalization has taken root to a certain degree and this universalization is technology with high probabilities of generalization. Unlike the past, architects are now perceived more as "Master Controllers" rather than "Master Builders". There is a need to view the realm of architecture more comprehensively, rather than viewing it from a narrow perspective

KYUNGHOON LEE The technology we are discussing refers to, more concretely, digital technology and the reason why this becomes an issue is that it has entered the realm of the design process. Forms unimaginable in the past have become possible visually through digital technology. However, currently there is lack of ability to technically realize this. It would be appreciated if anyone could discuss about what to do about this when one is in a position to teach students.

JOORYUNG KIM If we take the example of the Beijing Biennale, the places where the most odd and unrealistic forms are being constructed are in Asia. Countries like Japan are particularly so. I think that technology seen from engineering perspective is definitely different from technology in designing. Architects can be manifested because there exist excellent structure experts and the reason why German architecture has scored advancement is that it was supported by German engineers. However, what is unique about Korea is that in Korea architects act as engineers, coordinators and architects at the same time. The problem in Korea is not about shortages of engineers but lack of mindset that can incorporate several architectural elements.

NEIL LEACH What I want to discuss on the issue of digital education is that education is more conservative than actual practice. In the case of many programs like Maya or Max, there are many cases where they are more effectively utilized in practice rather than in education.

ILHYUN KIM There are several discussion points and please give us brief answers to each point. Mr. Neil Leach, you mentioned "new nature". Please tell us why we need to follow the norm of nature, while discussing autogenous, unpredicted, and changeable structures.

NEIL LEACH There is a reason why rely on nature, especially living creatures. What was understood as a style in the 1920s and 1930s is different from biological norms which are discussed nowadays. At that time, the focus was placed on style, the recent biological norm is more focused on the process. Lessons from nature are based on my belief that forms are originated from material calculations. Likewise, in architecture as well, digital calculations tend to follow material calculations.

ILHYUN KIM In terms of digital culture, what was termed "cyberspace" in the past is very different from what is termed "ubiquitous" today. Under the circumstances, Ms. Jooryung Kim, what do you think about whether architecture should be "a decorated barn or a duck."

JOORYUNG KIM I think architecture should be a duck. I am a person whose style is to try to control everything, including external aspects of buildings, even to the point of making myself fatigued.

ILHYUN KIM Mr. Chanjoong Kim, are you close to Toyo Ito or Frank Gehry?

CHANJOONG KIM Toyo Ito. I think this is about an issue of identity. My identity is about technology being integrated in design, which makes reproduction impossible, thereby elevating one's identity, unlike Frank Gehry who designs and then transfers it to engineers.

ILHYUN KIM Classification of styles until the times of modern architectural movements had been clear, yet it has become blurred after Post-modernism and Deconstructivism. Mr. Neil Leach calls the present trends as "digital tectonics" and please tell us about why you call it that and whether we can say this is a new kind of Avant-gardism.

Neil Leach It is true that since Post-modernism architecture is having difficulty in defining itself, most of the responsibility falls on Charles Jencks. Architectural trends after Modernism can be divided into something typical and something regressive. What is typical could be Deconstructivism and what is not could be Post-modernism which attempted to borrow historical theories as embellishments. Charles Jencks views Deconstructivism as Modernism. Initially De-constructivism was heavily reliant on Russia's Constructivism in terms of formal aspects and it was on the brink of being implemented in the name of "Neo-constructivism". However, I think instead, the word "Deconstructivism" was adopted as it sounded more sensational. The term "digital tectonics" is used by myself for the first time and I used this word to criticize the book titled *Tectonics in Architecture* authored by Kenneth Frampton. I intended to criticize ideas that digital architecture is invading external aspects of architecture. Regardless of digital or analog, I am basically interested in materials. Architects like Frei Otto took high interest in materials and I am also interested in the practical nature of digital architecture.

Young Kweon As for a question whether the digital environment is as intense as the changes that took place in 1968, Mr. Neil Leach said it was not so, and I agree on that point. Creating figurative forms by utilizing digital tools may not be regarded as a new species. Rather, I think architecture as new species can be created when things are created to which new sensibilities of architects can invade radically.

Kilyong Park There is something I felt since completing this exhibition. The feeling has become clearer after listening to today's discussion. Certainly, Korea's architecture has succeeded in terms of diversity of species. It has been achieved in only 30 years, which is truly marvelous. This is the first feeling and my second feeling is that Korea's architecture is being communicated with that of the world simultaneously. The biggest factor contributing to this simultaneity is the use of computer. This can be understood as another form of Internationalism. While wrapping up this event, I have personal questions on whether the Internationalism achieved through Esperanto and the current Internationalism through the Maya language have any difference and whether it is okay to accept this Internationalism without filtering. I hope that there would be more comments on this topic.

Neil Leach The lesson the 1968 revolution gave us is that it was critical of everything. To sum up, when I look back on what has happened so far, there is a need to be critical of being digital. Rather than excessively focusing on digital technology or rejecting it too critically, it is important to understand it and induce it to be utilized to benefit humans. I think this exhibition is all the more valuable as it highlights changes that took place over the three decades. However, what is more important is that this exhibition itself raises critical questions. I think this is the very value of this exhibition. The issue of generalization, internationalization and globalization are becoming very universal. However, in the process of such generalization there certainly exist differentiated regional differences.

PHOTO CREDIT

ⓒ 김용관 Yongkwan Kim
핀크스미술관 Pinx Museum
임진각 기념관 (모훈) Imjingak Memorial (model)
청계천문화관 Checnggyecheon Culture Center
첫마을 (모형, 온고당) The First Town (model, Ongodang)
예화랑 Gallery Yeh
서울대학교 미술관 Museum of Art, Seoul National University
삼성미술관 리움 Leeum Samsung Museum of Art
S-갤러리 S-Gallery
스피드 돔 Speed Dome
전시장 사진 View of Exhibition

ⓒ 김종오 Jongoh Kim
대전대학교 혜화문화관
Hyehwa Cultural Center for Daejeon University
평화누리+청소년수련원
PyoungHoa Nuri Peace Park+Youth Training Center
아시아출판문화정보센터
Asia Publication Culture and Information Center
허유재병원 Heryoo ae Women's Hospital

ⓒ 송재영 Jaeyoung Song
아쿠아아트브리지 Aqua-Art Bridge
열린책들 사옥 House of Open Books
보림출판사 Borim Publishing House and Marionette Theater
힐튼 남해 골프 & 스파 리조트 Hilton Namhae Golf & Spa Resort

ⓒ 박영채 Youngchea Park
밀레니엄 커뮤니티센터 Millennium Community Center
경희대학교 건축전문대학원
Graduate School of Architecture Kyunghee University
타워팰리스 III Tower Palace III

ⓒ 염승훈 Seunghoon Yum
타워팰리스 III Tower Palace III

ⓒ 이재성 Jaeseong E
안양 피크 Anyang Peak

ⓒ 이인성 Insung Lee
명필름 Myung Film Company Building

ⓒ 이기환 Kihwan Lee
광주디자인센터 Design Center of Gwangju

* 이 책에 사용된 대부분의 작품 사진 및 자료들은 사용 동의를 얻은 것입니다. 저작권자를 찾지 못한 일부 작품은 저작권자를 확인하는 대로 출판 동의를 구하겠습니다.

* The photographs and materials used in this book have been used with the copyright holders' consent. We have tried our best to contact all copyright holders. In indivicual cases where this has not been possible, we will seek their consent as soon as we get in touch with the copyright holders.

PROFILE *작품(아티클) 게재순

조성룡 _건축사사무소 조성룡도시건축
1944년 생 | 인하대학교 건축공학과, 동대학원 졸업 | 1996년~2003년 서울건축학교 교장, 현 조성룡도시건축 대표 | 수상-서울시건축상(1987), 한국건축가협회상(1993), 김수근문화상(2003), 아시아선수촌 및 기념공원 국제현상설계 1등(1983) 등 | 작품-인하대 학생회관, 양재287.3, 해운대빌리지, 의재미술관, 선유도공원 등

승효상 _이로재 종합건축사사무소
1952년 생 | 서울대학교 건축학과, 동대학원 졸업, 비엔나공과대학 수학 | 1989년 이로재 설립 | 수상_미국건축가협회 Honorary Fellowship(2002), 2002 올해의 작가 등 | 저서-빈자의 미학, 지혜의 도시 지혜의 건축 등 | 작품-수졸당, 수백당, 웰콤시티, 차오웨이 소호 상업업무시설 콤플렉스, 장성주거단지 2차 계획 등

박길룡 _국민대학교 건축대학
홍익대학교 건축학과, 석·박사학위 | 국민대학교 조형대학장, 박물관장, 건축대학장, 현 국민대학교 건축대학 교수, 김수근문화재단 이사 | 수상-한국건축가협회상(1994) 등 | 저서-건축이라는 우리들의 사실, 한국현대건축의 유전자, 시간횡단: 건축으로 보는 터키 역사

이타미 준(유동룡) _이타미준 건축연구소
1937년 생 | 무사시공업대학 졸업, 1968년 이타미준건축연구소 설립 | 수상-한국건축가협회상(2001), 프랑스예술문화훈장(2005), 김수근문화상(2006) 등 | 저서-돌과 바람의 소리, 조선의 건축과 문화 등 | 작품-핀크스뮤지엄, 학고재, 포도호텔, 핀크스-퍼블릭 골프하우스 등

아눅 레정드르
1961년 생 | 파리 라빌레트건축학교 졸업, 프랑스정부공인건축사, 건축환경조경학위. 도시개발학 박사준비과정학위 | 파리 빌멩건축학교 강연책임교수(2000-2001)
니콜라 데마지에르
1962년 생 | 파리 라빌레트건축학교 졸업
X-TU 아키텍츠
아눅 레정드르, 니콜라 데마지에르 설립 | 수상-전곡선사박물관 국제공모 당선(2006), 보르도시 건축상 대상(2003), 프랑스 건설부 선정 젊은 건축앨범상 대상(1999) 등 | 작품-릴르건축문화

관, 생데니스경찰청, 낭테르(파리)대 캠퍼스 재건축, 보르도 레므뉘고등학교 등

민현식
_한국예술종합학교 건축과, 기오헌 건축사사무소
1946년 생 | 서울대학교 건축학과 졸업, 런던 AA스쿨에서 수학, 1992년 민현식건축연구소 설립, 현 한국예술종합학교 건축과 교수, 기오헌 고문 | 수상-미국건축가협회 명예회원(2006), 김수근문화상(1992) 등 | 작품-대전대학교 마스터플랜 및 기숙사, 체육관, 신도리코 본사 및 공장, 동숭교회, 파주출판도시 도시설계 및 출판물유통센터, 인포품, 한국전통문화학교 등

김병윤 _홍익대학교 건축대학
1952년 생 | 한양대학교 건축공학과 석·박사학위, 런던 AA스쿨 수학 | 공간, 간삼, 토탈디자인, 영국의 콜린스와 조지 메이더스 사무소 근무, 건축연구소 스튜디오메타 개설, 백제예술대학 교수, 현 홍익대학교 건축대학 교수 | 수상-김수근문화상(2004) | 작품-파주 아시아출판문화정보센터, 정릉동 성당 등

안나 라노바 룬드스트롬
1973년 생 | 덴마크 왕립예술건축학교 졸업 | 보쉬허슬렛, KHR, 시그널 사무소 근무
쿠보타 토시히로
1976년 생 | 프랑스 크레아폴 실내건축학과 졸업 | 아틀리에 장 누벨과 함께 덴마크 코펜하겐 콘서트홀 등 당선
스테판 마티스
1973년 생 | 독일 RWTH 아헨 졸업 | 장 누벨 사무소에서 실무 | 독일 Euregio 상 수상(2002)
엘리펀트
2005년 룬드스트롬, 토시히로, 마티스가 파리와 코펜하겐에 설립

장 누벨 _아틀리에 장 누벨
1945년 생 | 파리국립미술대학교 졸업 | 수상-프랑스건축학회 금메달(1998), 메종오브제 디자이너(2006) 등 | 작품-아랍문화원, 아그바타워, 구트리극장, 브랑리미술관, 삼성미술관 리움 등

정영선 _조경설계 서안
1941년 생 | 서울대학교 농학과, 서울대학교 환경대학원 조경학과 졸업 | 1987년 조경설계 서

안 설립 | 수상-서울시건축상, 한국건축가협회상, 김수근문화상, 세계조경가협회 조경작품상, 미국조경가협회 프로페셔널상 등 | 작품-파리공원, 예술의전당, 인천국제공항 여객터미널, 선유도공원, 비전빌리지 등

안드레 페레아 오르테가
1940년 생 | 1968년부터 Escuela Tecnica Superior de Architecture de Madrid 겸임교수, The Initiative for an Architecture & Urbanism +Sustainable 창립멤버 | 수상-행정중심복합도시, 서울공연예술센터 당선, 아시아문화전당 국제공모 3등 등

MVRDV
뷔니 마스, 야콥 판 레이스, 나탈리 더 프리스가 1991년 로테르담에 설립 | 수상-Concrete Award, Dudok Award, Merkelbach Award(1997) | 작품-Wozoco, Mirador 아파트먼트, 더블하우스, Villa VPRO, 바르셀로나 리뉴스, 하노버세계박람회 네덜란드관 등

정영욱
1972년 생 | 성균관대학교 건축학과, 베를라헤 졸업 | 2003-2006 MVRDV 프로젝트 리더, 2006 델프트공대 건축학과 초빙강사 | 작품-웸블리 돔 호텔, 부산 시티 소파, 안양 피크, 롱탄 파크 주거단지 등

이공희 _국민대학교 건축대학
국민대학교 건축학과, 동대학원 졸업 | 원도시건축, 유창건축 근무 | 2002년부터 국민대학교 건축대학 교수 | 작품-강북구민회관, 일민미술관, 국민대 법학관 등

유걸 _아이아크 건축사사무소
1940년생 | 서울대학교 졸업 | 무애건축연구소, 김수근건축연구소 근무, 1979년 유걸건축연구소 설립, 경희대학교 건축전문대학원 교수, 현 아이아크 공동대표 | 수상-한국건축가협회상(1998), 김수근문화상(1996), 미국건축가협회 Honor Award 등 | 작품-밀레니엄 커뮤니티센터, 밀알학교, 전주대학교회, 배재대학교 국제교류관 등

다비드 피에르 잘르 콩 _디피제이앤파트너스
1968년생 | 파리 콩플랑건축학교, 팡테온 소르본대학교, 파리 빌멩건축학교 졸업 | 한국 주재 프랑스 상공회의소 사무총장, 디피제이앤파트너스 대표 | 수상-아카데미 보자르 건축대상, 피에르 가르뎅 건축부문 등 | 작품-서울 프랑스고등학교, 프랑스문화원 아쿠아아트 브리지, 센트럴포인트 브리지, 능평리 주택 등

윤웅원
1964년 생 | 연세대학교, 파리 라빌레트건축대학 졸업 | 연세대학교, 동국대학교 출강 | 전시-베니스비엔날레 국제건축전 한국관(1996), 건축과 나(2001), 롤링스페이스전(2004) 등

김정주
1968년 생 | 연세대학교 졸업, 파리 라빌레트건축대학 졸업 | 홍익대학교 건축학과 출강 | 전시-베니스비엔날레 국제건축전 한국관(1996) 등

제공 건축사사무소
2001년 설립 | 윤웅원, 김정주 공동대표 | 수상-명동성당 축성 100주년 기념 현상설계 우수작(1996), 서울공연예술센터 3등(2005) 등 | 작품-명필름 사옥, 정릉근생, 정릉주택, 마로니에 미술관 파빌리온 등

도미니크 페로 _도미니크 페로 아키텍처
1953년 생 | 파리사회과학대학원 졸업 | 수상-전기전자고등학교, 베를린산업관 설계공모 당선, 이화여대 지하캠퍼스 국제공모전 당선(2004) | 작품-프랑스 국립도서관, 올림픽사이클경기장 및 수영장, 해비타트호텔, 스페인 축구경기장 등

김종규 _한국예술종합학교, M.A.R.U.건축사사무소
1960년 생 | 연세대학교, 런던 AA스쿨 졸업 | 동우건축, 빌딩디자인파트너십, 플로리안 베이겔 사무소 근무, 한국예술종합학교 건축과 교수, M.A.R.U. 건축사사무소 대표 | 전시-베니스비엔날레 국제건축전(1996, 2002) 등 | 수상-김수근문화상(2001), 한국건축문화대상 등 | 작품-의재미술관, 카이스갤러리, 헤이리아트밸리(BB스튜디오, MA갤러리, 아고라뮤지엄) 등

마츠오카 사토시
1973년 생 | 교토대학교 건축학과, 도쿄대학교 건축공학석사, 컬럼비아대학 대학원 졸업, 델프트공과대학에서 도시론 연구 | 유엔스튜디오, MVRDV, SANAA 근무 | 전시-SD리뷰전(2002)

타무라 유키
1977년 생 | 니혼대학교 건축학과, 도쿄예술대학 건축석사 졸업 | 엔릭 미라예스 & 베네데타 탈리아부 사무소, SANAA 근무

마츠오카사토시타무라유키
2005년 설립 | 작품-Balloon Caught(벤쿠버), 분화구극장(서울공연예술센터 현상설계), 스톡하우스, 전곡선사박물관 현상설계 3등, Arex(나고야) 등

다니엘 바에
마드리드공과대학 졸업, 베를라헤 수학 | FOA, 에두아르도 아로요 사무소 근무, 한국예술종합학교 초빙교수, 서울시립대학교 출강 | 작품-임진각 기념비 현상설계안, 교문출판사 사옥 입면계획 등

정림건축 종합건축사사무소
대표이사 김정철, 김정식, 문진호 | 1967년 설립 | 수상-한국건축문화대상 대상(2006), 서울사랑시민상 대상(2006), 경기도건축문화상 대상(2006), 한국건축가협회상(2005), 부산다운건축상 금상(2005) 등 | 작품-국립중앙박물관, 청계천문화관, 할렐루야교회, 무악교회 등

희림 종합건축사사무소
대표이사 이영희 | 1970년 창립 | 수상-한국건축문화대상 대상(2006), 한국건축문화대상 우수상(2006), 대한민국토목건축대상 최우수(2005), 한국건축문화대상 우수상(2004) 등 | 작품-제주노형지구 공동주택, 천안시청사, 분당 벤처타운, 경기테크노파크 등

유석연 _hna온고당 건축사사무소
서울대학교 건축학과, 동대학원 졸업, 미국 펜실바니아대학교 건축학석사학위 | 2004 베니스비엔날레 국제건축전 한국관 초대작가, 현 hna온고당 대표, 고려대학교 건축학과 대학원 강사 | 수상-한국건축문화대상 특선(2005), 한국건축가협회상(2005) 등 | 작품-해남도 알로에관광광장 마스터플랜, 남양알로에 에코넷센터, 다음글로벌미디어센터, 강서구청사 별관 등

피에르 비토리오 아우렐리 _도그마 오피스
베니스대학교 도시계획 박사, 델프트공과대학 건축박사학위 | 멘드리시오대학교, 델프트공과대학 초빙교수

피터 아이젠만 _아이젠만 아키텍츠
1932년 생 | 코넬대학교, 컬럼비아대학교, 케임브리지대학교 졸업, 일리노이대학교에서 명예예술학박사, 1967년 건축도시연구소(IAUS) 개설 | 수상-베니스비엔날레 국제건축전 1등상(1985), AIA상(1989) | 저서-House X, Moving Arrows, House of Cards 등 | 작품-랩스톡공원 마스터플랜, 로미오와 줄리엣, 유태인추모공원, 콜럼버스컨벤션센터 등

해안 종합건축사사무소
대표이사 윤세한 | 1990년 설립 | 작품-향군2010사업 현상설계, 차세대융합기술연구원 TK, 오송생명과학단지, 일산 라페스타 등

장윤규 _국민대학교 건축대학, 운생동
1964년 생 | 서울대학교 건축학과, 동대학원 졸업 | 현 국민대학교 건축대학 교수, 운생동 운영 | 수상-신건축 타키로 국제현상, UIA 바르셀로나 국제현상, 이스라엘 평화광장 국제현상 13 최종작 등 | 작품-이집트대사관, 서울대학교 건축대학, 광주디자인센터, 갤러리정미소 등

최문규 _연세대학교 건축공학과, 가아 건축사사무소
1961년 생 | 연세대학교 건축공학과, 동대학원, 컬럼비아대 건축대학원 졸업 | 이토 도요 사무소, 한울건축, 시건축 근무, 1999년 가아건축 개소, 현 파주출판도시 섹터건축가, 연세대학교 건축공학과 교수 | 수상-한국건축가협회 엄덕문건축상, 한국건축문화대상, 서울사랑시민상, AIA상(2005) 등 | 작품-딸기테마파크, 정한숙기념관, 쌈지길, 독서지도회, 아름드리미디어, 서해문집, 태학사 등

조민석 _매스스터디스
1966년 생 | 연세대학교 건축공학과, 컬럼비아대 건축대학원 졸업 | 1998년 조슬레이드아키텍처 설립, 2003년 매스스터디스 설립 | 수상-일본 신건축국제도시주거 현상공모 당선(1994), 미국 젊은건축가상(2000), 프로그레시브 건축상(1999, 2003) | 전시-베니스비엔날레 국제건

축전 주제전(2004) | 작품-네이처포엠, 비틀린 집, 픽셀하우스, 딸기테마파크 등

FOA _Foreign Office Architecture
1992년 런던에 설립 | 파시드 무사비(비엔나예술아카데미 교수)와 알레한드로 자에라 폴로(베를라헤 학장)가 설립 | 작품-요코하마국제터미널, 동남아시아 연안 공원 & 오디토리엄, 번들 타워, ANY그룹 주최 잠재성의 집 현상설계안, BBC 화이트시티 등

장림종 _연세대학교 건축공학과
1961년 생 | 한양대학교 건축공학과 졸업, 하버드대학교 졸업 | 코닝아이젠버그 사무소 근무, 카이저스라우테른대학교 객원조교수 | 현 연세대학교 건축공학과 교수, 장림종디자인연구실 대표

김영준 _김영준도시건축
서울대학교, 동대학원, 런던 AA스쿨 건축대학원 졸업 | 현 김영준도시건축 대표, 한국예술종합학교 튜터, 서울건축학교 코디네이터 | 전시-2002 베니스비엔날레 국제건축전 | 수상-한국건축가협회상(2005), 김수근문화상(2005), 행정도시 첫마을 현상설계 2등(2006), 함부르크 건축올림픽 초청(2006) 등 | 작품-울산 프라우메디병원, 일산 허유재병원, 자하재 등

하태석 _아이아크 건축사사무소
성균관대학교 건축학과, 런던 AA스쿨 졸업 | 알솝, SOM, 아칸서스LW 사무소 근무, 경희대 건축전문대학원 겸임교수, 현 아이아크 공동대표

시로 나홀레 _GDB
아르헨티나 생 | 코넬대학교 초빙교수, 컬럼비아대학교, 베를라헤, 부에노스아이레스대학교 출강 | MID 근무, GDB 디렉터 | 수상-젊은건축가상 2등(2001) | 전시-프라하아트비엔날레, 베이징건축비엔날레(2004) 등

권영 _국민대학교 건축대학
국민대학교 건축학과 졸업, 런던 바틀렛건축학교 건축학석사학위 | 계원예술대학 실내건축학과 겸임교수 | 현 국민대학교 건축대학 강의전담교수 | 수상-UIA국제설계공모 입상(아시아, 2004)

김주령 _도시건축연구소 SaiA+
런던 AA스쿨, 로드아일랜드대학 졸업 | 포스터앤파트너스, 자하 하디드, 유엔스튜디오 근무, 2002년 도시건축연구소 SaiA+ 설립 | 수상-베이징건축비엔날레 한국대표작가(2006), 새건축사협의회 신인건축가상(2004) | 작품-아시아 에볼루션, 파주 덕은리 K씨 빌딩, PKM갤러리 등

손 머레이 _독립건축가
런던 바틀렛건축학교 건축학석사학위 | 2003 이스트사미미술관 현상설계 입선 | 전시-스테레오 모노, 영역교란, 광인의 다리 | 저서-영역교란, Reflexive Practice in Architectural Ecologies-Communicating Vessels 등 | 작품-주라기 해안을 위한 관문, 장에 내재한 본래의 건축, 사미미술관 등

신혜원 _로컬디자인
연세대학교 건축공학과 졸업 | 런던 AA스쿨 졸업 | 데이비드 치퍼필드, OMA아시아, 테리 파럴 사무소 근무, 2001년 로컬디자인 대표, 2004년 한국예술종합학교 겸임부교수 | 수상-베니스비엔날레 국제건축전 한국관 작가(2006), 타이완 치치 추모공원 국제건축공모 2등(2004), European 5 Competition Geesthacht 1등(1998) 등

김찬중 _시스템 랩
고려대학교 건축공학과 졸업, 하버드 건축대학원 건축학석사학위 | 현 시스템 랩 대표 | 작품-거제도 레지던스 신축설계, 한강 보트클럽2, 컨테이너 보트클럽, 경주엑스포 상징타워, 이건창호 이미지숍 계획 등

양수인 _리빙 아키텍처 랩
연세대학교 건축공학과 졸업, 컬럼비아대학교 건축학석사학위 | 프랫인스티튜트 객원교수, 현 컬럼비아대학교 부교수, 리빙 아키텍처 랩 공동대표 | 작품-더 좋게 더 싸게 더 빠르게, 리버글로우, 리빙 글래스 등

김기홍 _키오스크건축
1969년 생 | 국민대학교 졸업, 파리 라빌레트건축학교, 파리8대학교 수학, 경기대학교 설계튜터, 키오스크건축 멤버 | 전시-한국전통공간의 미(1997) | 수상-건축문화 영디자이너 100인 선정(2005), 부산 BIARC 제2회 국제건축공모 3

위(2003), 타이완 치치 추모공원 국제건축공모 우수상(2004)

이경훈 _국민대학교 건축대학
뉴욕 프랫인스티튜트 건축학석사학위 | 건축사(뉴욕, 한국), 현 국민대학교 건축학부 조교수

폴 프라이스너 _QuaVirarch
일리노이대학교, 컬럼비아대학교 졸업 | 피터 아이젠만, 필립 존스, SOM, 우드-자파타 사무소 등에서 근무 | 건축사무소 QuaVirarch 소장 | 2006년 네브래스카-링컨대학교 건축학과 초빙교수

론 콤
켄터키대학교, 컬럼비아대학교 석사학위 졸업 | 다니엘 리벤스킨트 스튜디오, SOM 근무

로나 이스턴
웨스트민스터대학교 졸업, 런던 바틀렛건축교 석사학위 | 영국, 독일, 중국, 미국 등지에서 교편

이스턴콤 아키텍츠
2001년 설립 | 작품-말라마교육센터, 휴스턴 제공항 비행장, 밀센터 등

마르코스 노박 _UCLA
오하이오주립대학교 건축학석사학위 | 오하이오주립대학교, 텍사스 오스틴대학교, 아트센터대학 출강, UCLA 건축 프로그램 및 디지털 미디어 프로그램 지도 | UCLA 건축 및 도시계획학과 교수 | 저서-Avatarchitectures: Fashioning Vishnu after Spacetime, Next Babylon: Algorithm to Play in 등 | 작품-파라큐브, 데이터연동 형태들, 변화하는 데이터 폼 등

김준성 _건축사사무소 hANd
1956년 생 | 맥켄지대학교 졸업, 프랫인스티튜트, 컬럼비아 건축대학원 수학 | 알바로 시자, 스티븐 홀 사무소 근무, 한국예술종합학교 객원교수, 현 hANd 대표 | 수상-한국건축가협회상(2003) 등 | 작품-열린책들사옥, 헤이리 커뮤니티센터, 아트레온 시네멀티플렉스 등

서혜림 _건축사사무소 힘마
컬럼비아대학교, 하버드 건축대학원 졸업 | 서혜림건축연구소(1995-97) | 1997년부터 건축사사무소 힘마 대표 | 수상-건축가협회상, 서

울시건축가상 | 작품-서울시청어린이집, 국립문화재연구소, 현대고등학교, 열린책들사옥, 보림출판사 등

나데어 테라니 _오피스 dA
페르시아계 영국 생 | 로드아일랜드대학교, 하버드대 건축대학원 졸업, 런던 AA스쿨 건축역사 및 이론 수료 | 하버드대 건축대학원 부교수, 오피스 dA 공동대표 | 수상-젊은 건축가상, 프로그레시브건축상, 실내설계상 등 | 작품-노스이스턴대학교 종교센터, RISD도서관, 뉴잉글랜드주택, 통산아트센터, 맥칼렌빌딩 등

봉일범 _국민대학교 건축대학
서울대학교 건축학과, 동대학원 졸업, 하버드대 건축대학원 건축학석사학위 | 현 국민대학교 건축대학 전임강사 | 저서-건축 지어지지 않은 20세기, 볼륨제로 등

문훈 _문훈건축발전소
1968년 생 | 인하대학교 건축공학과, MIT 건축대학원 졸업 | 2001년 문훈건축발전소 설립 | 전시-문훈전(2005) 등 | 수상-한국건축가협회상(2005) | 작품-포천주택, 묵동다세대, 현대고등학교, 상상사진관, 진주동물원 등

OMA
1975년 렘 콜하스 엘리아 & 조 젱겔리스, Madelon Vriesendorp 설립 | 수상-프리츠커 건축상(2000), 일본 Praemium Imperiale상(2003), 영국왕립건축학회 금메달(2004), 미스 반 데 로에 건축상, 유럽연합 현대건축상(2005) 등 | 작품-졸퍼아연역사 박물관 및 에센 지역 마스터플랜, 삼성미술관 리움, 서울대학교 미술관, 포르토 카사 다 무지카 음악당 등

켄 민 성진 _SKM 건축사사무소
캘리포니아대학교 건축과, 하버드대 건축대학원 졸업 | 작품-힐튼 남해 골프 앤 스파 리조트, 아크로비스타 주상복합, 파주 북시티 타운하우스, 숭실대 형남공학관 청담동 더 웨즈 등

자하 하디드 _자하 하디드 아키텍츠
1950년 생 | 런던 AA스쿨 졸업 | 컬럼비아대학교, 하버드대학교 객원교수, 비엔나응용예술대학 교수 | 1983년 더 피크 당선 | 작품-The Vitra Fire Station, Mind Zone at the Millennium Dome, Car Park and Terminus, Ski Jump, CCTV 등

안드레아 보체티
1969년 생 | 베니스대학교 건축과 졸업, 베니스 I.U.A.V와 컬럼비아대학교에서 도시학박사학위 | 튜린공과대학 건축디자인 초빙교수, 루카대학교 강의교수

알베르토 프란치니
1969년 생 | 플로렌스대학교 건축과 졸업, 밀라노 공과대학에서 도시학박사학위 | 밀라노공과대학 도시계획과 조교수

메트로그라마
1998년 밀라노에 설립 | 수상-카솔라 산업지역 마스터플랜 1등상(2006), 중국 광저우 행정지역 마스터플랜 및 신청사 현상 1등상(2002), 미스 반 데 로에 건축상(2002) 등 | 작품-레지오 에밀리아의 1가구 주택, 밀라노 로자노 D3 마스터플랜, 라베나 다르세나 마스터플랜, 네라노 지구 타당성 계획 및 주거건물 빌라 메라노 건설

공간그룹
대표이사 이상림 | 1960년 설립 | 수상-한국건축문화대상 대상(2005, 2002) 등 | 작품-부산아시안게임 주경기장, 달성군청사, 한성대학교 도서관, 양재동복합물류센터, 마포종합행정타운 등

최왕돈 _국민대학교 건축대학
서울대학교 건축학과, 동대학원 졸업, 맨체스터대학교 건축학박사학위 | 해안건축 근무, 현 국민대학교 건축대학 교수 | 저서 및 역서-아키텍처 리더, 모더니즘 이후의 현대건축 등

진아건축도시 종합건축사사무소
대표 부대진, 김무현, 부상훈 | 1969년 설립 | 수상-한국건축문화대상 입상(1991), 대전시건축상 금상(1996) 등 | 작품-대전엑스포 IBM 전시관, 이화삼성캠퍼스센터 건축기획, 드비스타워 등

건원 종합건축사사무소
대표이사 김종남, 함인선 | 1984년 설립 | 수상-서울시건축상 금상(1995), 한국건축문화대상 본상(1999), 한국건축문화대상 대상(2004) 등 | 작품-전쟁기념관, 한국무역센터, 고양국제전시장 등

삼우 종합건축사사무소
대표이사 손명기 | 1976년 설립 | 수상-한국건축문화대상 대상(2005) 등 | 작품-삼성미술관 리움, 서울시립미술관, 자갈치시장, 타워팰리스 등

무영 종합건축사사무소
대표이사 안길원 | 1985년 설립 | 수상-서울시건축상(2001), 한국건축문화대상 대상(2003) 등 | 작품-대한무역투자진흥공사, 경기중소기업종합지원센터, 시몬느 사옥 등

간삼파트너스
대표이사 김자호 | 1983년 설립 | 수상-인천시건축상 우수상(2005), 한국건축문화대상 우수상(2005/2006) 등 | 작품-포스코센터, 코오롱 과천사옥, 포스코역사관 등

원도시 건축사사무소
대표이사 변용, 장응재, 김석주 | 1969년 설립 | 수상-한국건축가협회상(1995), 서울시건축상 동상(2002), 대구광역시 건축상 금상(2003) 등 | 작품-한국종합무역센터 사무동, 서울대학교 체육관, 대법원청사, 사법연수원 청사, 기상청청사 등

원양 건축사사무소
대표이사 이종찬 | 1981년 설립 | 작품-국립남도국악원, 공군회관, 문화컨텐츠 콤플렉스 및 종합영상 아카이브센터, 오송생명과학단지, 은평뉴타운2지구 A공구, 동남권유통단지 이주 전문상가 등

토문엔지니어링 건축사사무소
대표이사 최기철 | 1990년 설립 | 수상-한국건축문화대상 본상(2004), 설계VE경진대회 최우수상(2004) 등 | 작품-용인신갈 주공 새천년 주거단지, 평내 주공 그린빌, 용인 구갈 써머트빌 등

창조 종합건축사사무소
대표이사 김병현 | 1984년 설립 | 수상-한국건축가협회상(1996), 서울시건축상 금상(1998), 한국건축문화대상(2003) 등 | 작품-SK생명보험사옥, 한국경제신문사사옥, 경기대 원격화상회의센터 등

PROFILE · in the order of appearance in the projects and articles

Sungyong Joh
_JOHSUNGYONG Architect Office
Born in 1944, Japan | Inha Institute of Technology, Master of Architecture, Inha University | President of Seoul School of Architecture(1996-2003) | Principal of JOHSUNGYONG Architect Office | Awards-1993 the KIA Award, 2003 the Culture Award of Kim Swoo-Geun, etc | Works-Athlete's Village and Memorial Park for Seoul Asian Games, Uijae Art Museum, Seonyudo Park, SOMA Art Museum, etc

H-Sang Seung _IROJE Architects & Planners
Born in 1952, Korea | Graduated from Seoul National University, studied at Technische Universität Wien | 1989 Established IROJE Architects & Planners | Awards-2002 Honorary Fellow of AIA, 2002 The Artist of Year, etc | Publications-Beauty of Poverty, City of Wisdom, Architecture of Wisdom, etc | Works-Sujoldang, Subaekdang, Welcomm City, Commune by the Great Wall, Chaowai SOHO Office_Commercial Complex(Beijing), 2nd phase plan(Beijing), etc

Kilyong Park _SAKU
PhD, Hongik University | Dean, College of Architecture and Design, Director, Kookmin University Museum, Dean, School of Architecture, Kookmin University | Professor at School of Architecture, Kookmin University | Awards-1994 KIA, etc | Publications-Architecture as Fact, The Gene in Korean Contemporary Architecture, Crossing Time: History of Turkey on Architecture

Itami Jun _ITAMIJUN Architects
Born in 1937, Japan | Graduated from Musashi Institute of Technology, 1968 Established ITAMIJUN Architects | Awards-2001 the KIA Award, 2005 Grade de Chevalier dans Lordre des Arts et des Lettres, 2006 the Culture Award of Kim Swoo-Geun | Works-Pinx Museum, Hakgojae, Podo Hotel, Pinx-Public Golf House, etc

Anouk Legendre
Born in 1961, France | Graduated from School of Architecture, Paris La Villette, Diploma of Architecture Dplg, DEA and theses [en cours] Urban Development Paris iv | Professor of Paris-Villemin Architecture School(2000-2001)

Nicolas Desmazieres
Born in 1962, France | Graduated from School of Architecture, Paris La Villette, Diploma of Architecture Dplg

X-TU Architects
Awards-2006 Gyeonggi-do Jeongok Prehistory Museum Competition 1st prize, 2003 Grand Prix d'architecture de la Ville de Bordeaux, 1999 Albums de la Jeune Architecture awarded by Construction Plan Ministry, France | Works-Lille Architecture Culture Center, Police Center in Saint-Denis, Nanterre University Campus Renovation(Paris), Bordeaux Les Menuts High School, etc

Hyunsik Min
_KNUA, KIOHUN Architects & Associates
Born in 1946, Korea | Graduated from Seoul National University, studied at the Architectural Association School of Architecture in London | Professor at the Korean National University of Arts, advisor to KIOHUN Architects & Associates | Awards-2006 the AIA Bestowed Hon. FAIA, 1992 the Culture Award of Kim Swoo-Geun, the KIA Award, etc | Works-master plan, Daejeon University and several buildings, Sindoricoh Headquarters and Factories, Dongsoong Presbyterian Church, Paju Book City design and several buildings, The Korean National University of Cultural Heritage, etc

Byungyoon Kim
_School of Architecture, Hongik University
Born in 1952, Korea | Phd, Hanyang University, AA school, London, established Studio METAA, professor at School of Architecture, Hongik University | Awards-2004 the Culture Award of Kim Swoo-Geun, etc | Works-Asia Publication and Information Center, Jungreungdong Catholic Church, etc

Anna Ranova Lundström
Born in 1973, Sweden | Graduated from the Royal Academy of Art and Architecture, Denmark | Worked at Bosch Haslett Architects, KHR Architects, and Signal Architects

Toshihiro Kubota
Born in 1976, Japan | Received an Architecture-Interior degree from Creapole-Esdi in Paris | the 1st prize of Competition Copenhagen Concert Hall with Ateliers Jean Nouvel

Stephen Matthys
Born in 1973, Germany | Received a diploma in Architecture from RWTH Aachen | Worked at Ateliers Jean Nouvel | Nominated for the European Architecture Award Euregio

ELEPHANT Office
Founded in 2005, Copenhagen and Paris

Jean Nouvel _Ateliers JEAN NOUVEL
Born in 1945, France | Graduated from the Nationale Superieure des Beaux Arts, 1994 Ateliers JEAN NOUVEL | Publication-Ateliers Jean Nouvel, etc | Awards-1998 gold medal from the French Academy of Architecture, etc | Works-the Arab World Institution, AGBAR Tower, Guthrie Theater, Mus?e du Quai Branly, Leeum Museum, etc

Youngsun Chung
_SEOAHN Landscape Architects
Born in 1941, Korea | Graduated from Agriculture in Seoul National University, SNU Graduate School of Environment | Established in 1987 SEOAHN Landscape Architects | Awards-the Culture Award of Kim Swoo-Geun, Landscape Award from the Eastern Region of the IFLA, the Professional Award from the American Society of Landscape Architects, etc | Works-Paris Park, The Seoul Arts Centre, Incheon International Airport, Seonyudo Park, Aloemime Vision Village, etc

Andrés Perea Ortega
Born in 1940, Spain | Professor of Escuela Tecnica Superior de Arquitectura de Madrid, a founding member of the Initiative for an Architecture & Urbanism +Sustainable | Awards-The Multi-functional Administrative City, International Competition, prize, International Ideas Competition for the design of the Seoul Performing Arts Center, prize, Asia Culture Center International Competition 3rd prize, etc

MVRDV
Winy Maas, Jacob van Rijs, Nathalie de Vries, principal | Established in 1991, Rotterdam | Awards-1991 the Berlin European Competition 1st prize, 1997 the Concrete Award, the Dudok Award, 1997 the Merkelbach Award, etc | Works-WOZOCO, Mirador Apartment Building, Double House, Villa VPRO, Barcelona Renews, the Dutch Pavilion at the Hannover World Exhibition Expo 2000, etc

Youngwook Joung
Born in 1972, Korea | Graduated from Sungkyunkwan University, Berlage Institute, Master in Architecture | 2003-2006 MVRDV Project Leader, 2006 TU Delft University, guest teacher | Works-Wembly Dome Hotel, Busan City Sofa, Anyang Peak, Longtan Park Housing, etc

Gonghee Lee _SAKU
Graduated from the College of Architecture and Design, Kookmin University | Worked at WONDOSHI Architects Group, YUCHANG Architects | Professor at School of Architecture, Kookmin University

Kerl Yoo _iArc Architects
Born in 1940, Korea | Graduated from Seoul National University, worked for Swoogeun Kim, 1979 established KERL YOO Architects, professor at Graduate School Architecture of Kyunghee University | Partner of iArc Architects | Awards-1998 the KIA Award, 1996 the Culture Award of Kim Swoo-Geun, etc | Works-Millennium Community Center, Milal School, Chunjoo University Chapel, Paichai International Center, etc

DAVID-PIERRE JALICON _D.P.J. & Partners
Born in 1968, France | Graduated from Paris-Conflans School of Architecture, University of Paris Panthéon-Sorbone, Paris-Villemin School of Architecture | President of D.P.J. & Partners and David-Pierre Jalicon Architect | Awards-the Beaux-Arts Academy of the French Institute Award, Pierre Cardin Prize, etc | Works-Seoul French High School, French Cultural Center, Aqua Art Bridge, Central Point Bridge, Nungpyongri House, etc

WOONGWON YOON
Born in 1964, Korea | Graduated from Yonsei University, the School of Architecture at La Villette in Paris | Teaching at Yonsei University and Dongkuk University

JEONGJOO KIM
Born in 1968, Korea | Graduated from Yonsei University, the School of Architecture at La Villette in Paris | Teaching at Hongik University

JEGONG ARCHITECTS
2001 established | Awards-Competition for the 100th Anniversary of MyungDong Cathedral (2nd prize), 2005 Competition for Seoul Performing Arts Center, 3rd prize | Works-Myung Film Company Building, Jungnung Building, Jungnung Residence, etc

DOMINIQUE PERRAULT
_DOMINIQUE PERRAULT Architecture
Born in 1953, France | Graduated from Ecole Nationale des Ponts et Chaussees | Awards-the Engineers School, the Hotel Industrial Berlin in Paris, 2004 Ewha Campus Center Competition prizewinner | Works-the French National Library, the Olympic Velodrome and Swimming Pool, etc

JONGKYU KIM _KNUA, M.A.R.U. Architects
Born in 1960, Korea | Graduated from Yonsei University, AA Diploma at Architectural Association School of Architecture | Worked at Dongwoo Architects, Building Design Partnership, Florian Beigel Architects, president of M.A.R.U. Architects, professor at the Korean National University of Arts | Awards-2001 the Culture Award of Kim Swoo-Geun, the Korean Architecture Award, etc | Works-Uijae Museum, CAIS Gallery, Heyri Art Valley(BB Studio, MA Gallery, Agora Museum), etc

SATOSHI MATSUOKA
Eorn in 1973, Japan | Graduated from Kyoto University, Master of Engineering in Architecture from the University of Tokyo, Mater of Science in Advanced Architectural Design from Columbia University, Graduate School of Architecture, Planning and Preservations, researched urbanism in Delft University | Worked at UN Studio, MVRDV and SANAA | Exhibition-SD Review 2002

YUKI TAMURA
Born in 1977, Japan | Graduated from Nihon University, Master of Architecture from the Graduate School of Tokyo National University of Fine Arts and Music | Worked at Enric Miralles and Benedetta Tagliabue Arquitecto, SANAA

MATSUOKASATOSHITAMURAYUKI
Established in 2005, Tokyo | Works-Balloon Caught(Vancouver), Crater Theatre(The Seoul Performing Arts Center Competition, Stack House, Gyeonggi-do Jeongok Prehistory Museum 3rd prize(Korea), Arex(Nagoya), etc

DANIEL VALLE
Graduated from the E.T.S. Architecture in Madrid, Spain, educated at the Berlage Institute | Worked for Foreign Office Architects and Eduardo Arroyo(NOMAD), visiting professor at the Korean National University of Arts, teaching in the University of Seoul | Works-Imjingak Memorial Competition, Kyomunsa Offices facade development, etc

JUNGLIM ARCHITECTURE
Jungchul Kim, Jungsik Kim, Jinho Moon, principal | 1967 established | Awards-2006 KIA Design Award(Cheonggyecheon Culture Center), 2006 City of Seoul Award Program(National Museum of Korea), 2002 City of Seoul Award Program (World Cup Korea Main Stadium), etc

HEERIM ARCHITECTS & PLANNERS
Younghee Lee, principal | Founded in 1970 | Awards-1st prize, Korean Architecture Awards(Apartment at Nohyeong, Jeju, block 2), 3rd prize, Korean Architecture Awards(Cheonan City Hall), 1st prize, Korean Civil & Architecture Awards (Bundang Venture Town 1), etc

SUKYEON YOO _hna On-go-Dang
Graduated from Seoul National University, University of Pennsylvania(M.Arch) | Exhibitor of Korean Pavilion in Venice Biennale 2004 | Principal of hna On-go-Dang, lecturer in Korea University | Awards-2005 the Korean Architecture Award, highest honors, 2005 the KIA Award, etc | Works-master plan of Aloe Tourist Farm at Haenamdo, Namyang Aloe Econet Center, Daum Global Media Center, Gangseo-gu Ministration Building, etc

PIER VITTORIO AURELI _DOGMA Office
Graduated from Master's degree in Urban Design Univesita Ca Foscari d Venezia, Architecture from Delft University | Visiting professor at the Academy of Architecture, Mendrisio and Delft University

PETER EISENMAN _EISENMAN Architects
Born in 1932, USA | Studied at Cornell University, Columbia University and the University of Cambridge, an honorary Doctor of Fine Arts Degree from the University of Illinois Chicago, established IAUS | Awards-1985 the 3rd International Architectural Biennale in Venice, 1st prize, 1989 Award from AIA | Works-Rebstock Park Master Plan, Romeo and Juliet Castles, Memorial to the Murdered Jews in Europe, etc

HAEAHN ARCHITECTURE
Sehan Yoon, principal | Founded in 1990 | Works-KOR.V.A. 2010 Project, Osong Bio-Health Technopolis, Ilsan Lafesta, etc

YOONGYOO JANG
_SAKU, UnSangDong Architects
Born in 1964, Korea | Graduated with a M.A. in Architecture from Seoul National University | Professor at School of Architecture, Kookmin University, Partner of UnSangDong | Awards-Takiron International Architectural Design Competition, UIA International, Barcelona, nominated one of 13 finalists in the Rabin International Peace Forum, Israel | Works-Egypt Embassy, Architecture Department of Seoul National University, Gwangju Design Center, etc

MOONGYU CHOI
_Yonsei University, Ga.A Architects
Born in 1961, Korea | Graduated from Yonsei University, Master of Architecture from Columbia University, worked in TOYO ITO Architects, HANUL Architects and Group SEE, 1999 founded Ga.A Architects, professor at Architectural Engineering of Yonsei University and a sector architect of Paju Book City | Awards-2005 KIA Special Award, 2005 Korean Institute of Registered Architects Award, 2005 Seoul Architecture Award, etc | Works-Dalki Theme Park, Cheong Hansook Memorial, Ssamziegil, Taehaksa Publishing, etc

MINSUK CHO _MASS STUDIES
Born in 1966, Korea | Graduated from Yonsei University, the Graduate School of Architecture at Columbia University, 2003 the principal of MASS STUDIES | Awards-1994 Shinkenchiku International Residential Architecture Competition, 2000 the Architectural League of New York's Young Architects Award, Progressive Architecture Awards(1999, 2003) | Works-Nature Poem, Torque House, Pixel House, Dalki Theme Park, etc

FOA _Foreign Office Architecture
Established in 1992, London | Founder-Farshid Moussavi(professor at the Academy of Fine Arts Vienna), Alejandro Zaera Polo (dean, Berlage Institute) | Works-Yokohama International Port Terminal, South-East Coastal Park and Auditoriums, The Bundle Tower, Virtual

House Proposal for ANY Corporation, BBC White City, etc

Leemjong Jang _Yonsei University
Born in 1961, Korea | Graduated from Hanyang University, Harvard University | Worked at Koning Eizenberg Architecture, assistant Professor of Kaiserslautern University | professor at Architectural Engineering of Yonsei University, representation of ATStudio

Youngjoon Kim _yo2 Architects
Graduated from Seoul National University, studied at AA School in London, director of yo2 Architects, coordinator of SA School of Architecture and Paju Book City, 2002 Venice Biennale, 2006 Architectural Olympiad Hamburg | Awards-2005 Korea Institute of Architects Award, 2005 the Culture Award of Kim Swoo-Geun, First Town Competition(2nd prize), etc | Works-Fraumedi Womens Hospital, Heryoojae Womens Hospital, Jahajae Residence, etc

Tesoc Hah _iArc Architects
Graduated from Sungkyunkwan University, Architectural Association School of Architecture in London, RIBA, worked at Alsop Architects and Skidmore, Owings & Merrill | Adjunct professor at the Graduate School of Architecture, Kyunghee University, partner at iArc Architects

Ciro Najle _GDB
Architect from Buenos Aires | Visiting professor at Cornell University, taught at Columbia University, Berlage Institute and the University of Buenos Aires, director of GDB, General Design Bureau and previously at MID, Met Infrastructural Domain | Award-2001 Young Architect of the Year 2nd prize | Exhibitions-Prague Biennale of Art, Beijing Bienniale of Architecture 2004

Young Kweon _SAKU
Master of Architecture, Bartlett School of Architecture, UCL | Professor, Dept. of Architecture & Interior Design, Kaywon School of Art & Design | Professor, School of Architecture, Kookmin University | Award-an Asian Finalist of UIA International Competition Celebration of City, 2004

Jooryung Kim
_Urbano Architectural Research Group, SaiA+
Graduated from Architectural Association School of Architecture in London, Rhode Island School of Design in Providence | Worked at Foster & Partners, Zaha Hadid and UN Studio, 2002 the founder of SaiA+ | Awards-2004 KA Young Architects Award | Works-Noon Pictures Photo Studio Interior, PKM Gallery, etc

Shaun Murray _Independent Architect
An Illustrated Primer, M. Arch, University College London, Bartlett School of Architecture, UK | 2003 commended in East Sámi Museum Competition | Exhibition-Stereo-Mono, Disturbing Territories, The Leg of a Maniac, etc | Publication-Disturbing Territories, Reflexive Practice in Architectural Ecologies-Communicating Vessels, etc | Works-Gateway to the Jurassic Coast, Herent Architecture in the Field, Sami Museum, etc

Haewon Shin _Lokaldesign
Graduated from Yonsei University, AA School of Architecture, RIBA Part2, worked for David Chipperfield Architects, OMA Asia, Terry Farrell & Partners, 2001 established Lokaldesign, 2004 visiting professor of Korean National University of Arts | Awards-2006 Venice Biennale, 2003 ChiChi Earthquake Memorial Competition, 2nd prize, 1998 European 5 Competition Geesthacht, 1st prize, etc

Chanjoong Kim __System Lab
Graduated from Architecture Engineering of Korea University, M.A of Graduate School of Design at Harvard University | Principal of _System Lab | Works-Geoje Island Residence, Boat Club2, Container Boat Club, Gyeongju Expo Tower Competition, Eagon Window Image Shop, etc

Sooin Yang _The Living Architecture Lab
Graduated from Architecture Engineering of Yonsei University, Master Degree at Columbia University GSAP | Visiting professor of Pratt Institute, N.Y., assistant professor of Columbia University | Co-founder and principal of THE LIVING Architecture Lab | Works-Better_ Cheaper_Faster, River Glow, Living Glass, etc

Kihong Kim _Architect KYOSKS
Born in 1969, Korea | Graduated from Kookmin University, studied at School of Architecture of Paris La Villette, University Paris 8, Studio | Tutor Kyonggi University, member of Architect KYOSKS | Exhibition-1997 Korean Traditional Space | Awards-2005 Selected 100 Young Designers by ANC, 2003 Busan International Architectural Competition 3rd prize, 2004 Chichi International Competition finalist, etc.

Kyunghoon Lee _SAKU
Master of Architecture, Pratt Institute, Registered Architect(New York, Korea), professor at School of Architecture, Kookmin University

Paul Preissner _QuaVirarch
Graduated from the University of Illinois, studied at Columbia University | Worked at Eisenman Architects, offices of Philip Johnson, Skidmore Owings and Merrill and Wood-Zapata, principal of QuaVirarch | 2006 visiting hyde chair professor at the University of Nebraska-Lincoln School of Architecture

Lonn Combs
Graduated from the Architecture Dept. at University of Kentucky, Master Degree for Architecture at Columbia University | Working experience at Daniel Libeskind Studio and SOM

Rona Easton
Trained in the United Kingdom, a Professional Degree from the University of Westminster, Master Degree from Bartlett School of Architecture, licensed in the USA and the United Kingdom | Broad professional experience in the UK, Germany, China and the USA

Easton + Combs Architects
Established in 2001 | Works-Malama Learning Center, Airpark IAH and the Mill Center, etc

Marcos Novak
The Masters of Architecture at Ohio State University, taught at Ohio State, University of Texas Austin, the Architecture program at UCLA, the Digital Media program at UCLA, Art Center College of Art & Design, professor of UCLA Dept. Architecture and Urban Design | Publications-Avatarchitectures: Fashioning Vishnu after Spacetime, Next Babylon: Algorithm to Play in, etc | Works-Paracube, Data-Driven Forms, Variable Data Forms, etc

Junsung Kim _hANd Architecture
Born in 1956, Korea | Graduated from Mackenzie University, studied at Pratt Institute and Columbia University | Worked at architectural firms such as Alvaro Sizas and Steven Holls, adjunct professor of Korean National University of Arts, 2006 principal of hANd Architecture | 2003 the KIA Award | Works-Open Book Publishing, Heyri Art Valley Community House, Artreon Cine-multiplex, etc

Hailim Suh _Architecture Studio HIMMA
Born in Korea | Graduated from Columbia University, Harvard University Graduate School of Design | HAILIM SUH Architecture Studio(1995-97), principle at Architecture Studio HIMMA(1997-) | the KIA Award, the Seoul Architectural Work Awards | Design Vanguard of Architecture Record 2005 | Works-Borim Publishing House, Open Book Publishing, National Research Institute of Cultural Properties, Childcare and Children's Educational Center for Seoul Metropolitan Government, etc

Nader Tehrani _Office dA
Born in England | Rhode Island School of Design 1986 | Architectural Association in London 1987 | Harvard University Graduate School of Design

1991 | Principle at Office dA 1991-present | Awards-8 Progressive Architecture Awards, the Harleston Parker Award, Academy Award from the American Academy of Arts and Letters | Works-Tongxian Arts Center in Beijing, Rhode Island School of Design library, Macallen Building Condominiums in South Boston, Northeastern University's Multi-faith Spiritual Center, etc

ILBURM BONG _SAKU
Graduated from Seoul National University, studied at Harvard Graduate School of Architecture, MArch | Professor at School of Architecture, Kookmin University | Publications-Unbuilt 20th Century, Volume Zero, etc

HOON MOON _MOON HOON Design Lab
Born in 1968, Korea | Graduated from Inha University, M.Arch from MIT | Awards-2005 KIA Awards | Works-Mookdong Housing, Jeonju Zoo, Hyundai High School, Sangsang Museum, etc

OMA
Founded in 1975 by Rem Koolhaas, Elia and Zoe Zenghelis and Madelon Vriesendorp | OMA Rotterdam, New York, Beijing | Awards-2000 the Pritzker Architecture Prize, 2003 the Praemium Imperiale Japan, 2004 the RIBA Gold Medal, the Mies van der Rohe Prize(Neterland Embassy in Berlin), 2005 European Union Prize for Contemporary Architecture, etc | Works-Zeche Zollverein Historical Museum and master plan in Essen, Leeum Museum, Seoul National University Museum of Art, the much acclaimed Casa da Musica in Porto, etc

KEN MIN SUNGJIN _SKM Architects
Graduated from University of Southern California and Harvard University Graduate School of Architecture, American Institute of Architects AIA | Works-Hilton Namhae Golf & Spa Resort, Daesang Acrovista Residential Tower, Paju Book City Town House, Soongsil University School of Engineering, The Wedge Building, etc

ZAHA HADID _ZAHA HADID Architects
Born in 1950, Iraq | Studied at Architectural Association School of Architecture in London, taught at Harvard Graduate School of Design, professorship at University of Applied Arts, Vienna, visiting lecturer at Royal College of Arts | Works-The Vitra Fire Station, Mind Zone at the Millennium Dome, Car Park and Terminus, Ski Jump, CCTV, etc

ANDREA BOSCHETTI
Born in 1969, Italy | Architecture from Venice University, PhD in Urbanistics at I.U.A.V., Venice, 1995 Cooperation with OMA, 1998-2001 PhD in Urbanistics at I.U.A.V. and Columbia University | 2002-today visiting professor of Architectural Design

ALBERTO FRANCINI
Born in 1969, Italy | Architecture from Florence University, PhD in Urbanistics at Polytechnic of Milan | 2002-today Assistant Professor in Urban Design at Politecnico di Milano

METROGRAMMA
Founded in 1998, Milano | Awards-2006 master plan Cassola Industrial Area : 1st place, 2006 master plan of Administrative Area and new Headquarters of Ministry of Guangzhou, Nansha China: 1st place, 2002 Mies van der Rohe, etc | Works-One family villa in Reggio Emilia., Masterplan Area D3, Rozzano, Milano, master plan Area Darsena, Ravenna, Feasibility planning and construction of a residential building Villa Merano in Merano-Total area, etc

SPACE GROUP
Sangleem Lee, principal | Awards-2005/2002 Grand Prize for Korean Architecture & Culture Awards, etc | Works-Government Building in Dalseonggun, Library in Hangsung University, Yangjae Merchandising Complex, Mapo Integrated Government Building, Busan Asian Games Main Stadium, etc

WANGDON CHOI _SAKU
Seoul National University, University of Manchester(Ph.D.) | Partner, HAEAHN Architects & Engineers | Professor at School of Architecture, Kookmin University | Publications-Architecture Reader, Architecture after Modernism(Korean Translation), etc

JINA ARCHITECTS
Daejin Bu, Moohyun Kim, Sanghoon Bu, principal | Established in 1969 | Awards-1991 Korean Architecture Award, 1996 Daejeon Architecture Award, etc | Works- Daejeon Expo IBM Exhibition Hall, Ehwa Samsung Campus Center Architectural Consulting, Devis Tower, etc

KUNWON ARCHITECTS
Jongnam Kim, Insun Hahm, principal | Established in 1984 | Awards-1995 Seoul Architecture Award, 1999 Korean Architecture Award, 2004 Korean Architecture Award 1st prize, etc | Works-War Memorial of Korea, Korea Trade Center, Goyang International Exhibition, etc

SAMOO ARCHITECTS & ENGINEERS
Myunggi Sohn principal | 1976 founded | Works-Samsung Museum of Art, Leeum, Seoul Museum of Arts, Jagalchi Market, Tower Palace, etc

MOOYOUNG ARCHITECTS & ENGINEERS
Gilwon Ahn, principal | Established in 1985 | Awards-2001 Seoul Architecture Award, 2003 Korean Architecture Award Residential Prize, etc | Works-Korea Trade Information Center, Gyeonggi Small Business Center, Simone Headquarters, etc

GANSAM PARTNERS
Jaho Kim, principal | Established in 1983 | Awards-2005 Incheon Architecture Award, 2005/2006 Korean Culture and Architecture Annual Award, etc | Works-Posco Center, Kolon Tower, Posco Historic Museum, etc

WONDOSHI ARCHITECTS GROUP
Yong Byun, Eungjae Jang, Seokju Kim, principal | Established in 1969 | Awards-1995 Korea Institute of Architecture Award, 2002 Seoul Architecture Award, 2003 Daegu Architecture Award, etc | Works-Office Building Tower Korea Trade Center, Sports Center Seoul National University, Judical Research And Training Institute, Korea Meterologic Administration, etc

WONYANG ARCHITECTURAL DESIGN GROUP
Jongchan Lee, principal | Established in 1981 | Works-The National Center for Korea Traditional Performing Art, Air Force Club, Korea Culture & Contents Agency+Korean Film Archives, Osong Bio-Health Science Park, EunPyung Newtown 2 district, Dongnam District Shopping Mall Turn Key Competition, etc

TOMOON ENGINEERING ARCHITECTS
Gichul Choi, principal | Established in 1990 | Awards-2004 Korea Architecture Award Housing Part Prize Award, 2004 VE Design Competition 1st prize, etc | Works-Yongin Shingal New Millennium Apartment, Namyangju Pyeongnae District Apartment, Yongin Gugal Summitest Vill Apartment, etc

CHANGJO ARCHITECTS
Byunghyun Kim, principal | Established in 1984 | Awards-1996 Korean Institute of Architecture Award, 1998 Best Building Design of Seoul Awards, 2003 Korean Culture and Architecture Annual Award, etc | Works-SK Life Insurance Headquarters, Korea Economic Daily Headquarters, Kyonggi University Teleconference Center, etc

통섭지도: 한국건축을 위한 아홉 개의 탐침
CONSILIENT MAPPING: NINE PROBES FOR ARCHITECTURE KOREA

초판 1쇄 인쇄 2007년 3월 3일		First Published in Korea March 2007
초판 1쇄 발행 2007년 3월 12일		Copyright ⓒ 2007 Space Publishing Co.

엮은이	국민대학교 건축대학
데이터 작업	한상일 봉일범 이정우 이혁찬 김형술
펴낸이	이상림
펴낸곳	(주)공간사
출판등록	1978년 4월 25일 제1-18호
편집총괄	박성태
마케팅총괄	오현준
책임편집	권순주
편집	김도언 김정옥
디자인	정진주
마케팅	김천세 김미경 한경화
종이	이지포스트
출력	오비케이시스템
제작	본디자인

Compiled by School of Architecture, Kookmin University
Contributed by Sangil Han, Ilburm Bong, Jeongwoo Lee, Hyukchan Lee, Hyungsul Kim
Publisher Sangleem Lee
Directed by Seongtae Park
Marketing Directed by Hyunjun Oh
Edition Directed by Soonjoo Kwon
Edited by Doeon Kim, Jeongok Kim
Designed by Jinjoo Jeong
Marketing by Cheonse Kim, Mikyung Kim, Kyunghwa Han
Paper Supplying _E.G Post
Color Separation _OBK Color Process System
Printing _Bon Design

(주)공간사 Space Publishing Co.
110-280 서울시 종로구 원서동 219
219 Wonseo-dong, Jongno-gu, Seoul, 110-280 Korea
Tel 82-(0)2-3670-3633(Editing), 82-(0)2-3670-3627(Marketing)
Fax 82-(0)2-747-2894
Web site http://www.vmspace.com
E-mail webmaster@vmspace.com

35,000 won
ISBN 978-89-85127-25-7 03600

* 이 책은 국민대학교 개교 60주년 기념 건축전 '통섭지도: 한국건축을 위한 아홉 개의 탐침'을 바탕으로 구성되었습니다.
* 저작권법에 따라 보호를 받는 저작물이므로 무단전재나 복제, 광전자 매체 수록 등을 금합니다.
* 파본이나 잘못된 책은 교환해 드립니다.
* This book is based on the contents of "Consilient Mapping: Nine Probes for Architecture in Korea" exhibition.
* All right reserved. No part of this publication may be reproduced, stored in a retrieval system, or transmitted by any means, electronic, photocopying recording or otherwise, without the prior permission of the copyright owner.
* Damaged or incorrect books may be exchanged.